D0976775

HEINERMAN'S NEW ENCYCLOPEDIA OF FRUITS & VEGETABLES

REVISED & EXPANDED

JOHN HEINERMAN

PARKER PUBLISHING COMPANY
West Nyack, New York 10994

10 9 8 7 6 5 4 3 2 1

This book is a reference work based on research by the author. The opinions expressed herein are not necessarily those of or endorsed by the publisher. The directions stated in this book are in no way to be considered as a substitute for consultation with a duly licensed doctor.

Library of Congress Cataloging-in-Publication Data

Heinerman, John.
[New encyclopedia of fruits and vegatables]
Heinerman's new encylopedia of fruits and vegatables / John Heinerman.—Rev. and expanded.
p. cm.
ISBN 0-13-209222-0 (cloth).—ISBN 0-13-209230-1 (paper)
1. Materia medica, Vegatable—Encyclopedias. 2. Vegetables--Therapeutic use—Encyclopedias. 3. Fruit—Therapeutic use--Encyclopedias. I. Title.
RM236.H45 1995
615'.321'03—dc20 95-19016
CIP

ISBN 0-13-209222-0 (ppc)
ISBN 0-13-209230-1 (p)

PRINTED IN THE UNITED STATES OF AMERICA

Other Books by the Author

Approximately 40 books on food therapy, folk medicine, general health, and nutrition, including the following recent titles:

Double the Power of Your Immune System

Heinerman's Encyclopedia of Healing Juices

Heinerman's Encyclopedia of Nuts, Berries & Seeds

Heinerman's New Encyclopedia of Fruits and Vegetables

Heinerman's Encyclopedia of Healing Herbs and Spices

Fondly dedicated to

My Guardian Angel, Briant Stringham Stevens, and my dear father and brother, Jacob and Joseph Heinerman

My conclusion is that disease can be cured through the proper use of correct foods.

—Henry G. Bieler, M.D., author of *Food Is Your Best Medicine* (New York: Random House, 1966). The quote appears on the page entitled "To the Reader."

FOREWORD

In 1987 my good friend Lendon Smith, M.D., wrote an appealing foreword to the original version of this book when it included medicinal plants. The final line of his remarks was a genuine expression of appreciation to me for all the work and effort I had put into that book: "Thank you, John Heinerman, for opening up a whole new world of health [care] to us, the uninitiated and naive."

Toward the end of 1994 my publisher asked that I split the contents into two separate reference works. Nearly all the medicinal plant information was transferred to a brand new *Encyclopedia of Healing Herbs and Spices*. With a somewhat reduced contents, it was a challenge to come up with other material to replace what had been omitted.

The result has been an almost entirely new book all together. To begin with, there are over 20 *additional* entries, ranging from bamboo shoots to vinegar. Some sections, for example, garlic, mushrooms, and peppers, were greatly expanded to include new information, while others, such as that for seaweed, appear here for the very first time. Other sections, though, remained untouched, such as those for nuts and seeds.

In addition, this revised edition includes a number of surprising incidents that never appeared in the first printing. The fantastic "mushroom cure" for breast cancer that came through a near-death experience in a hospital operating room from a patient's deceased relatives is one example. The amazing story of how a Russian man survived on nothing but mustard greens for many weeks during the terrible German siege of Leningrad in 1941–1942 while tens of thousands starved to death around him is another.

I felt it was essential to include *more* produce items that are quite common on a regional or ethnic basis. This is why, for instance, that the kudzu vine shows up here in spite of its nuisance status throughout much of the South. While difficult to eradicate, it has proven to be a valuable curative as well as tasty culinary treat. Or, because of the huge Hispanic population in this country, why jicama—a popular food and medicine for that culture—is included.

Sometimes the popularity of one or more items demands that sufficient pages be devoted to them. Vinegar is so common in most American households that a list of its many medicinal applications was warranted. The same thing may be said for chile peppers in the American Southwest or the general appeal of garlic: exhaustive data on both may be found under their appropriate headings. And in the increased interest shown by more of the public in cuisines and remedies from the Orient justified a special section on seaweeds.

I've introduced over 87 new remedies and added 62 more recipes to this edition than appeared in the old version. This is data gleaned from trips made to a number of new countries since 1988. And for easier cross-referencing, there are not one but now two table of contents matching symptoms with food remedies and vice versa.

So here's to better health, better appetite, and better reading all around.

—John Heinerman, 1995

CONTENTS

I. HEALTH SYMPTOMS AND THEIR FOOD REMEDIES

NOTE: Some of the food remedies appearing here with each problem may not always be found within the general text; nevertheless, their presence is justified for that particular malady. Also several problems with different remedies may be related, such as **BLEEDING** and **HEMORRHAGING**; therefore, cross-referencing is encouraged for a broader range of potential treatments.

II. Food Remedies and Health Symptoms

III. Appendix

Amaranth

(Amaranthus hypochondriacus, A. cruentus)

Brief Description

Not a true cereal, but rather a fruit, it belongs to the same family (chenopodium) as the edible weed, lamb's-quarter. Looking like a sesame seed, it has a pleasant, nutty flavor and can be popped like corn or steamed and flattened into a flake. It requires very little water and fertilizer, growing almost anywhere the common weeds do. It is thought that amaranth was brought to this hemisphere by those first migrants from the Tower of Babel, who traveled eastward across China and launched their barges on the Pacific, eventually reaching what is now western Mexico around 2000 B.C.

Food of the Future

Amaranth seed has been described as the "perfect protein food of the past for meals of the future," by *U.S. News & World Report* for November 25, 1985. Robert Rodale, publisher of *Prevention* and *Organic Gardening* magazines, first reintroduced amaranth to this country in the early 1970s after the U.S. Deptartment of Agriculture showed a lack of interest in it.

Diarrhea and Bleeding

Amaranth seed and leaves have been used effectively as an astringent for stopping diarrhea, bloody stools and urine, and excessive menstruation. It also makes a good wash for skin problems ranging from acne and eczema to psoriasis and hives. It's an excellent douche for vaginal discharges of purulent matter, a nice gargle for sore mouths, gums, teeth and throats, and a fantastic enema for colon inflammations and rectal sores.

AMARANTH TEA

To make an amaranth tea for all these purposes, simply bring 3 cups of water to a rolling boil. Then add 2 tsp. of seeds, cover and simmer on a very low heat for about 5 minutes. Remove from heat and add 1 tsp. of leaves (if available) or else just let steep for 30 minutes. Drink 2 cups daily for internal problems.

Chinese Recipe

AMARANTH WITH RICE STICK NOODLES

Needed: 2 oz. rice stick noodles; 1 $^1/_2$ tbsp. soy sauce; 1 tsp. sesame seed oil; 2 tbsp. pale dry sherry; $^1/_8$ tsp. hot sauce; 3 tbsp. water; 3 tbsp. peanut oil; 1 large clove garlic, finely chopped; 2 scallions, including some green, finely chopped; 1 small zucchini, ends removed, scrubbed and sliced into julienne strips; 1 cup amaranth leaves and stems, washed and chopped medium fine (about 1 $^1/_2$ oz.).

Soak the rice stick noodles in hot water to cover for 15 minutes. Cut them into 2″ lengths. In a small bowl combine the soy sauce, sesame seed oil, sherry, hot sauce and water. Heat a 12″ wok or iron skillet over high fire for about 30 seconds. Add the peanut oil and heat for another 30 seconds. Add the garlic and scallion, mix, and then add the zucchini and amaranth. Stir-fry for 3–5 minutes until the vegetables are tender. Mix in the rice noodles and then add the bowl of seasoning mixture. Bring to a boil and serve. Serves 4 at a Chinese meal or 2 at a Western-style meal.

Apple
(Pyrus malus)

Brief Description

The wild apple or crab apple grows throughout Europe and as far as Central Asia, where it seems to have originated. The people of Asia Minor took no notice of it for a long time, but the Greeks had already cultivated it. In Rome at the time of the Emperor Augustus there were no less than 30 different varieties recorded. Today there are more than 1,400 varieties of apples worldwide. Although tradition holds that the forbidden fruit in the Garden of Eden was supposed to have been the apple, apocryphal literature strongly suggests that it was the grape instead.

An Apple a Day Keeps the Doctor Away

This very old rhyme had become a polite way universally of explaining that apples are an ideal preventative of both constipation *and* diarrhea. The July 1978 number of *The American Journal of Clinical Nutrition* reported that apples helped to decrease the time it took to have a bowel movement, by increasing the stool weight, which in turn increased the number of trips to the bathroom during a 24-hour period.

In her book on herbs, Mary Quelch recommends "a baked apple at night to be followed by another at breakfast [as] one of the most efficacious remedies" known for constipation. So the next time you begin to reach for the Ex-Lax, think instead in terms of munching a Rome Beauty, Delicious, Jonathan, or Winesap.

Conversely, apples are also very good for treating acute diarrhea or "Montezuma's Revenge," the bane of so many tourists traveling to foreign countries where cleanliness is definitely a low-priority item. Take one ripe apple and grate it, allowing the pulp to stand at room temperature for several hours until considerably darkened before eating. The oxidized pectin present in the fruit is the same basic ingredient in Upjohn Pharmaceutical's Kaopectate brand diarrhea medicine.

An Infection Fighter

You probably never imagined that an apple could possess some penicillinlike properties. Well, Canadian scientists in the December 1978 *Applied & Environmental Microbiology* demonstrated that fresh apple juice or fresh apple sauce can knock the heck out of stomach flu and polio viruses.

And how about those nasty little germs that can cause tooth decay? Two British doctors gave a group of kids one or two thin slices of raw apple after each meal or snack and discovered in a while that they had a substantial decrease in cavities as a result.

Cider Vinegar Eliminates Gallstones

A woman from Napanee, Indiana who had been bothered with gallstones for a number of years went to a clinic for a total physical. The X-rays showed a lot of stones of varying sizes. A naturopathic physician advised her to drink nothing but apple cider vinegar for four days straight, in 1/2 cup amounts 5 times a day. And on the second, third, and fourth days, she drank equal parts (1/4 cup each) of apple cider vinegar and pure virgin olive oil mixed together. The stones passed on the fifth day, never to come back again.

This shouldn't come as much of a surprise. According to Volume 31 (1987) of *Annals of Nutritional Metabolism*, "Apple fiber extracts containing a high level of pectins decreased the level of cholesterol in hamsters." Therefore by eating more apples, cholesterol-induced gallstones can be prevented from forming.

Other Uses for Cider Vinegar

Apple cider vinegar has still a wider range of applications as the following list shows:

- Helps heal burns when soaked gauze is applied to injured areas.
- Relieves pain and itchiness when rubbed on insect bites and stings.
- Removes dandruff when used as a hair rinse after washing hair.
- Eliminates body odor when used in place of an underarm deodorant.
- Cures athlete's foot when sore feet are soaked daily in strong solution.

Also see under VINEGAR for additional information.

Massage Away Your Headaches

Have you ever tried massaging away a headache? This method works even better when apple juice is used. Finely grate an apple and squeeze out the juice. This can be easily accomplished by putting the grated pulp into a double-layered cheesecloth, gathering the four corners together, and twisting it tightly until all the liquid has been extracted. Soak the tips of the forefingers of each hand and gently massage into the scalp in back-and-forth and sideways motions.

Bringing Down the Temperature of Fever

A poultice made of apple helps to reduce fevers. Finely grate an apple and spread it 1/2″ thick on a piece of double-layered cheesecloth about 6″ in length. Apply this over the forehead. Make a similar poultice for the throat area or chest if necessary. Leave on for an hour.

Unique Baked Apple Recipes

The first recipe that follows is adapted with the permission of La Rene Gaunt and the publishers of her book, *Recipes to Lower Your Fat Thermostat*, while the second is of my own creation when I was in the restaurant business years ago.

Baked Apples with Date Stuffing

Needed: 6 washed, cored and halved Rome Beauties; 6 tbsp. chopped, pitted dates; $^1/_3$ cup cranberry juice; dash each of cinnamon and nutmeg. Put apples in glass baking dish. Stuff them with 1 tbsp. chopped dates and sprinkle with cranberry juice. Sprinkle them with cinnamon and nutmeg next. Cook, covered, in an oven at 400° F. until they're tender, 45–60 min. Serve lukewarm, topped with a little plain yogurt and a smidgen of pure maple syrup.

Baked Beans with Apples

Golden Delicious apples add a juicy note to this rich-flavored bean bake. Core and slice one Golden Delicious apple. Layer slices on bottom of a 1 $^1/_2$ quart casserole greased with lecithin, reserving several slices to put on top. To 1 quart of baked beans, stir in $^1/_4$ cup packed brown sugar, 1 tbsp. each orange juice and apple cider vinegar and 2 tbsp. prepared mustard. Mix these ingredients thoroughly and pour into the casserole dish. Arrange apples on top. Cover and bake at 350° F. for about 1 hour. Uncover and bake 30 min. longer. Yields 4–6 servings.

APRICOT
(Prunus armeniaca)

Brief Description

Botanists have characterized this sweet-sour fruit as more of a plum, although it does belong to the same family as peaches and almond nuts. This brass- or copper-colored fruit originated anciently in Central Asia. It is said that when the first body of emigrants to leave the Tower of Babel in central Iraq crossed over the Caucasus Mountains then turned westward towards the Caspian Sea, they brought with them young apricot tree seedlings, some of which were planted along the way.

Luxurious Beauty Agent

A two-step program I've recently introduced to selected audiences around the country has been drawing praise from many of the people who've tried it themselves, reporting back how much softer and wrinkle free their skin was afterward. The method originated with my Hungarian grandmother, Barbara Liebhardt Heinerman, whose skin was as soft and unlined as a baby's behind when she passed away in her eighties.

Step 1: In a blender put enough chopped fresh apricots with a smidgen of water to equal 1 cup, or else pour in a cup of concentrate, without the water.

Next, coarsely chop or slice a peeled, halved avocado (without the pit) into the blender. Thoroughly blend until the mixture is a smooth, even consistency. Add a tad of pure virgin olive oil, and blend again for a minute or two.

Apply this mask in a thin, uniform layer all over your face, neck and throat, leaving it on for 45 minutes. Rinse off with water. The ideal time for doing this is several hours before retiring to bed.

Step 2: Put a little heavy dairy cream or half-and-half into a dish and add the juice of half a lemon: blend well. Use only enough to cover those areas you deep-cleansed with the fruit-vegetable mask. With fingertips or cotton balls, gently massage in with a rotating action. This may take a while to do, but be sure to get all the areas previously covered with the mask. Then retire for the night.

In the morning gently bathe the face with a barley or oatmeal soap bar, which can be obtained at local health food stores or specialty cosmetic outlets. I don't recommend the use of rouge or other cosmetics while on this program.

The apricot-avocado mask can be used two or three times per week, but the heavy cream–lemon juice combination should be applied every single night before retiring. Within a matter of weeks, your skin will look and feel more luxurious and should elicit comments from those around you who are wondering where you got your "face lift" from.

Two Delicious Recipes

The fantastic fruit salad has been adapted from better Homes & Gardens' *Eating Healthy Cook Book*. The dried fruit–based bonbons are quick and easy to prepare. They also make perfect gifts or last-minute bake-sale contributions for worthy charity causes.

Apricot Fruit Salad

Chill the unopened can of apricots several hours before serving so they'll be refreshingly cold. Needed are one 16-oz. can unpeeled apricot halves (juice pack), chilled; $^1/_2$ cup halved seedless green grapes; 4 Romaine lettuce leaves; 2 oz. Neufchatel cheese (softened); $^1/_4$ tsp. ground ginger and ground nutmeg.

Drain apricots, reserving 1 tbsp. of the juice. Cut apricot halves in half to form quarters. Arrange apricots and grapes on 4 individual lettuce-lined salad plates. Stir together Neufchatel cheese and ginger. Gradually stir in reserved apricot juice to make of drizzling consistency. Drizzle over fruit. Sprinkle with nutmeg. Makes 4 servings.

Apricot Rum Balls

Needed: 1 $^1/_2$ cups dried apricots; $^2/_3$ cup hazelnuts, toasted; $^2/_3$ date sugar; $^1/_4$ cup light or dark rum; 2 tsp. grated orange zest; and 3 oz. bittersweet (not unsweetened) chocolate.

In a Vita-Mix whole food machine or equivalent unit, combine the apricots and hazelnuts; pulse just until finely chopped.

Transfer to a medium bowl. Stir in the date sugar, rum, and orange zest. Roll the mixture into 1″ balls, arranging them close together in rows on a baking sheet; set aside. In the top of a double boiler over hot, not boiling, water, melt the chocolate. Remove the top pan from the heat and let stand for 1 minute to cool slightly. Dip a table knife into the melted chocolate and drizzle it decoratively over the tops of the candies. (Alternatively, spoon the chocolate into a plastic sandwich bag and cut a tiny hole in one corner. Pipe the chocolate over the candies.) Refrigerate until the chocolate has set, at least 30 minutes. The rum balls can be stored in an airtight plastic container in the refrigerator for up to 1 week. Makes about 2 dozen rum balls.

To inquire about the Vita-Mix whole food machine, see the Appendix.

ARTICHOKE
(Cynara scolymus)

Maurice Messegue, Europe's greatest herbalist, says that the part of the artichoke we are in the habit of eating is the least active, while all the rest of it, which is unbelievably bitter, is actually the most nutritious and therapeutic for you. "For myself," he states quite emphatically, "I use every bit of the artichoke and encourage others to do the same!"

Two vegetables are called artichokes, but have absolutely no relation to each other. We distinguish them as the globe artichoke and the Jerusalem artichoke. The former is a green vegetable somewhat like a tiny cabbage, except that its leaves are smaller and thicker, while the latter isn't even an artichoke, and has nothing to do with Jerusalem. It came here from South America, and first was called "girasole" from its likeness to the sunflower. Later this was corrupted into "Jerusalem." The tubers are pleasant enough to consume, but have no medical value.

A Terrific Cholesterol Manager

To make your own special leaf tincture for better managing the problems of too much cholesterol, slightly crush and soak about 5 1/4 cups of artichoke leaves in 2 pints of alcohol for 10 days. Strain, and take 1 tbsp. twice daily in between meals. This should help keep cholesterol from accumulating in fatty globs within the body.

Volume 5 of *Experimental Medicine & Surgery* for 1947 confirmed artichoke's cholesterol fighting properties very well. Laying hens and human subjects manifesting early signs of atherosclerosis had their blood cholesterol contents lowered by administration of artichoke powder. Cynarin is the compound within artichokes that protects man and beast alike from hardening of the arteries and keeps serum triglyceride levels very low.

Brain Food to Make You More Alert

To increase your mental powers, pull an artichoke to pieces, leaf by leaf, and put into a jar with barely enough water to cover. Set a saucer on the jar and stand it in a pan of boiling water for 2 hours, adding more water to that in the pan as it boils away. Remove the jar from the pan and strain the contents, squeezing the leaves well. Take 3–4 tbsp. of this infusion should be taken 3 times a day.

Artichoke leaves also seem to be pharmacologically active in the brain and portions of the central nervous system. According to *Nutrition Reviews* for April 1978, the leaves contain "several active compounds similar to caffeine" in some ways. An herbal combination called Artichoke/Garlic may be obtained through the mail from Old Amish Herbs out of St. Petersburg, Florida (see the Appendix). Recommended intake has been 2–4 capsules per day with meals as necessary.

Remedy for Liver Problems

Certain acids in artichokes definitely help to activate liver function. Scientific literature indicates a definite improvement in liver problems when artichoke is regularly used.

A Three-Minute Salad

Ever in a hurry and don't have much time to fix yourself something to eat? How about whipping up an instant artichoke salad? Prior to this, when you have more time on your hands, cook some artichoke hearts until tender; drain, then marinate in apple cider vinegar in the refrigerator until needed. Cut a plump, ripe tomato in quarters or slices. Arrange around marinated artichoke hearts on pieces of chilled Romaine lettuce. Spoon some cold yogurt over them, sprinkle with a dash of kelp, and eat with delight.

See also JERUSALEM ARTICHOKE.

A Trio of Exciting Recipes

HOT ARTICHOKE-PARMESAN SPREAD

This spread is delicious when put on pita crisps or toasted slices of French bread.

Needed: 1 14-oz. can artichoke hearts, drained; 1 cup plus 1 tbsp. freshly grated Parmesan cheese; $^{1}/_{4}$ cup low-fat mayonnaise; 1 clove garlic, finely chopped; 1 tsp. grated lemon zest; pinch of cayenne pepper; granulated kelp to taste.

Preheat the oven to 400° F. Place the artichoke hearts in the center of a clean kitchen towel; gather up the ends and twist firmly to extract all the juice. In a Vita-Mix whole food machine or equivalent food processor, combine the artichokes, 1 cup of the Parmesan cheese, mayonnaise, garlic, cayenne pepper, and lemon zest. Process until smooth, scraping down the sides of the bowl. Season with kelp. Spread the mixture in an even layer in a small ovenproof gratin dish. Sprinkle with the remaining 1 tbsp. Parmesan cheese and bake for 15 minutes, or until the top is golden. Serve warm. Makes about 1 $1/3$ cups.

Vegetable-Artichoke Paella

Here is a nice takeoff on the classic Spanish dish with saffron-flavored rice, minus the shellfish, chicken, and sausage. If you want to stick with tradition, however, include some grilled shrimp or fish in the menu.

Needed: Vegetable cooking spray; $2/3$ cup chopped onion; $2/3$ cup diced red bell pepper; 2 garlic cloves, minced; 1 cup frozen artichoke hearts, thawed; 1 $1/2$ cups tightly packed torn fresh spinach; $1/2$ cup mineral water; 2 (10 $1/2$ oz.) cans low-salt chicken broth; 1 $1/4$ cups uncooked jasmine rice; $3/4$ tsp. salt; $1/2$ tsp. Hungarian sweet paprika; $1/4$ tsp. saffron threads; 1 cup frozen baby lima beans, thawed; $1/3$ cup frozen green peas, thawed.

Coat a large saucepan with cooking spray, and place over medium-high heat until hot. Add onion, bell pepper, and garlic, and sauté 3 minutes. Add the artichokes; sauté 2 minutes. Next add the spinach, mineral water, and broth; bring to a boil. Stir in the rice and next 3 ingredients on the list. Cover, reduce the heat, and simmer for 15 minutes. Then stir in the lima beans and peas; cover and cook an additional 10 minutes or until the liquid has been entirely absorbed. Remove from the heat; let stand, covered, 5 minutes. Then fluff up with a fork. Serves 7.

Artichoke and Sun-Dried Tomato Pizza

Needed: 1 package dry yeast; $1/2$ cup warm water (105° to 115° F.); 1 $1/2$ cups all-purpose flour, divided; 1/4 tsp. salt; vegetable cooking spray; 3 tbsp. sun-dried tomato tidbits; $1/4$ tsp.

oregano; 1/4 tsp. dried basil; 1/8 tsp. salt; 1/8 tsp. coarsely ground pepper; 1 (14 1/2 oz.,) can (no-salt-added) stewed tomatoes, undrained and chopped; 1 garlic clove, minced; 1 1/2 tsps. cornmeal; 3/4 cup drained, coarsely chopped canned artichoke hearts; 1 cup (4 oz.) shredded Italian provolone cheese.

Dissolve the yeast in warm water in a large bowl, and let stand for 5 minutes. Stir in 1 1/4 cups flour and salt to form a soft dough. Turn the dough out onto a lightly floured surface. Knead until smooth and elastic (5 min.), and add enough remaining flour, 1 tbsp. at a time, to prevent dough from sticking to hands. Place dough in a large bowl coated with cooking spray, turning to coat top. Cover and let rise in a warm place (85° F.), free from drafts, 1 hour or until doubled in bulk. Combine tomato tidbits and next 6 ingredients in a small saucepan; bring to a boil. Reduce heat, and simmer, uncovered, 20 minutes or until reduced to 1 1/3 cups, stirring occasionally. Remove from heat; let cool. Punch dough down, and roll into a 12" circle on a lightly floured surface. Place dough on a 12" pizza pan coated with cooking spray and sprinkled with cornmeal. Crimp edges of dough with fingers to form a rim. Cover; let rise in a warm place 30 minutes. Spread tomato mixture evenly over prepared crust, and top with artichoke hearts and cheese. Bake at 500° F. for 12 minutes on bottom rack of oven. Remove the pizza to a cutting board, and let stand for 5 minutes. Yield: 6 servings of 1 wedge each.

Asparagus
(Asparagus officinalis)

Brief Description

Asparagus was cultivated in ancient times by the Romans. The vegetable is a member of the lily family, and grows like weeds on the seacoasts of England and in the southern parts of the USSR and Poland, where the tundra steppes are literally covered like a carpet with this garden delicacy. Cattle and horses graze on it with delight.

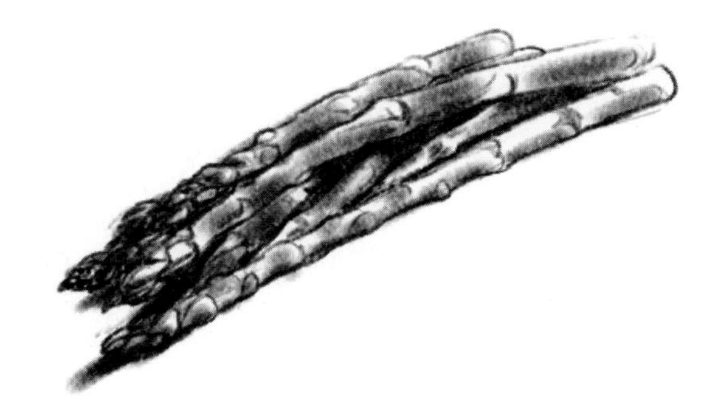

Blemish Remover

For those bothered with blackheads, pimples and general facial and lip sores, this simple preparation might do the trick in getting rid of these problems. Tie 24 large spears into two separate bundles of 12 each. Trim even. Stand butts down in preheated boiling water up to about 1 $^1/_2$" below the tips. Simmer for half an hour uncovered until tender. Store cooked spears in refrigerator and use in the recipe below. Save the asparagus water, however, and cleanse the face morning and night with it.

Remedies Kidney Problems

Cooked asparagus and its watery juices are very good for helping to dissolve uric acid deposits in the extremities, as well as inducing urination where such a function might be lacking or only done on an infrequent basis. Asparagus is especially useful in cases of hypertension where the amount of sodium in the blood far exceeds the potassium present. Cooked asparagus also increases bowel evacuations.

Handy Casserole

This has been adapted from *Recipes to Lower Your Fat Thermostat* with permission of the publisher.

Luscious Layered Casserole

Needed: 1 cup sliced onion; 1 cup chopped green pepper; 1 cup sliced mushrooms; $^{1}/_{2}$ lb. sliced potatoes; 1 cup thinly sliced carrots; $^{1}/_{3}$ cup raw brown rice; 1 $^{3}/_{4}$ cups short parboiled asparagus spears (use those in remedy above); 3 $^{1}/_{2}$ cups stewed, mashed tomatoes (you can replace the tomatoes with onion soup and turn this into a different casserole).

Sauté the onion, pepper, and mushrooms in a lightly oiled frying pan set on medium heat. Use either olive oil or lecithin from your local health food store to oil pan and baking dish. Next place sliced spuds in a 2 $^{1}/_{2}$ quart baking dish. Then alternate layers of carrots, rice, onion mixture and asparagus. Finally, stir kelp into mashed tomatoes and pour over the vegetables. Cover and bake in preheated oven at 350° F. for 2 hours. Serves about 8 people.

Hot Asparagus–Broccoli–Pea Soup Drink

This recipe is courtesy of the Vita-Mix Corporation, makers of the world's most popular, economical, and reliable juicer/whole food machine.

Needed: $^{1}/_{2}$ cup peas, frozen or fresh, steamed; $^{1}/_{2}$ cup broccoli, steamed; $^{1}/_{4}$ cup fresh asparagus, steamed or canned; 1 tbsp. onion; $^{1}/_{2}$ tbsp. Kyolic liquid garlic extract; $^{1}/_{4}$ tsp. chicken bouillon or soup base; $^{1}/_{2}$ cup chicken broth, hot; 1 tbsp. low-fat cheddar cheese. Place all the ingredients in the Vita-Mix whole food machine container in the order given. Secure the complete two-part lid by locking under both tabs. Move the black speed control lever to HIGH. Lift the black lever to the ON position and allow the machine to run for 2 $^{1}/_{2}$ minutes, until smooth. Makes 1 $^{1}/_{2}$ cups. (See the Appendix for more information on wakunaga's kyolic garlic and vita-mix).

Avocado
(Persea americana)

Brief Description

The avocado tree, which is related to the laurel, grows in semi-tropical climates. Orchards occur from Santa Barbara, California, all the way to Lima, Peru. Today, southern California harvests about 600 million avocados each year. Giant prehistoric ground sloths feasted on ripe avocados, rapidly packing masses of oily flesh into their mouths, and later defecating seeds with hardly a sense of their passing. The famous Amazon plant specialist, Richard Spruce, wrote that he was acquainted with wild jaguars deep in the rain forest, that would sometimes gather around an avocado tree, "gnawing the fallen fruit and snarling over them as so many cats might do."

Lowers Blood Cholesterol

Patients at the V.A. Hospital in Coral Gables, Florida ranging in age from 27 to 72 were given $1/2$ to $1\ 1/4$ avocados per day. Twice a week blood samples were taken. Fifty percent of them showed a definite decrease in serum cholesterol from 8.7–42.8%. Eating half an avocado every other day would probably help your own cholesterol drop some.

Avocado-Chamomile for Psoriasis

A rather remarkable twofold approach towards relieving the itchy misery of psoriasis is by eating half of an avocado daily and applying an extra-rich cream of chamomile flowers extract to the skin. The oils in the avocado will work internally toward the surface of the skin, soothing deep muscle inflammation. The oils in CamoCare Soothing Cream help the skin to literally repair itself from the damage done by psoriasis. CamoCare is available at most health food stores.

An Ancient Mayan Beauty Secret

While working at an archaeological site several years ago near the Honduran-Guatemalan border, I noticed that all the Chorti women (descendants of the ancient Maya) rubbed their hair and bodies with an oil to keep them soft and resilient.

Through our interpreter, I learned that they were using avocado oil to keep their skin from getting burned by the hot, glaring sun and the rough elements of wind and rain. They even rubbed some on their lips to keep them nice and moist.

Some the Chorti women seemed to be in their late twenties or early thirties. Imagine my utter astonishment when my interpreter told me that most of them were in their mid to late *fifties!* Now I'm a pretty good judge of age because of my training in anthropology, but their constant use of avocado oil sure fooled me about how old I *thought* they were.

You too, can experience near ageless beauty again simply by using avocado oil in place of other lotions and creams.

A Quick Laxative Recipe

My father, Jacob Heinerman, uses ripe avocados regularly as a fast-acting laxative. He'll peel two of them and mash the meat up in a dish, adding a little kelp, 3 tbsp. apple cider vinegar, and 1 tsp. lemon juice. After mixing them together, he'll spread the mixture on some sprouted cracked wheat or pumpernickel bread and eat it.

Not only does it make incredibly delicious sandwiches, but usually within just a couple of hours or less it will promote a pretty vigorous bowel movement. He seldom has constipation as a result of this in spite of being 74.

Two More Avocado Recipes

THE ULTIMATE GUACAMOLE DIP

Needed: 4 large peeled, pitted avocados; 7 tsp. peeled, grated onion; $1/_8$ tsp. cayenne (optional); $1/_2$ tsp. kelp; two 8 oz. cans of peeled tomatoes; 4 tbsp. plain yogurt; $1/_2$ tsp. lemon juice; $1/_2$ tsp. Worcestershire sauce. Mash all the ingredients together in a large mixing bowl and whip until well blended and smooth. Chill before serving. Use natural corn chips from your local health food store for dipping.

AVOCADO AND GRAPEFRUIT SALAD WITH POMEGRANATE SEEDS

Needed: 2 cups loosely packed torn romaine lettuce; 2 cups loosely packed torn chicory; 2 cups grapefruit sections (about 2 large grapefruit); $1/_2$ small red onion, sliced and separated into rings; 1 medium avocado, peeled and sliced; 1 cup pomegranate seeds; 2 tbsp. honey; 2 tbsp. fresh lime juice.

Place $1/_2$ cup romaine and $1/_2$ cup chicory on each of 4 salad plates. Divide the grapefruit, onion, and avocado evenly among the plates; top with pomegranate seeds. Combine the honey and lime juice, stirring with a wire whisk. Drizzle 1 tbsp. of the dressing over each salad. Serves 4.

BAKING SODA

Brief Description

In a world contaminated by dangerous chemicals and in a society thoroughly disillusioned by expert studies on everything from sunshine to hot dogs, at least one substance remains pure, wholesome, and always dependable. I'm referring to sodium bicarbonate, of course, which the public is rediscovering in droves. You would probably know it better by the stubby, school–bus–colored box it comes in labeled "baking soda."

It is emerging as a modern-day miracle substance. New product labels proclaim to consumers that baking soda can whiten your teeth, freshen your cat's litter box, deodorize your sweaty armpits, clean your laundry, tame your smelly shoes, shine your appliances, maintain your household plumbing, extinguish your small fires, and battle your gorilla breath.

But, then, it seems it always could do these things and a lot more. Our grandmothers knew the virtues of baking soda when our own parents were still in diapers. Put simply, it's cheap and it works. Sodium bicarbonate is an inorganic salt that occurs naturally in ground water and mineral deposits, including vast deposits in southwestern Wyoming. It is also manufactured by the human body. A white, crystalline solid or powder, sodium bicarbonate acts as a mild abrasive for cleaning, but it also buffers acidity and alkalinity to make it effective as a deodorizing and leavening agent.

Allergy Relief Guaranteed

There are many types of allergies that evoke different symptoms in those who are very sensitive to their environment. One of the more common consists of unpleasant odors that tend to make people sneeze. A little bit of baking soda goes a long ways in bringing immediate relief to the situation.

Remove onion, garlic, and other food odors from your kitchen cutting board by rubbing it with baking soda. Rinse with clear water. Swish baking soda in vacuum bottles and plastic containers—or let the solution stand in them overnight—to remove lingering odors. Sprinkle in your cat's litter box to reduce offensive urine odor. Also, sprinkle on pet bedding, let sit 15 minutes and vacuum. Clean pet accidents on your carpet with club soda; allow to dry thoroughly. Then sprinkle on baking soda, let sit for 15 minutes, and vacuum. Soak the diapers of infants or the enfeebled elderly in a solution of 2 quarts water and $^{1}/_{2}$ cup baking soda, or sprinkle liberally in the container holding disposable diapers. These and other offensive odors that may aggravate existing asthma, bronchitis, or hayfever and induce watery eyes, plugged sinuses, and a runny nose can all be helped by using baking soda liberally, which will purify the air around you.

Skin Sores and Body/Mouth Odors Cleared Up

Apply a little bit of baking soda to canker sores or chickenpox sores for relief of pain or itch. First moisten the area, and then apply some of the powder on the surface and leave for awhile.

Sprinkle some baking soda on your wet toothbrush to whiten teeth and freshen breath. Dissolve 1 tsp. baking soda in half a glass of water and swish between teeth to neutralize mouth odors without leaving a "cover-up" scent. You can even brush your dog's teeth with baking soda and water solution to eliminate that familiar "doggy breath."

Rash and Insect Bite Pain Relieved

Make a paste of baking soda (3 parts baking soda to 1 part water) or add $^{1}/_{2}$ cup baking soda to lukewarm bath to reduce itch or pain from insect bites and stings, poison ivy, poison oak, or prickly heat.

BAMBOO SHOOTS
(Bambusa arundinacea)

Brief Description

Bamboo is a member of the grass family and is especially abundant in the monsoon area of East Asia. Bamboos are the largest grasses, sometimes reaching 100 feet in height. The stalks are nearly always rounded, jointed, occasionally thorny, and hollow or solid with evergreen or deciduous leaves. Some types die after fruiting, and some don't flower until they are about 30 years old.

In many places bamboo is used as wood for construction work, furniture, utensils, fiber, paper, fuel, and innumerable small particles. When I've been in places like Manila, Bangkok, and Jakarta, I've noticed that the erected scaffolding around new building projects is always bamboo lashed securely together with plenty of ropes. I've slept on beds and drank out of cups made from bamboo.

Bamboo is an important food for man and beast alike. The pandas of mainland China subsist almost entirely on bamboo. In Asian cooking, the fibrous shoots are highly prized for their mild, unpretentious flavor and their crunchy texture. In Japan, bamboo shoots are available on a year-round "fresh-cooked" basis. But here in America they are usually only obtained in canned form from Oriental food stores and some larger supermarkets.

Before using canned bamboo, be sure to wash very thoroughly after slicing each shoot in half lengthwise. There is a grainy white residue in the interior ridges of the shoots—a result of the commercial preparation process—which has a sour flavor and must be removed. Once opened, canned bamboo will keep for about 10 days if stored in a covered container in the refrigerator. Be sure to change the water every other day.

Bamboo is an extremely hardy plant. To illustrate just how tough this grass is, consider the following true incident. On August 6, 1945 an atomic bomb was dropped on the Japanese city of Hiroshima with an estimated equivalent explosive force of 12,500 tons of TNT. The bomb caused widespread death, injury, and destruction. A tremendous shock wave accompanied this terrific explosion, followed by intense neutron and gamma radiation, both of which were very damaging to living tissue. About 90% of the city was instantly leveled and some 130,000 people either killed, injured, or declared missing.

Much of the city has since been reconstructed, but a gutted section was set aside as a "Peace City" to illustrate the devastating effects of this horrible weapon. A large museum details the horror and history of this explosion. Two very unforgettable displays held my attention longer than the rest. The first was a large color photograph of a stand of bamboo about a half mile from the epicenter of the blast, which somehow managed to survive albeit in a rather frazzled condition. The other one was the partial remains of a concrete staircase with the carbonized form of what had once been a human being, driven with tremendous impact into the cement by the terrific heat that instantaneously vaporized the individual.

Bamboo Shoots for Peptic Ulcers

Throughout Japan, there are many medical doctors who practice a form of alternative health care known as *kampo* in addition to their regular medical practices. Kampo medicine relies a great deal on more natural folk remedies which are safer for the body. One such remedy often prescribed to patients suffering from peptic or duodenal ulcers is bamboo shoot tea. Simmer $^{1}/_{4}$ cup diced bamboo shoots in 2 $^{1}/_{2}$ cups water on low heat for 20 minutes. Cool to lukewarm, strain, and drink on an empty stomach twice daily.

Bamboo Broth for Hiatal Hernia

When I went to Japan some years ago with others as guests of Wakunaga Pharmaceutical Co. (makers of Kyolic garlic), we had a chance to visit the city of Sapporo (famous for the beer of the same name) on Hokkaido Island in the very north. One of our number suffered from a hiatal hernia; every time she ate something she would experience sort of a "reverse regurgitation" that would leave an unpleasant burning sensation deep in her esophagus.

One of our official guides recommended that she try some *Takenoko Suimono*, which translated simply means "bamboo broth or soup." She was soon served some in a pretty, decorated porcelain cup, which she was told to sip slowly. Within minutes she pronounced herself as feeling much better and went on about her usual business of sight-seeing with no further discomforts.

The Japanese make this soup by boiling several bite-sized pieces of fish fillet or part of the fish head (eyes included) and $1/2$ cup chopped bamboo shoots in 1 pint of water, flavored with a pinch of granulated kelp. It can be consumed before regular meals or sipped lukewarm for medicinal purposes.

Delicious Recipe

Takenoko Gohan (Bamboo Rice)

Needed: $1/2$ lb. bamboo shoots; $1/2$ lb. boned chicken leg, with skin; 3 $1/3$ cups short-grain rice, washed.

Needed (for boiling the rice): 4 cups chicken stock; 1 tsp. granulated kelp; 2 tbsp. light soy sauce; 2 tbsp. sake; $1/2$ sheet toasted *nori* seaweed.

Halve the canned bamboo shoots lengthwise and wash thoroughly. Cut them into half-moon slices. Wash again and drain in a colander. Parboil them for 2 minutes, then rinse again in cold water. Drain. Next debone the chicken leg, cut the meat into $1/2$" cubes, parboil for 2 minutes, and wash in cold water to eliminate the excess fat. Drain also.

Next put the washed rice into a heavy, tight-lidded pot. Mix the liquid for boiling it and pour over the rice. Stir in the bamboo and chicken. Cover and bring to a boil over medium heat, then raise heat to high and cook till most of the liquid is absorbed. Reduce the heat to very low and cook, covered, until all the liquid is absorbed; this takes about 25 minutes. Turn off the heat, line the underside of the pot lid with a kitchen towel, cover, and let rest about 15 minutes before serving. To serve, gently fluff the rice with a wooden spoon and serve in individual bowls. Garnish with crumbled toasted *nori* seaweed available from any Oriental food store. Yield: Serves 4.

BANANA
(Musa sapientum)

Brief Description

Banana is an herb that grows up to 20 feet or more in height. It has a stout, cylindrical, succulent pseudostem arising from a large, fleshy corm. This corm sends up a series of suckers, forming clumps. Bananas are native, in various forms, from India and Burma through the Malay Archipelago to New Guinea, Australia, Samoa, and tropical Africa. It's universally cultivated in tropical regions.

Curious Uses for Banana Peel

A familiar trick used in silent screen comedies was for someone to slip on the proverbial banana peel.

But more sober and serious uses for banana peels appear in folk medical literature throughout the world. Green banana peels are grated and dried or else burned to ash in Curacao and then applied to cancerous sores, herpes lesions, and diabetic leg ulcers with some good effects. In Trinidad a poultice of ripe banana peel is applied to the forehead and back of the neck to relieve excruciating migraines. In the Bahamas a decoction of fresh green peel is taken as a remedy for hypertension. Also the inner surface of the ripe banana skin may be applied directly to burns, rash, and boils for healing relief. Curiously enough, the inside of the peel also makes a dandy shoe polish for scuff marks.

M.D. Cures over 200 Wart Cases with Ripe Peel

One of the most remarkable folk treatments I've ever encountered has to do with the successful removal of plantar warts.

Matthew Midcap, M.D. shared with me some of his firsthand experiences with ripe banana peel, which he uses in his clinical practice in Morgantown, West Virginia.

Dr. Midcap says this cure will work on all kinds of warts and "has always been 100% effective thus far."

> I first read about this banana treatment in the Dec. 1981 issue of the *Journal of Plastic Reconstructive Surgery*. A short article by an Israeli physician from the city of Safad related how he treated a single patient—a 16-year-old girl with painful plantar warts—with just ripe banana peels. The doctor in Wheeling with whom I was working in a clerkship capacity while still a medical student then, decided to join me in conducting further experiments based on this one recorded episode. His name is Dr. Phillip Polack.
>
> One of our very first patients was a male Caucasian, 48 years old. He's a prominent banker in Wheeling and loves to play golf. But a cluster of plantar warts on the bottom of one foot, about 2 inches in diameter, prevented him from enjoying his favorite sport. By the time he came to us, he had tried all the standard medical therapies available: acid therapy, cryotherapy where they freeze it off, surgery and even radiation. But nothing seemed to work, they just kept coming back. I cut a piece of ripe banana skin and applied the inside white mushy part against the warts, taping it down good with some adhesive tape.
>
> I instructed him to keep it on, removing only when bathing or showering. After which, he was to dry his foot well and apply another ripe peel. He came in each week to have us scrape away the old, dead wart tissue. Within less than a month we could see his warts were definitely getting smaller. In six months he was not only completely cured, but has never experienced another recurrence of them since then. He has been golfing a lot since then without any pain.

A Healing Fruit

In parts of Central America, cooked banana pulp is given as a remedy for diarrhea. In other places the mashed fruit pulp is bound around the neck to relieve sore throat and to reduce swelling of the adenoids. It's helpful in some cases of stomach ulcers, diverticulitis and colitis, but not in every instance. The oil in banana makes it a very difficult food to digest for some adults, especially the elderly. It needs to be *well* chewed and consumed slowly in small portions in order to digest properly.

A diet of bananas, according to an early 1950–1960 issue of *Postgraduate Medicine*, is ideal for treating celiac disease, an allergic sensitivity to the gluten in some grains and usually found in young children. Physicians at the Vanderbilt School of Medicine fed as many as 10 bananas a day to some of their young patients with remarkably good results. When treating a young child with this disease, be sure to consult with a proper medical or health authority regarding his or her entire diet, since things like candy and sugar are also restricted as a rule.

Healthy Snack Food for Adults

Several health benefits for older adults from banana consumption have been reported in the medical literature. In one study when a banana was given as a snack, either alone or with milk during the daily rest period, reported illness among industrial female workers decreased. The fruit was an ideal midmorning pickup for them. In a second report, improvement in morale and decreased absenteeism among clerical workers were observed when employees were given supplements of bananas. Workers who were given these natural fruit snacks were more cheerful and attentive and less tired on the job than those who didn't receive any bananas. The last experiment was conducted at a retirement home. 117 residents received a ripe banana each day for periods of 16–30 days. After awhile these residents no longer suffered any digestive disorders. Banana is also good to offset insulin shock in diabetics.

Building Incredible Strength and Size

Bananas are good for putting on extra pounds in order to acquire more muscle and strength for very physical types of competitive sports such as football or wrestling. Sports medical doctors who work with athletes usually prescribe foods like bananas, potatoes and rice along with the usual weight lifting and other muscle-building exercises in order that the extra weight is added as usable muscle rather than turned to mere flab. Japanese Sumo wrestlers eat tremendous amounts of rice and bananas in order to maintain their incredible girths (which is *all* muscle, by the way)!

For those needing to put on an extra 40 pounds or so, it's recommended that a couple of bananas each day either in a liquid drink or pudding or just plain straight, will help toward building additional muscle protein when accompanied with the usual exercises.

Scrumptious Banana Bread

A delicious banana–date bread recipe has been modified from one appearing in the Better Homes & Gardens' *Eating Healthy Cook Book*, which I recommend.

BANANA DATE BREAD

Needed: 1 1/2 cups all-purpose flour; 1/2 tsp. baking soda; 1/2 tsp. ground cinnamon; 1/4 tsp. baking powder; 1/4 tsp. ground nutmeg; 2 egg whites; 3/4 cup dark honey; 1/4 cup blackstrap molasses; 1 cup mashed ripe bananas; 1/4 cup olive oil; 1/2 cup chopped, pitted dates and some liquid lecithin from any health food store.

Stir together all dry ingredients. Next blend well the egg whites, honey, molasses and bananas; add oil and keep stirring. Then turn flour mixture into it. Add the dates. Lightly oil an 8 × 4 × 2 loaf pan with lecithin. Put batter into pan. Bake at 350° F. for 55 minutes. Cool in pan 10 minutes, then remove and cool on a wire rack.

A Delicious Dessert Treat

SPICED BANANAS WITH RUM SAUCE

Needed: 2 small, firm ripe bananas (about 1/2 lb.); 1 tbsp. butter; 1 tbsp. brown sugar; 2 tbsp. thawed apple juice concentrate; 1 tsp. vanilla extract; 1/4 tsp. ground cinnamon; 1/8 tsp. ground allspice; 2 tbsp. dark rum; 1 cup vanilla low-fat frozen yogurt.

Cut bananas in half crosswise, then lengthwise; set aside. Melt the butter in a large nonstick skillet over medium-high heat. Add the brown sugar, apple juice concentrate, vanilla, cinnamon, and allspice, and cook for 1 minute. Turn the banana pieces over, and cook an additional minute. Heat the rum in a small saucepan. Pour over the banana mixture. Arrange 4 banana pieces on each of 2 dessert places; spoon sauce over the banana pieces, and top each serving with 1/2 cup frozen yogurt. Yield: Serves 2.

Barley

(See under GRAINS)

Beans
(Phaseolus genus)
(See also under GREEN BEANS)

Brief Description

BLACK BEANS. These are small, oval beans with a tender texture and a mushroom-flavored, somewhat earthy taste to them.

BLACK-EYED PEAS. These oval. medium-sized beans have a nutty crunchiness to them and possess a fine-flavored taste reminiscent of a meatless vegetable stew.

CHICKPEAS. These round beans have a distinctive nutty flavor and chewy firmness to them.

FAVA BEANS. Such beans are large, flat and oval with a firm texture and a dainty taste. They are very popular in Europe, especially in Great Britain.

KIDNEY BEANS (Dark Red, Light Red, and White). All three colors of beans are oval shaped with a somewhat soft, bland taste to them. The dark red ones are mostly sold in cans and used in salads, while the light red are sold dry and made into chili, refried beans and creole dishes. An interesting item on kidney beans appeared in the October 30, 1994 issue of the *New York Daily News.* A man described as "a career criminal who loves kidney beans" was apprehended in the middle of a burglary at an upscale Fire Island, New York, home. The residents had arisen to check out noises in the house but found no one. However, in the vicinity of a closet door, they heard repeated flatulence and discovered Richard Magpiong, age 56, hiding in a closet. They quickly locked him in until the police arrived. As he was led away in handcuffs, he was heard to mutter beneath his breath; "Damned beans!"

LENTILS. These small, disk-shaped beans have a subtle mildness characterized with a distinctive flavor and rather firm texture. Under certain extreme situations they can supply a person with just about everything the body needs in the way of nutrients. Debi Holmes from backwoods Alabama knows all about this. She left her self-imposed wilderness exile in the west Utah desert in mid-December 1994 to return to civilization. This woman, who has been to all American states except Hawaii and to five foreign countries, decided to do a 40-day exodus into complete solitude in early November. She selected a site high in a west-facing canyon in Snake Valley about 10 miles from Fish Springs National Wildlife Refuge near the Utah-Nevada border (about 70 miles south of the gaming town of Wendover). She told me during her 40-day "soul quest" that she only ate 4 of 12 pounds of lentils and 6 of the 25 pounds of rice she had brought with her, besides cooking 6 pounds of flour. Holmes had less than a gallon of water left when her stay ended. She confessed that the rice and flour weren't enough to sustain her energy levels for very long. "Luckily for me I had those lentils," she said. Out of curiosity I asked her, "Why just 40 days?" Her prompt response: "Moses, Christ, Mohammed, all experienced retreats—and 41 days would have killed them," she said with a laugh. This woman with a degree from Cal Poly San Luis Obispo in California plans to write a book on her isolation odyssey "for a big-name publisher somewhere so it will become a national bestseller." Then pausing with a hungry look, she suddenly asked me, "By the way, you wouldn't know where I could get some French fries in a hurry?"

LIMA BEANS. These large, white oval beans date from 7000 to 5000 B.C. in Peru and yield a mild taste and a soft texture when cooked. Quite often, though, they are dried, canned and marketed under the names of wax or butter beans.

MUNG BEANS. Also known as green or golden gram in India, they are highly esteemed for their tiny seeds, which become rather sticky on cooking, but are accounted both wholesome and nourishing. These are dried and boiled whole or split, or else parched and ground into flour. In China they are added to green noodles and used for bean sprouts, a use to which they are also put here in America.

NAVY, WHITE and *GREAT NORTHERN BEANS.* All three of them are firm, mild beans in varying sizes, ranging from small to medium and large.

PINTO BEANS. These are small, oval beans with a mild flavor and texture to them.

SPLIT and *WHOLE PEAS* (Yellow and Green). Both varieties are small and possess a soft, grainy texture marked with a certain distinctive flavor.

SOYBEANS (*Glycine* or *Soja max*). These firm, round, bland-tasting beans are not of the genus *Phaseoulus* as the others happen to be.

Food to Make You Healthier Looking

Very few of us these days possess what might be correctly termed "a radiant glow of health." In a large part, its absence in many of our countenances is due to our poor dietary and social habits, and unwise ways of living. But by consuming varieties of beans more often, we can regain a healthier and more fuller look in the course of time.

The best direct evidence for this comes from the opening chapter in the Book of Daniel. We are told that "Daniel purposed in his heart that he would not defile himself with the portion of the king's meat, nor with the wine he drank." But instead asked Melzar, the prince of the eunuchs, "Give us pulse to eat, and water to drink." At the end of nearly a week and a half, Daniel and his three friends' "countenances appeared fairer and fatter in flesh than all the children which did eat the portion of the king's meat." "Pulse" is believed to have been either lentils or beans.

If you want to maintain a vibrant look of radiant health, then beans and lentils should be a frequent part of your diet.

Legumes Promote Vigor and Vitality

In ancient times beans and lentils were often associated with men and women of strength, or considered to be the standard fare of consumption in activities requiring great feats of strength and vigor. The Bible provides us with several remarkable accounts to this effect. Esau's "mess of pottage" made for him by Jacob is one example and is believed to have been lentil soup. II Samuel 23:11–12 is another instance in which lentils were associated with valor and courage.

Some clinical evidence exists to back up the vigor and vitality claims which have been assigned to various legumes. Chickpea (garbanzo bean) is a very important food staple in India where it goes by the common name of Bengal gram. According to Volume 7 of the *Journal of Ethnopharmacology* for 1983, the sprouted seeds of chickpea are "extremely nourishing and constitute a regular item of diet for athletes and professional wrestlers in India" and "is used as a food for horses which gives them an untiring stamina." This is due to the high content of pangamic acid or vitamin B15, which has been sold in health food stores nationwide as a stamina-builder of sorts. Additional studies concerning the strength-giving properties of B15 in other legumes and vegetables have appeared in Soviet, Hungarian, and Indian medical journals in the past.

Bean Juice for Constipation and Hyperactivity

When I was at a recent scientific symposium held at the University of Rhode Island in Kingston during July 1987, a chemist from Japan whom I was introduced to, explained to me how black bean juice is used in his country to correct constipation caused by eating too much white bread and refined foods and to calm hyperactive children. According to him, 2 tbsp. of cleaned black soybeans are boiled in 2 quarts of water for 10 minutes, then simmered until just 1 quart of water remains. Some kelp is added to season before the broth is strained. One cup of juice three times per day is recommended.

Possible Cancer Preventative

Certain enzymes called proteases break down proteins and also play multiple roles in the production and development of various cancer tumors. Beans and grains contain protease inhibitors (PIs), a variety of substances which block protease activity. When ingested as part of the diet, PIs interfere with these cancer-producing enzymes. These same PIs also prevent the growth of tumor cells. Additionally, they prevent the release of deadly oxygen radicals, thereby protecting against possible DNA damage and subsequent cancer. Finally, PIs prevent radiation-induced cancer and enhance tissue resistance to invasion by tumor cells. If people consume adequate amounts of PIs in the form of beans and grains, they can then be protected against cancer at a variety of sites.

Lowers Cholesterol and Triglycerides

The American Journal of Clinical Nutrition for October 1983 cited a number of recent studies indicating that all varieties of bean can definitely lower serum cholesterol and trigylceride levels in the body substantially. Enough so, in fact, to suggest they be reintroduced into our diets again on a more frequent basis in order to keep too much fat from accumulating in the circulating blood.

A possible dietary pattern to follow to bring this about, would be to have a bowl of oatmeal for breakfast three times a week and bean soup (without ham or sausage) several times at lunch. This combination of both is an ideal grain-legume mixture to fight cholesterol buildup. Variations to this theme might be oatmeal cookies or muffins for breakfast or snacks and baked beans or bean casserole for dinner.

A Diabetic's Delight

Published dietary studies show that beans can be a diabetic's delight. The *Indian Journal of Medicinal Research* for February 1987 reported that three types of legumes had a very significant effect on lowering blood sugar levels in diabetic patients. These reductions were directly related to the dietary fiber content of each legume. Because of this legumes ought to be included in the diet at least twice a week.

Help for Hypertensives

Dr. Louis Tobian, chief of the hypertension section of the University of Minnesota Hospitals, has been a busy man the last few years. You see, he's been conducting some very important studies concerning the role of potassium in treating high blood pressure.

Using stroke-prone rats that were bred to have dangerously high blood pressure, he fed half a regular diet and half a diet with potassium supplements. Only 2% of the rats on the high-potassium diet died from strokes, compared to 83% on the regular diet. His findings led him to speculate that potassium actually protects against strokes and kidney disease, and that his rat studies suggest the way a high-potassium diet may act in humans.

The October 1986 issue of *Sports Fitness*, which reported Dr. Tobian's work, recommended navy beans (1 cup = 750 mg.) or lima beans (1 cup = 1163 mg.), among other foods, as giving the body the potassium power it needs to prevent and reduce hypertension.

Potassium estimates for other varieties of cooked beans in 1 cup portions are as follows: red kidney, 629 mg.; pinto, 670 mg.; black-eyed peas, 573 mg.; chickpeas or garbanzos, 570–590 mg.; lentils, 500 mg. and soybeans, 972 mg.

"Meat" Protein Without the Meat

For those who are quite health minded and wish to reduce their intake of red meat somewhat without experiencing any loss of energy or strength, they might want to consider increasing their intake of beans instead. A registered dietician with the Stanford Heart Disease Prevention Program described beans as being "a power-packed, nutrition food." Elsewhere chickpeas, lentils and other legumes have been described as "ideal meat substitutes" and "unexcelled meat stretchers."

Boston Baked Beans Classic

Needed: 2 cups Great Northern or navy beans, soaked for 10 hours in 7 cups water; some ham hocks; 1 chopped white onion; 1 tbsp. lemon juice; 3/4 tsp. kelp (powdered seaweed from health food stores); 1 tbsp. dark honey; 2/3 cup blackstrap molasses; 1 tsp. pure vanilla; 1/2 tsp. pure maple syrup; 1 1/2 tbsp. catsup; 1/8 tsp. dry mustard; 1 chopped garlic clove.

Preboil ham hocks until tender. Remove meat from bones, adding it with a little of the juice to a large pan. While the ham hocks are cooking, simmer the soaked beans for 1 1/2 hours until tender but not broken. Skim off any foam on the beans prior to cooking. Put beans and half the juice into the large pan with the deboned ham pieces and their juice. Add the diced garlic and onion.

Mix together in a separate bowl the kelp, honey, molasses, vanilla, maple syrup, catsup and dry mustard, adding 1 1/2 cups boiling water. Stir well and pour over the beans. Cover and bake at 275° F. for about 7 hours, adding additional water if necessary. Stir with a wooden ladle every couple of hours. Uncover during the last 45 minutes to brown beans and meat a little. Be careful not to burn them. Serve warm with pumpernickel bread.

To get rid of the gas effect which beans produce, just soak them for at least 24 hours prior to cooking, adding $^1/_2$ tsp. ginger root for every 2 cups of beans. Another alternative is to take an antigas herb product called Ginger-up (2 capsules) with any meal that includes beans or may cause heartburn. It's available from Great American in St. Petersburg, Florida (see Appendix). Drinking acidophilus or buttermilk also helps to disperse gas.

Homemade Tofu

Tofu resembles a soft cheese and is a custardlike food made from soybeans in much the same way that cottage cheese is made from milk. It's quite mild tasting, and has been called the "food of 10,000 flavors" because it tends to borrow the flavor of the foods, sauces, and marinades it's prepared with. Tofu is now widely available in most supermarkets, having been popularized by the Japanese.

I am grateful to Mishio Kushi and his publisher (St. Martin's Press) for their generous use of his instructions for making homemade tofu as found in his excellent treatise.

Diet for a Strong Heart

Needed: 3 cups organic yellow soybeans; 6 quarts spring water; 4 $^1/_2$ tsps. natural nigari. Soak beans overnight, strain and grind in an electric blender. Place ground beans in pot with 6 quarts of water and bring to a boil. Reduce flame to low and simmer for 5 minutes, stirring constantly to avoid burning. Sprinkle cold water on beans to stop bubbling. Gently boil again and sprinkle with cold water. Repeat a third time. Place a cotton cloth or several layers of cheesecloth in a strainer and pour this liquid into a bowl. This is soy milk. Fold corners of the cloth to form a sack or place cloth in a strainer and squeeze out remaining liquid. Pulp in sack is called *okara* and may be saved for other recipes. In a blender, grind the nigari, a special salt made from sea water and available in many natural or health food stores nationwide.

Sprinkle powdered nigari over soy milk in a bowl. With a wooden spoon carefully make a large, X-shaped cut with two deep strokes in this mixture and allow to sit 10–15 minutes. During this time it will begin to curdle. The next step calls for a wooden or stainless steel tofu box (available in many natural food stores) or a bamboo steamer. Line box or steamer with cheesecloth and gently spoon in soy milk. Cover top with layer of cheesecloth and place lid on top of box or steamer so it rests on cheesecloth and curdling tofu. Place a brick or weight on the lid and let stand for an hour or until tofu cake is formed. Then gently place tofu in a dish of cold water for half an hour to solidify. Keep the tofu covered in water and refrigerate until used. Tofu will stay fresh for several days in the refrigerator. However, it's best to change the water daily.

Tofu is extremely versatile and has a variety of textures depending upon how it's cooked. It can be sliced, diced, cubed, pressed, or mashed and boiled in soups, sautéed with vegetables or grains, or baked in casseroles. It may also be used in dips, sauces, dressings, and desserts. The *okara* or pulp mentioned earlier can be added to soups or cooked with vegetables. A simple salad dressing can be made by mixing together in a food blender some tofu, sesame seed oil, lemon juice, kelp and tarragon. You've got a creamy dressing that's less expensive than commercial dressings made from dairy products, and has *only one-third* their calories.

A Meal That Stays with You

Curried Chickpeas and Black Beans

Needed: 2 tsp. olive oil; 1 cup chopped onion; 1 tbsp. minced peeled ginger root; 2 tsp. curry powder; 1 (14.5 oz.) can diced tomatoes, undrained; 1/2 tsp. granulated kelp; 1 (15 oz.) can black beans, rinsed and drained; 1 (15 oz.) can chickpeas (garbanzo beans), rinsed and drained; 1/2 cup chopped fresh parsley; 1 tbsp. lime juice.

Heat the olive oil in a large nonstick skillet over medium heat. Add the onion and ginger root; sauté 3 minutes or until tender. Stir in curry powder; cook an additional minute. Then add the tomatoes; cook another minute or until the mixture is slightly thickened, stirring occasionally. Add the kelp, black beans, and chickpeas; stir well. Cover, reduce heat, and simmer

5 minutes. Remove from heat; stir in fresh parsley and lime juice. Serve warm. Yield: 4 main dish servings in 1 cup amounts each.

Japanese-Style Salad

Bean Sprouts Salad with Soy Sauce Dressing

Needed: 1 $^1/_2$ cups bean sprouts; 1 cup carrots; 3 green bell peppers.

Needed (for soy sauce dressing): 1 tbsp. vegetable oil, 1 tbsp. sesame oil, 3 tbsp. soy sauce, 2 tbsp. diluted apple cider vinegar.

Mix the ingredients for the dressing first of all. Then wash the bean sprouts under running water and drain in a colander. Cut the carrots into 1″ long thin strips. Quarter the green peppers lengthwise and remove their seeds and ribs. Next, slice each quarter crosswise into long fine strips. Bring to boil 3 $^1/_2$ cups water containing $^1/_4$ tsp. salt. Put in the carrots and cook 1 minute, then add the bean sprouts and green bell peppers. When the water simmers again, turn off the heat. Drain off water through a colander, and quickly cool by fanning. Squeeze slightly to remove the excess water. Combine all ingredients and mix with dressing just before serving. Arrange in mounds in small salad bowls. Yield: Serves 3.

Beets and Swiss Chard
(Beta vulgaris, B. vulg. cicla)

Brief Description

Edible beet roots are related to the sugar beets (*Beta vulgaris saccharifera),* grown for the making of sugar, the extraction of which began in the eighteenth century. The edible roots of beets are of two basic colors, red and yellow. The English pioneered the red variety, but elsewhere in Europe the yellow-rooted kind was once much preferred because of its sweeter taste and greater suitability for pickling.

There is also a white-rooted beet which is cultivated chiefly for its leaves and stalks. The tops are used in place of spinach and the ribs are like asparagus in some ways. In the Middle Ages, no meal was considered complete without a soup made from the leaves of this Swiss chard, as it is often called.

Hospital Beet Therapy for Cancer

One of the most remarkable and tremendously successful programs for treating many different kinds of cancer tumors was commenced in the late 1950s by Alexander Ferenczi, M.D., at the Department for Internal Diseases at the district hospital at Csoma, Hungary, using nothing but raw, red beets. Portions of his intriguing medical success were recently translated from Hungarian and reprinted in the Australian *International Clinical Nutrition Review* for July 1986.

Dr. Ferenczi's clinical report included methods of administering the beets and several very important case studies:

> In D.S., a man of 50 years of age, a lung tumor was diagnosed by me, and subsequently confirmed in a Budapest hospital and also in a country hospital, which corresponded clinically to lung cancer . . . I started the treatment with beetroot in the described manner. After 6 weeks of treatment the tumor had disappeared . . . After 4 months of treatment he gained 10 kg. (22 lbs.) in weight, the erythrocyte (mature red blood cell) sediment rate (e.s.r.) was reduced from 87 millimeters/h to 77 mm/h. Thus he represented the symptoms of a clinical recovery.

A side-by-side comparison of two cancer patients, one on beet therapy and the other not, further demonstrates the efficacy of this marvelous treatment.

> We received simultaneously two patients for treatment. One suffered from cancer of the prostate and the other from cancer of the uterus. The body weight of both was the same. The patient with cancer of the prostate was treated with beetroot. The patient with cancer of the uterus could not take it, but remained in our ward. The condition of the man started to improve. When admitted, he was bedridden with a permanent catheter. After one month the catheter was removed. The patient walked around and put on weight, whereas the female patient lost weight. After 3 months, there was a difference in weight of 10.5 kg. (23.15 lbs.) between the two.

> Experience gained up to now points to the fact that beetroot contains a tumor inhibiting (anti-cancerous) active ingredient. However for the present no clue has been found as to the nature of this active substance. One thing is certain, that it is not very unstable because it also acts when taken orally; therefore digestion does not harm it. The very apparent red color may suggest that the active substance is the coloring matter. Treatment with beetroot presents several advantages over the rest of the medication used in the treatment of cancer. Firstly, because it is nontoxic and one can administer red beetroot in unlimited quantities. Also there are unlimited supplies of beetroot at our disposal. We have therefore endeavored to administer to the patient this active substance in the most concentrated form and in the largest quantities possible, because the beetroot or rather the juice could not be given in larger quantities.

Now beet root is available to consumers several different ways. One Lawrence, Kansas firm, Pines Int'l, makes a very nice organic red beet root concentrate. This beet powder is available at most local health food stores.

One, however, has to be careful with the amount of beets consumed at any given time—certainly not because they're harmful, but rather due to their incredibly strong ability to quickly break up cancer in the body. A woman in her thirties who was treated with beet root for breast cancer contracted a fever of 104° F due to the rapid breakdown of the tumors. In instances such as this, beets clean up the cancer faster than the liver is capable of processing all of the wastes dumped into it at any one time. Consequently, the internal administration of beet root needs to be staggered somewhat, and closer attention given to detoxifying the liver and colon at the same time the beet root therapy is commenced.

Dr. Ferenczi concluded his medical report with this undeniable fact: "The results achieved with beetroot are no worse than those with well-known chemical preparations, such as those with Tetramin (an experimental anti-neoplastic)." He attributed the anticancer strength in beets to their natural red coloring agent, betaine.

Iron Fortification for Women

Several very good sources of iron for women, besides dessicated liver tablets, egg yolks, legumes, and iron-fortified cereals, are red beet root and Swiss chard. One to 2 level teaspoons of Pines' beet powder added to an 8 oz. glass of water or juice supplies a lot of iron. The balance may come from a variety of other foods, including cooked Swiss chard, dark Romaine lettuce, parsley, poultry, and fish. Beets, along with carrots and parsnips, are also good for hypoglycemia.

Anticancer "Quack Salad"

I'm indebted to Dr. James Duke, head of the USDA's Germplasm Resources Laboratory in Beltsville, Maryland, for letting me feature here a slightly revised version of the cancer preventative salad that he calls Quack Salad. You'll need the following ingredients: 1 cup washed, unpeeled, raw, grated red beet; handful of chopped walnuts (unpackaged, unsalted); 3/4 cup coarsely diced celery; 1/2 cup washed, snipped endive; 1 medium to large washed, unpeeled and sliced cucumber; 1/4 tsp. cumin; 1 tbsp. flaxseed; 1 peeled, chopped garlic clove; pinch of powdered cayenne pepper; 1/2 peeled, chopped white onion; handful of shelled, chopped peanuts (not canned, salted or fried); 1/2 tsp. sage; 2 medium-sized, washed, quartered, ripe tomatoes.

Lightly toss everything together in a large wooden salad bowl until thoroughly mixed. To make the dressing, add 1/4 tsp. kelp and 1 finely minced garlic clove to 2 1/2 cups of lemon juice. Mix well and pour over the salad.

Swiss Chard Plaster for Relieving Coughing, Glandular Swelling, and Sore Throat

Swiss chard makes a wonderful plaster for relieving persistent coughing, glandular swelling, and sore throat. It also helps to remove accumulated toxins from the body through the skin. It works best when used in combination with other vegetables.

To make an effective drawing poultice, first cut the leaves of Swiss chard very fine. If not in season or readily available, the leaves of spinach or mustard greens or even kale will do nicely.

Next add a little grated Pontiac (red) potato. Figure on using about one quarter the amount of the Swiss chard leaves. Mix the cut leaves and grated potato with $1/4$ teaspoon fresh grated ginger root. Or the equivalent amount of black pepper with a little water added to give it the consistency of grated ginger root, can be substituted instead. Add just enough whole wheat flour to give the mixture a nice, even consistency. Then spread this mixture $1/2$" thick on a piece of clean cotton flannel and apply directly to the skin. Secure in place with a gauze bandage. Leave on overnight or for 6 hours.

Russian Beet Soup

During my visit to the Soviet Union in the summer of 1979, I was fortunate enough to meet an old gentleman in Leningrad whose father had been the personal chef of the last of the great Russian czars, Nicholas Romanov. Through an interpreter I learned a marvelous beet-cabbage soup (called borscht in Russian) that his father prepared for the Romanovs.

Palace Borscht

Needed: $3/4$ celery; 1 $1/4$ cups raw beets; $3/4$ cup carrots; $3/4$ cup finely chopped green onions; 1 cup shredded cabbage; 3 tsp. grated ginger root; 12 oz. of roast duck (or chicken or turkey) carcass; 2 $1/2$ cups beef stock; $1/4$ tsp. kelp; 3 tsp. caraway; $3/4$ tsp. finely shredded orange peel; $3/4$ tsp. basil; $3/4$ tsp. thyme; $1/2$ cup tomato purée; 1 $1/2$ cups apple cider, $1/2$ cup vodka.

Place poultry carcass in oven and brown for a while. Bring beef stock to a boil. Add the poultry carcass. Reduce to low heat and simmer for one hour before draining. Cut the celery, beets and carrots julienne style, that is, into strips about 2" long and $1/4$" square, and add to the stock, with the chopped green onions, spices and tomato purée. Stir together well with a wooden ladle, then simmer for another 30 minutes Add the finely shredded cabbage, apple cider, and vodka. Simmer another 15 minutes. Serves about 6.

BERRIES

Brief Description

The following is a selected list of berries known for both their nutritional as well as medicinal values.

BLACKBERRY (*Rubus villousu*). Known by their deep purple-black fruits.

BLUEBERRY (*Vaccinium gaylussacia, V. corymbosum*). The cultivated kind of a blue-black color used principally for its food value.

BOYSENBERRY (*Rubus* species). A huge blackberrylike fruit with a raspberrylike flavor. Named after Rudolph Boysen, their originator.

CRANBERRY (*Vaccinium macrocarpum*). A low, creeping shrub common to New England states and boggy areas. Distinguished by its bright red berry. Most commonly used around holidays like Thanksgiving and Christmas.

CURRANTS (Black, Red) (*Ribes nigrum, R. rubrum*). Small red, black, and even white berries closely related to gooseberries, with a tart flavor.

DEWBERRIES (*Rubus canadensis*). Part of the blackberry family and regarded as one of the tastiest of the entire *Rubus* species. The hybrid youngberry was developed from it.

ELDERBERRY (Sweet, Black, Red) (*Sambucus canadensis, S. nigrum, S. racemosa*). A small shrub or tree yielding red-brown to shiny black berries.

GOOSEBERRY (Garden) (*Ribes grossularia*). The small fruit looks like a little green basketball with a stem because its skin has striated lines that appear to divide the berry into uneven sections. Has a tartness to it.

HAWTHORN BERRY (*Crataegus oxyacantha*). Many species found around the world. Some are tastier than others. Berries can be red, purple or nearly black, depending on the species, and are found on either trees or large bushes.

HUCKLEBERRY (*Vaccinium myrtillus*). Same as the blueberry, only more noted for its medicinal uses. Fruit can be either blue-black or red. Both kinds of blueberries are more closely related to the cranberry.

JUNIPER BERRY (*Juniperus connumis*). An evergreen shrub found in dry, rocky soil throughout North America, Europe, Asia and even the Arctic Circle. Fruit is a berrylike cone which is green the first year and ripens to a bluish-black or dark purple color in the second year.

LOGANBERRY (*Rubus loganobaccus*). Developed in California by a Scotsman. A hybrid from raspberry and blackberry. The fruit is large, long, dark red in color with a strong tart flavor to it.

OLLALLIEBERRY (*Rubus species*). Originally from Oregon but grown extensively in California, it's a cross between the black loganberry and youngberry, and is bright black, medium size, firm and sweeter than loganberry is.

MULBERRY (*Morus alba, M. rubra, M. microphila*). The white mulberry occurs throughout New England, the red in the Appalachias and the Texas variety in that state, with many hybrids as well. Tart, but juicy and tasty.

RASPBERRY (Red, Black) (*Rubus idaeus, R. crataegifolius, R. occidentalis*). Red raspberry produces a spring and fall crop, with the latter being sweeter on account of the cooler weather (unless the spring is cool, too). The red raspberry has less seeds and is juicier than the black variety. The black is darker and its shape is more odd, being that of a skull cap rather than the ball shape of the red kind. Its season is only four weeks (June–July).

ROSE HIPS (*Rosa* species). Over 100 kinds in the world. Fruit resembles a berry, but is actually a ripened hypanthium (an enlargement of the torus below the calyx). Noted for its strong vitamin C content.

SALMONBERRY (*Rubus spetabilis*). Its name comes from its yellow color. Common to the northwestern United States. Has a strong, sour taste. Better cooked than raw.

STRAWBERRY (Cultivated, Wild) (*Frangaria ananassa, F. vesca*). Well-known enough that it needs little or no description to speak of.

THIMBLEBERRY (*Rubus parviflorus* or *R. nutkanus*). Common throughout the Pacific Northwest, its raspberry-red berries are shaped like a thimble, hence the name. It has an especially soft texture like juicy velvet that almost seems to literally melt in your mouth.

General Health Benefits

One thing nearly all berries are good for are as tonics for rejuvenating both the heart and blood. A major medical journal noted that an extra serving of fresh fruit, such as berries, each day may decrease the risk of stroke by as much as 40%, regardless of other known risk factors due to their high potassium content.

For another, they are remarkable cleansing agents serving as effective stimulants for the bladder and colon. Then too they apparently seem to make ideal accompaniments to heavy meals consisting of fatty meats.

Finally, it can be said that most berries have varying degrees of antiviral activity to them. Volume 41 of the *Journal of Food Science* for 1976 carried an interesting article, "Antiviral Activity of Fruit Extracts," in which it showed that poliovirus was inactivated by strawberry extract. Several other fruits such as raspberries, blueberries and wild cranberries helped to inactivate other intestinal viruses, including herpes simplex virus. Berries are also very purifying for the blood, cleansing for the skin and increase the beauty of your complexion. Following are some entries for most of the berries listed here and several health benefits to be expected from each, although some of them may contain a dozen or more uses.

Blackberry for Bowel Regulation

Blackberry is one of those remedies which seems to work both ways in correcting bowel disorders. A lady from Costa Mesa, California, related her personal experience with blackberry:

> When my youngest daughter was about 6 months old she had diarrhea. I took her to a doctor and he prescribed a medication for her which didn't help. Grandmother came to visit and she told me to put about

1 tsp. of allspice in a cheesecloth bag, and simmer it in unsweetened blackberry juice for a few minutes. When cool, I gave my daughter a teaspoonful about every 4 hours. Within 24 hours the complaint was checked and in 48 hours she was completely over it. I've used this same remedy on the rest of the family whenever they have diarrhea, only in larger doses, and have found it works miracles.

Blueberry for Noninsulin Diabetes

There is a substance in blueberries called mytrillin that reduces blood sugar as insulin would. To make an antidiabetes tea for the nonuser of insulin, steep 1 tsp. of the cut, dried leaves in a cup of hot water until lukewarm. Drink 1 cupful four times daily. Clinical evidence shows that several *Rubus* species (blackberry, blueberry, raspberry, etc.) have produced a noticeable decrease in blood sugar levels in diabetic rabbits and humans. But the tea made from the berries' leaves needs to be used two to three times daily on a regular basis or else glucose levels soar. This hypoglycemic effect may be due to an increase in the liberation of insulin in the beta cells of the pancreas. Blueberries and the leaves also exhibit antidiarrhea properties as blackberries do.

Cranberry for Kidneys

If you have any kind of kidney problems, then you should be drinking cranberry juice every single day! At least this is what a number of major medical journals and some doctors have to say about the subject. One report noted that 60 patients with acute urinary tract infection were given 2 cups or 16 fl. oz. of cranberry juice per day for 3 weeks with over 70% of them showing moderate-to-excellent improvement in their conditions. *The Journal of Urology* for 1984 revealed "that cranberry juice is a potent inhibitor of bacterial adherence" in the urinary tract.

Cranberry juice is also good for dissolving kidney stones. So said a U.S. Navy doctor in the January 3, 1963, *New England Journal of Medicine*. He wrote then, "I have found that an 8 oz. glass four times daily for several days followed by 1 such glassful twice daily is valuable therapy in stone-forming patients." A personal testimony to this effect comes from a friend of mine, Charles Eady of Harahan, Louisiana (a suburb outside of New Orleans).

> Four years ago (1981) in carrying a heavy display case in an awkward position, I managed somehow to dislodge a kidney stone I'd apparently had for a long time. I began urinating blood and having severe pains as a result of this. I began drinking cranberry juice from the health food store and in 3 days the problem was completely cured—no more blood, burning or pain.

Recently there has come onto the market a heavily concentrated cranberry formula for cleansing the kidneys. The powdered, encapsulated product is based in part, on an old Pennsylvania Dutch remedy from the Amish country around Lancaster, Pennsylvania. It may be obtained by mail order from Old Amish Herbs in St. Petersburg, Florida. Or an equally nice dried cranberry juice product under the Nature's Way label is available from your local health food store. Three capsules of either on an empty stomach twice daily is recommended.

Currants Work Health Miracles

Black, red, and white currants all manifest strong antiseptic properties—enough so, in fact, that they can be used in the treatment of Candida yeast infection, some forms of cancer, whooping cough, multiple sclerosis, and various skin diseases. They're also an excellent antidote for any kind of ptomaine food poisoning, especially from meat.

Black currant fruit and the berry seeds both contain the rare and badly needed gamma linolenic acid (GLA), which only occurs in mother's milk and evening primrose. Black currant constitutes one of the richest natural sources of GLA yet discovered.

What kind of health benefits then can we expect from it? Well, stronger immune and central nervous systems for one. And, for women, a relief from possible premenstrual syndromes, which include migraines and menstrual cramps. Also for both sexes, stronger hearts and improved circulations, with considerably less bad cholesterol in the blood that would clog major arteries. Probably this is why Eskimos eat a lot of berries such as currants with their varieties of fatty meats. Hypertension and arthritic inflammations also receive definite improvements when GLA-rich currants (either fresh or frozen) or currant juice are consumed on a more frequent basis.

Elderberries for Inflammation and Coughing

Fresh elderberry juice evaporated into a syrup (on low heat) and mixed with lard or a creamy base, makes a good ointment for burns. A tea made by lightly cooking the berries is a soothing lotion for sore eyes. An extract is available from health food stores.

To make a good tonic for the throat, bring to a boil 1 quart of elderberry juice with 1 tbsp. each of cloves, nutmeg, and cinnamon. Replace the evaporation loss with water. Thirty minutes later, strain, then add $1/2$ cup each dark honey and blackstrap molasses, boil and skim, cool and refrigerate. Can be diluted or used straight for coughs, sore throats, and lung irritations.

Gooseberry Remedies from Pioneer Days

Pioneer women in the old American West made a tea from gooseberry fruit and leaves to help cure any uterine difficulties incurred from too many childbirths. Some pioneer women's answer to a cold was to add a little red current jelly to a glass of whiskey and to give this to the patient just before sleep. The berry juice from either black currants or gooseberries mixed in with a little honey was regarded on the frontier as an almost infallible remedy for throat irritation. Rocky Mountain Indian squaws had a hankering for gooseberries during early stages of their pregnancy, much as modern women might experience unusual cravings for pickles and ice cream. Any kind of acute skin inflammation, ranging from erysipelas to poison ivy rash could be treated in those days by making an infusion from the ripe gooseberries, straining them well and then rubbing that lotion on for immediate relief.

Hawthorn Is Good Heart Medicine

Hawthorn is a valuable drug for the treatment of various heart ailments and circulatory disorders, including angina. You can get hawthorn berries from your local health food store. About 2–3 capsules per day is sufficient for adequate maintenance of the heart. A tea can also be made by soaking 1 tbsp. of crushed berries in 1 1/2 cups cold, distilled water for 8 hours, then quickly boiling and straining. Sweeten with honey and drink when lukewarm. As always with serious ailments of this sort, seek and follow the advice of a trusted health care professional of your choice.

Nothing Like Juniper for Wounds and Sores

Juniper berries are used to flavor gin and alcoholic bitters. A strong tea made of the berries (8 tbsp. berries in 1 quart boiling water steeped 1 hour) makes a great remedy for scalds, burns, sores, and all kinds of infected wounds when they're washed thoroughly with the tea several times a day.

A few years back on our farm in the wilderness desert of southern Utah, I accidentally cut my hand badly on some rusty barbed wire. Like lightning, the thought flashed into my mind to have someone pick some juniper berries from a nearby tree, pound them into a poultice, and apply it to my wound. This was done and my hand bound up with a clean, wet cloth. By that night, the throbbing pain had ceased—and come the next morning, the laceration was on its way to healing very nicely.

Hawthorne tea can be inhaled when lukewarm to relieve nose, throat, and lung congestion. Just drape a heavy towel over your head and sit with your face held about 8″ above the pot the tea is in.

Raspberry Leaves for Easy Births

A Utah Mormon mother took raspberry leaf tea to make her deliveries a lot easier. Prior to this, a great deal of pain had attender her labors. But when she switched to using the tea throughout her pregnancy, little or no pain attended her. She relates that in the recovery room with her "were several other young women who had just given birth also, moaning and groaning," but she felt just fine.

She took a cup each day during her 9-month pregnancy, and about 4 cups of strong, *hot* tea prior to entering the hospital. Contractions started in just a couple of hours and her delivery was a snap and virtually pain-free. To make the tea, bring 4 cups of water to a boil. Remove from heat and add 6 tbsp. to dried raspberry leaves. Steep 40 minutes Drink cool twice daily to help curb morning sickness and very hot just prior to entering labor. This tea is also excellent for curbing nausea and morning sickness.

Rosehips for Pneumonia

Most of us recognize that the vitamin C from rosehips is good for fighting colds and flu. But it's outstanding for pneumonia, bronchitis, and other respiratory problems as well.

If fresh rosehips aren't readily available to you, then make your own tea with dried hips from any health food store. Keep in mind, however, that a lot of the vitamin C has disappeared during the drying process and more escapes during the tea making. One company that has been apparently able to retain a goodly portion of ascorbic acid in its Old Fashioned Rosehips capsulated product is Old Amish Herbs out of St. Petersburg, Florida (see Appendix.)

But hips still have quite a bit of minerals in them and other antibiotic principles to knock out a cold with, which aren't affected by heat. Heat a quart of water to lukewarm. Remove from heat and add 6 tbsp. of rosehips. Steep for 20 minutes or so. After straining, add some honey, stir well, then drink right away. Or you can take 3–4 rosehip capsules from any health food store several times a day with some warm juice or water.

Strawberries for Clean Teeth and Beautiful Skin

Nothing else quite cleans the skin so well as strawberries. A California herbalist I know, Kathi Keville, shared with me her strawberry facial cream:

Kathi's Strawberry Facial Cream

Needed: 1/2 cup fresh strawberry juice; 1 cup almond oil; 1/2 oz. beeswax; 1 tbsp. lanolin. Melt the beeswax and lanolin into the oil. Cool to body temperature or just a little warmer. Warm the juice to the same temperature and whip the two together until cool. Pour into a jar and let set. The strawberry scent is easily lost under the lanolin, so a few drops of essential oil may be desired to fragrance the cream.

Both strawberries and raspberries make excellent dentifrices—an almost perfect means of preventing tartar from settling on the teeth. When either berry is cut in half, the fruit is then rubbed over the tartar-covered teeth or mashed to a pulp and gently applied with a wet soft-bristled toothbrush. For best results, the juice should stay on the teeth for as long as possible and then only be rinsed with a little warm water.

To make a good berry mouthwash, simply add 1/4 cup honey to a cup of berry juice and bring to a boil for a few seconds. Cool then add 1 1/2 cups distilled water and 1/2 cup of thyme tea and 4 drops of peppermint oil. Gargle and rinse the mouth with this. Kills all kinds of infection and leaves a nice, sweet taste and odor.

Blackberry
(See under BERRIES)

Vita Drinks From Vita-Mix

The Vita-Mix Corporation makes the world's finest whole food machine. The following recipes are taken from their *Recipes and Instructions* book and used here by permission.

Berries with Honey and Milk

Needed: 1/2 cup fresh strawberries; 2 tsp. honey; 1/2 cup low-fat milk; 1/2 cup ice cubes.

Place all ingredients in the Vita-Mix Total Nutrition Center container in the order given. Secure the complete two-part lid by locking it under both tabs. Move the black speed control lever to HIGH. Lift the black lever to the ON position and allow the machine to run for about 15 seconds. Makes 1 1/4 cups of delicious drink.

Orange-Cranberry Yogurt Drink

Needed: 1/2 orange, peeled, seeded and cut in half; 1/4 cup fresh or frozen cranberries or whole berry cranberry sauce; 2 tbsp. low-fat vanilla yogurt; 1/2 cup skim milk; 1/2 cup ice cubes.

Place all the ingredients in the Vita-Mix container in the order given. Secure the complete two-part lid by locking under both tabs. Move the black speed control lever to HIGH. Lift the black lever to the ON position and allow the machine to run for 20 seconds. Makes 1 3/4 cups of a smooth yogurt shake.

Blueberry Surprise

Needed: 1/2 cup fresh or frozen blueberries; 1/4 cup low-fat blueberry yogurt; 1/2 cup lemon sherbert; 1 tsp. honey.

Place all the ingredients in the Vita-Mix container in the order given. Secure the complete two-part lid by locking under both tabs. Move the black speed control lever to HIGH. Lift the black lever to the ON position and allow the machine to run for 25 seconds. Makes 1 1/4 cups of good-tasting drink.

BLACK (WHITE) PEPPER
(Piper nigrum)

Brief Description

Black pepper is the dried full-grown but unripe fruit, while the white kind is the dried ripe fruit with the outer part of the pericarp removed by soaking in water and then rubbing it off. It's less aromatic than black, but has a more delicate flavor. Major producers of both kinds include India, Indonesia, Malaysia, and mainland China. They should not be confused with red or cayenne pepper which comes from the *Capsicum* species.

Nifty Frostbite Remedy

During my 1979 sojourn to mainland China, I happened to get up into the cold parts of Inner Mongolia. There I picked up a nifty remedy for treating frostbitten fingers, toes, ears, and nose I'd like to pass on to you.

Well in advance of the coming of winter, soak 2 tbsp. of whole black peppercorns and 1 tbsp. each of coarsely grated horseradish and ginger roots in 1 1/4 cups of white wine for a week. Then filter and strain, storing away in a tightly stoppered bottle in a cool, dark place until needed. At which time, using either a basing or artist's brush, paint afflicted body parts generously with this solution to bring immediate comfort and relief from stinging pain.

Great Insecticide

Black pepper has been proven toxic against a number of agricultural and household pests, including ants, potato bugs, silverfish, some roaches and moths. Sprinkle ground pepper in those areas where such insects frequent. A strong tea makes a good spray for keeping aphids and cut-worms off tomato and cabbage plants. In 2 quarts boiling water, put 5 tbsp. peppercorns and 2 tbsp. chopped garlic cloves. Simmer down to 1 quart for 2 hours. Cool, then use on plants.

Remedy for Montezuma's Revenge

When traveling abroad in foreign countries where the purity of the water is somewhat questionable, chronic diarrhea and abdominal cramps often result when such unpurified water is drunk by unsuspecting visitors. A handy remedy for this is to carry along with you some black pepper and kelp. Then whenever such intestinal discomforts suddenly hit you, just mix 3 tsp. of black pepper and 1 $^1/_2$ tsp. of kelp in 2 cups of *boiled* water that has been allowed to cool first. Repeat as often as necessary.

BLUEBERRY

(See under BERRIES)

BOYSENBERRY

(See under BERRIES)

BRANS

Wheat Bran (Triticum aestivum)
Sorghum Bran (Sorghum vulgarie)

Brief Description

Bran is a relatively inexpensive and abundant source of dietary fiber, being the coarse outer coat or hull of the grain of wheat, separated from the meal or flour by sifting or bolting.

Sorghum bran, on the other hand, comes from sorghum, a major cereal crop in many Third World and developing nations. In fact, it's the world's third largest cereal food grain. Sorghum is a coarse kind of grass with a stalk very similar to that of corn.

A Bran the Celiac Patient Can Tolerate

Celiac disease is an acute sensitivity to certain fractions of wheat gluten that may be expressed in such symptoms as chronic diarrhea, constant intestinal problems, possible anemia, and the like. Patients with celiac disease are placed on a totally wheat-free or gluten-free diet.

An interesting study reports that sorghum is able to increase defecation and stool softness as much as wheat does, yet lacks the gluten that's harmful to those suffering from celiac disease. It could serve as an ideal laxative bran for them in place of wheat. It's also of value in colitis or other chronic gastrointestinal inflammations which might be aggravated by the increased harshness of bran particles. Sorghum bran may be obtained from some larger health food stores or specialty outlets that import particular food items from other countries.

Free from Western Diseases

In the early 1970s South African rural blacks eating more than 50 grams of dietary fiber per day were relatively free from appendicitis, colon diseases like diverticulosis, polyps, hemorrhoids or cancer, coronary artery disease, diabetes, and hiatus hernia. But when they moved to the big cities and adopted Western eating preferences for more refined foods, then they began to suffer from these same

Western diseases. Bran is a valuable addition to your diet to help keep you from getting many of these ailments.

How Bran Works as a Laxative

Here is a rather novel way of describing bran's effects in the colon in nonscientific jargon for lay people to easily comprehend. Fiber passing through the intestines is somewhat like a wet sponge, absorbing and holding not only water and toxicants but such compounds as bile acids, which in turn might modify cholesterol metabolism. The sponge, due to its great bulk, also increases the size of the stool and decreases the emptying time of the colon.

Bran Prevents, Heals Ulcers

Wheat bran eaten by healthy volunteers on a regular basis has a greater buffering effect on gastric acid juices in the gut than refined carbohydrate foods do. Which suggests some protection against the development of duodenal ulceration.

In another experiment, 21 patients aged 15–70 years diagnosed with solitary rectal ulcers consumed at least 6 tbsp. of coarse wheat bran on a daily basis for an average of nearly a year (10.5 months). Fifteen patients were completely healed, indicating a 71% effectiveness for bran in regard to this type of ailment.

Curbing Hunger with Bran

Have a problem with the munchies? Always nibbling and getting fatter as a consequence? How about trying a little bran every day to reduce your appetite for those sinfully delicious but naughty no-no's? A 1983 Swedish experiment tested bran on 135 members of a weight loss club in Stockholm, discovering that if they took an extra helping of bran just before mealtime, it greatly reduced their hunger feelings.

You might try this either with a slice or two of stone-ground whole wheat toast (lightly buttered) or with 1 tbsp. of bran stirred into a glass of some kind of juice, taken before your next meal. You'll find out that you'll be eating *a lot less* as a result. And *less* fatty, sugary foods means *less* pounds put on.

A Good Cancer Preventative

No less authority than Dr. Varro Tyler, head of the Schools of Pharmacy, Nursing and Health Services at Purdue University, believes that bran can help prevent certain kinds of cancer. In his book, *Honest Herbal*, he explains how it probably works: "(1) Bran dilutes, in the large amount of water held by the fiber, any carcinogens which might be present and (2) decreases the contact time with potentially damaging substances, since the larger stool is expelled more rapidly."

Bran Good for Diabetes

Both adult and juvenile-onset diabetes mellitus can be helped by the frequent consumption of either wheat or sorghum brans. One clinical study noted that wheat bran can exert a mild reduction in the insulin needs of young, insulin-dependent diabetic subjects. An improvement of glucose tolerance was noted in diabetic patients after ingestion of bran, according to another report.

Based on the evidence just given, it's recommended that diabetics take about 1 $^1/_2$ tbsp. of bran per day, either mixed in with something like plain yogurt or vegetable juices such as V-8 or carrot-and-mixed-greens. Over an extended period of time, a definite improvement of some kind should become noticeable.

Best Tonic for Irregularity

There's no two ways about it—bran *is* the best tonic you could ever use for correcting the problem of irregular bowel movements. In one fascinating medical study, "before and after" results were given of stool weights in those who ate refined food, then added about 1 tbsp. of bran twice a day to their diets a week later. Below is a simple table illustrating the dramatic changes in the stool weights of eight healthy male doctors and medical students between the ages of 25–43 without and with bran in their weekly diets.

	Experiment A		Experiment B	
	Without Bran	With Bran	Without Bran	With Bran
Wet weight	107 ± 44	174 ± 51	126 ± 47	215 ± 22
Dry weight	26 ± 9	41 ± 9	31 ± 10	47 ± 6

The physicians and surgeons from Western General Hospital in Edinburg, Scotland who compiled this report, estimated that without bran intake, a 176 lb. man passing only 3 $1/_2$ oz. of stool per day would excrete fecal material equivalent to his body weight about every 2 years. But by the ingestion of about 2 tbsp. of bran daily this output would be achieved in only 12 months. And the coarser the bran particles are, the more water-holding capacity or laxative properties it's going to have.

A superb FarmLax used by some horse-and-buggy Amish in Ohio contains a unique bran blend and is available through Old Amish Herbs (see Appendix.)

Two Guaranteed Laxative Cereals

There are many prepared cereals on the market today claiming to be high in fiber and so forth. But why take the manufacturer's word for it? Be on the safe side and know what you're getting. Go ahead and make your own! Here are two recipes to choose from.

FRUIT/NUT GRANOLAX

Needed: 2 cups rolled oats; 3 cups wheat germ; $1/_2$ cup shredded wheat; $1/_2$ cup bran; $1/_2$ cup chopped dates; $1/_2$ cup raisins; 1 cup chopped dried apples; $1/_2$ cup shelled cashews; 3 cups dry milk; $1/_2$ cup sunflower seeds; $1/_2$ cup pumpkin seeds; $3/_4$ cup chopped dried papaya. Get a good cutting board and either a Chinese vegetable cleaver or a sharp French knife and start chopping and slicing everything up into tidbit portions (the size depends on your own preference).

Needed: $1/_2$ cup honey; $1/_8$ cup pure maple syrup; $1/_8$ cup blackstrap molasses; 1 tbsp. pure vanilla; 1 tsp. cinnamon; 2 tsp. cardamom. Mix well together by hand with a wooden ladle in a large bowl.

Combine the dry ingredients together with the liquid. Stir until uniformly mixed. Rub a shallow pan with some lecithin (from health food store), then spoon out mixture in an even layer. Bake for 1 $1/_2$ to 2 hours at 225° F. Bake a little longer if a dry, crunchy consistency is desired. Makes approximately 13 cups. Have a bowl of this every morning for breakfast and for a midnight snack.

Basic Granola

Needed: 8 cups rolled oats; 2 cups wheat germ; 1 cup shredded coconut; $^1/_2$ cup dark honey; 1 cup water; 2 tbsp. pure vanilla.

Stir the dry oats, wheat germ and coconut together in a large mixing bowl with a wooden ladle. Next mix together in a separate bowl the honey, water and vanilla. Then add it to the dry ingredients and mix well. Pour contents into a large, greased cake pan. Bake at 275° F. for about 1 hour and 15 minutes, depending on how crisp you like it. You can add raisins and chopped, pitted dates to the hot granola after taking it out of the oven if you wish.

Dried Fruit Granola

Needed: 3 cups regular oats, uncooked; $^1/_2$ cup sliced almonds; 1 $^1/_4$ cups honey; $^1/_2$ cup nutlike cereal nuggets; 1 tbsp. butter, melted; 1 tsp. ground cinnamon; 2 $^1/_2$ cups flaked whole-grain wheat cereal with almonds and raisins; $^1/_2$ cup dried cranberries; $^1/_2$ cup dried tart cherries; $^1/_2$ cup dried banana chips.

Combine the oats and almonds, and spread the mixture evenly in a 15″ × 10″ jelly roll pan. Bake at 325° F. for 15 minutes or until toasted, and set aside. Combine the honey, cereal nuggets, butter, and cinnamon in a large bowl. Add the oat mixture, and stir well. Spread this honey-oat mixture evenly into the jelly roll pan. Bake at 325° F. for 20 minutes or until dry. Let it cool, then break into large pieces, and set aside. Combine the whole grain wheat cereal and dried fruits in a large bowl, and stir in the honey-oat mixture pieces. Store the granola in an airtight plastic container. Serve with goat's milk. Yield: 20 one-half cup servings.

BREAD
(See also GRAINS)

Brief Description

One slice of pumpernickel, rye, or wheat bread is about equal to 8 slices of white bread for increasing stool output. "The best way to increase intake of fiber-rich food," writes Denis Burkitt, M.D., a prominent British physician, "is to increase bread consumption, making sure that white bread is replaced by bread made from flour that is as near as possible to wholemeal. Whole wheat bread has three times the amount of dietary fiber compared to white bread."

Satisfying the "Munchies" with Bread

Ever get the hungries and feel like snacking on foods that put on more weight? Well, two slices of whole wheat bread will fill you up for several hours over a lot of other refined foods. White flour products are virtually worthless. This can be illustrated by the fact that it takes eight loaves of white bread to have the same laxative effect on the colon as does just one loaf of whole wheat bread. So having a peanut butter sandwich on two pieces of whole wheat or two slices of whole wheat toast lightly buttered with some cinnamon and cardamom sprinkled across them can be a nice way of satisfying a growling stomach without adding too many extra calories.

Bread Poultice Checks Tetany

A case I recall some years ago in Lander, Wyoming, involved a young girl whose ankle was bitten by a pit bull terrier. An elderly gentleman down the street happened to hear of the incident and called on the distraught parents. He asked them for permission to treat her leg. They knew of his reputation as a local healer of sorts and agreed to it, not fully convinced themselves that they should subject their young girl to the further trauma of a hospital or doctor's office.

He immediately went to work and made a poultice by soaking some slices of slightly moldy bread in condensed milk and applying the same to their daughter's ankle with some strips of clean gauze taped on the ends with adhesive tape. The next morning he changed the dressing, and later that night all redness and inflammation had disappeared for good.

The Healing Aroma of Freshly Baked Bread

A therapeutic aspect seldom or never discussed in any self-care book that I'm aware of concerns the psychological and emotional benefits to be derived just from the aroma of freshly baked bread. While Freud or your favorite analyst may not have recommended this for mental problems, I do simply because I've witnessed the success of it on several occasions for myself.

Some cultures actually believe that the smell of fresh bread has medicinal importance. The January 1983 issue of *Natural History*, for instance, notes that "freshly baked bread, piled high on the body of a malaria sufferer, is believed to have the power to relieve malaria attacks," on the Italian isle of Sardinia. And in the Sardinian village of Esporlatu, people on the street will swear to the fact that "the smell of fresh bread is so potent it can keep sickness and death far away."

Some years ago I had an opportunity to spend some time interviewing several psychologists and therapists at the Utah State Hospital in Provo. One case, in particular, which was called to my attention then, compelled me to investigate this matter further in later years to my own satisfaction. In the case described to me, a middle-aged woman who had been committed for severe reactive depression wasn't responding at all to any of their given therapies. And the medication she was getting each day didn't really help that much either. She finally began to come out of her doldrums when they assigned her to work in the hospital kitchen with the baker, helping to make bread. When asked to report to her therapists what was making her feel better, she responded by telling them that the aroma of the freshly baked loaves gave her an indescribable exhilaration of sorts—so much so in fact, that she began to feel and think there was hope for herself after all. About 2 $^1/_2$ months later she was discharged, apparently helped by the smell she was exposed to for several hours each day, as well as working her frustrations out in the kneading and punching of the dough. After this she never required further therapy again.

Making a Good Loaf of Bread

The following recipe appeared in the Dec.–Jan. 1978 issue of *Quest* magazine.

Rye Bread

Needed: 2 packages active dry yeast; $^1/_2$ cup warm water 115°–125° F; 1 tbsp. honey; 1 $^1/_2$ cups dark malt beer; 2 tbsp. butter; 1 tbsp. salt; 1/4 cup plain yogurt; 2 $^1/_2$ cups dark rye flour; 1 cup pumpernickel flour (or $^1/_2$ cup rye and $^1/_2$ cup whole wheat); 1 cup gluten flour; 1 cup all-purpose white flour; 1 tbsp. egg white mixed with 1 tbsp. water; cornmeal.

In a large bowl dissolve yeast in warm water and stir in honey. Heat beer until warm enough to melt the butter in it; add salt and yogurt. Cool to lukewarm, then mix with yeast liquid. Add all the flours, except for about $^1/_2$ cup of the white. Mix well, then turn out on a floured surface and knead, adding the reserved flour as necessary, for 5–10 minutes.

When dough is smooth, although it may still be slightly sticky, put it in a buttered bowl, turn it once, cover it, put it in an 85° place (inside a cool oven with a pan of steaming water underneath works well), and let it rise until double in size (about 2 $^1/_2$ hours). Shape into 2 oval or round loaves and place on a baking sheet sprinkled with cornmeal. Let rise again, lightly covered, until double in size, about 1 $^1/_2$ hours. Preheat oven to 375° F.; bake for 45–50 minutes. It will sound hollow if hit with a knuckle. Cool on a rack. Makes 2 loaves.

BREADFRUIT/JACKFRUIT
(Artocarpus communis; also called A. altilis or A. heterophyllus)

Brief Description

Artocarpus is the large genus of tropical trees and shrubs to which breadfruit and jackfruit (a variety of breadfruit) belong. This genus is a member of the mulberry family, which is one of four families of the flowering plant order Urticales, or nettle order. This morphologically distinctive and evolutionarily advanced group of woody flowering plants include such important and well-known herbs as slippery elm, stinging nettle, hops, and marijuana, and other fruits like figs.

The breadfruit itself is a tall evergreen having milky juice and bearing fleshy fruit. It is indigenous to the Malay Archipelago and spread from this region throughout the tropical South Pacific region in prehistoric times. Its introduction into the New World was connected with the memorable voyage of Captain William Bligh in *H.M.S. Bounty*, a voyage recommended by Captain James Cook, who had seen the breadfruit in the Pacific islands and believed that it would prove highly useful as a foodstuff for slaves in the West Indies. After the failure of Bligh's first voyage, a second expedition resulted in the successful establishment of the tree in Jamaica, where it failed to live up to expectations because the slaves preferred bananas and plantains. The breadfruit native to tropical Africa is *Treculia africana*.

The breadfruit is not a fruit in the popular sense of the word; it contains considerable amounts of starch and is seldom consumed raw. It can be roasted, baked, boiled, fried, or dried and ground into flour. It is one of the highest yielding of food plants, with a single tree able to produce more than 800 grapefruit-sized, seedless fruits every season. Breadfruit constitutes one of the important staple foods in many tropical regions. In the West Indies and on the American mainland from Mexico to Brazil, this handsome evergreen is grown in dooryards, and the fruit is sold in local *mercados* (marketplaces).

Paleontologists have unearthed fossils of the genus to which breadfruit belongs in the Cretaceous rock formations of frigid Greenland. By comparing these extremely ancient remains of breadfruit with living descendants in tropical regions of the world, scientists were able to determine clearly that there was a much warmer climate for that glacial land some 100 million years ago.

Jackfruit is a popular variety of breadfruit now grown in most tropical zones. It can get very large, acquiring a weight of over 20 pounds, but smaller ones are also marketed. Jackfruit is irregularly shaped but roughly oval, and the densely spiny skin ripens from green to brown. The rather musty sweet aroma intensifies as the fruit ripens.

Unlike regular breadfruit, which is seldom eaten uncooked, jackfruit can be consumed raw when ripe and sweet. It usually accompanies curries and other meat dishes. The numerous edible white seeds can be dried and ground to provide a type of flour or roasted and eaten like chestnuts, in which case they should be boiled first (and the water discarded).

Wonderful Paste for Herpeslike Skin Disorder

Between 1984 and 1986 I traveled extensively throughout Indonesia. At various times I worked with faculty members of Airlangga University, in the city of Surabaya, and the University of Indonesia, in the capital of Jakarta. I also coordinated some of my research efforts with government scientists. My mission was to investigate the wonderful plant lore accumulated over many generations from a number of diverse ethnic groups in the country. I painstakingly recorded most of these remedies for inclusion years later in some of my best-selling food and herb encyclopedias. One remedy worth relating here involves the use of unripe breadfruit for treating a skin condition which slowly spreads like herpes.

The prickly brown skin is peeled away with a sharp knife and the inner starchy pulp then grated as one might do with a peeled potato. A little coconut oil and some powdered turmeric root and then mixed with the breadfruit to form a paste of even consistency.

This is then spread over the herpeslike sores and left in place for 24 hours, before being washed away and treated again in similar fashion. The sores on several patients whom I observed undergoing this treatment at different stages appeared to me to be clearing up very nicely. In larger U.S. cities such as Los Angeles, New

York, or Miami, there are certain markets specializing in such exotic fruits. But in the rest of the country where breadfruit isn't readily available, a potato from Idaho or Maine will serve as an effective substitute for the same remedy, provided the other two ingredients remain the same.

Severe Constipation Easily Remedied

As I've traveled the world over, the single most frequent category of remedies I find are those designated by different cultures as effective laxatives. Constipation, while more prevalent in technologically advanced Western societies, still seems to exist no matter where you go. In the United States alone, over 50% of *all* herbal products sold in health food stores are for solving infrequent or difficult evacuation of the feces.

While that may be good news for those who rely on such things to assure regularity, these products are not without their problems. In the course of time, the body can become dependent upon these natural laxatives. A more reasonable alternative employed in other societies is the consumption of foods that promote bowel movements *without* becoming habit forming. Dates, figs, and prunes are some of these.

So is jackfruit. In July 1994 I went to the Philippines in the role of a medical anthropologist for additional folk medicine research. One outstanding remedy there for successfully inducing a gently efficient bowel movement is the consumption of some jackfruit. The pulp (called *lamûkot*)) surrounding the numerous seeds is a golden yellow, richly sweet, and pleasantly aromatic. It is perhaps one of the most delicious laxatives I've ever encountered.

Anyway, I tried some for myself on several different occasions when a few days had elapsed without any bowel movements on my part (probably due to the stress of my trip and indiscriminate eating habits). And I can report that within *fewer than* 45 minutes, the sudden urge to find a toilet came upon me without warning. Jackfruit works quite well for constipation, thank you!

BROCCOLI
(Brassica oleracea italica)

Brief Description

Broccoli is a member of the cabbage family and an annual that produces flowers and seeds the same year. It produces bunches of flowering buds on the terminal stems as well as on branches that continue to be formed throughout the fall months. These buds are a great delicacy. Because of the high vitamin content, even the young leaves are eaten as well as the tender parts of the stem. Broccoli is probably the forerunner of cauliflower and was developed to its high degree of perfection by Danish gardeners. The plants are coarse and grow to a height of 4 feet. It does better in cool weather, particularly cool nights, after the heads begin to form. It is usually grown as a fall crop, following an early crop of spinach or snap beans.

Ideal Food for Recuperation from General Illness

Broccoli is a vegetable that is becoming more popular every year. It has a very high food value and has been recommended by some alternative-minded doctors to patients who are slowly recuperating from general illness. One physician in northern California told me awhile back that both the vegetable *and* the water in which it has been cooked are wonderful for helping the body to mend rapidly from long-term debilitating diseases or surgery.

When the stems and buds are cooked, they become tender in just a few minutes. Many people make the mistake of over-cooking broccoli, which causes the heads to fall apart and much of its flavor to be lost. It is better to undercook it and serve with a little bit of melted butter and a dash of granulated kelp—a seaweed obtainable from health food stores.

Broccoli for Breakfast, Lunch and Dinner

A James Bond Breakfast

The world's most famous secret agent prefers breakfast to any other meal of the day, according to information scattered throughout all the Ian Fleming and John Gardiner novels. And it seldom varies—orange juice, strong coffee without sugar, scrambled eggs, toast and frequently bacon. Occasionally, though, Bond has indulged himself in a little breakfast delight consisting of steamed broccoli and soft-boiled eggs.

Steam 4 stalks of broccoli until tender, and cut into thin strips. Soft-boil 4 eggs for 4 minutes. Lay the broccoli strips on a plate. Shell the eggs and chop them in a bowl. Blend in 4-6 tbsp. of lime juice with kelp to taste. Spread this over the broccoli, then garnish with paprika and a little finely chopped parsley. Serves 4.

Spicy Broccoli Salad

1 1/2 lb. broccoli; 3 tbsp. chopped red bell pepper; 3 tbsp. chopped red onion; 3 tbsp. rice wine vinegar; 1 tbsp. sesame oil; 2 tsp. light brown sugar; 1 tsp. red pepper flakes; granulated kelp to taste.

Cut off broccoli florets. Trim and peel stems; cut into 1/2" thick slices. Place broccoli florets and stems in a steamer basket over boiling water; cover and steam for 2 to 3 minutes, or until cooked but still crisp. Refresh under cold water. Drain well. In a serving bowl, stir together red peppers, onions, vinegar, oil, brown sugar, and red pepper flakes. Just before serving, add the broccoli and toss to combine. Season with kelp. Serves 4.

AGENT 007'S BROCCOLI HARLEQUIN

Albert R. Broccoli gained fame for 15 James Bond films produced over the past 25 years. In addition to that he belongs to that famous Italian family which brought the seeds of this now famous vegetable named after them, from Italy to America at the turn of the century.

Broccoli, himself, at the age of 78 offers this recipe.

Needed: 1 small head cauliflower; 2 small bunches broccoli; 4 tbsp. butter; 2–3 tbsp. plain yogurt; 2 tbsp. grated Parmesan cheese; 1/4 tsp. salt; freshly ground black pepper to taste (I recommend substituting kelp for the salt and pepper); 3/4 cup rye or pumpernickel breadcrumbs.

Break the cauliflower into florets and remove most of the white stalks. Steam the cauliflower for 3-4 minutes or until still firm. Steam the broccoli for about 6 minutes or until soft enough to purée. Preheat oven to 350° F. Purée broccoli with butter and yogurt. Lightly grease a 6 cup, ovenproof dish with olive oil, mound the cauliflower in it and sprinkle with the cheese, salt and pepper (or kelp). Season the broccoli purée with salt and pepper (or kelp); then spoon it over the cauliflower and sprinkle the top with the breadcrumbs. Bake for 20 minutes. Serves 6–8.

BRUSSELS SPROUTS
(Brassica oleracea gemmifera)

Brief Description

Brussels sprouts is another member of the cabbage family. It is grown for its miniature buds, produced in the axils of the leaves. It is a biennial and is grown in many sections of the United States. It is one of the older types of vegetables, having been grown in this country for a century and in Europe probably several centuries. It is a tall, erect, nonbranching plant with rather large, long, slender, and ruffled leaves. The buds are 1″ to 2″ across.

Headaches due to Exhaust Fumes Treated with Cooked Brussel Sprouts

In the last quarter of 1994 the National Park Service was investigating a link between the exhaust fumes of snowmobiles and heightened reports of illness by employees at the west entrance to Yellowstone National Park. Employees had been complaining of headaches and said they could taste oil in the air from exhausts of the two-cycle machines that came into the park regularly.

In fact, it isn't uncommon to find up to 1,200 snowmobilers visiting the park's west entrance on a typically busy winter day. At times, 50 to 70 snowmobiles are waiting in line with engines running.

One of the park rangers complained to me of recurring headaches he had been having on account of the excessive exhaust fumes from so many idling snowmobiles waiting to get into the park. I suggested brussels sprouts for this, recommending that he not only eat the cooked buds but also drink the water in which they had been *lightly* cooked. I thought doing this once or twice daily for a week should help.

He reported back a month later that the headaches had disappeared, even though the problem of air pollution still existed. He asked how the vegetable worked for him. I explained that the exhaust from one snowmobile can produce the same volume of hydrocarbons and nitrous oxides as 1,000 cars, as well as carbon monoxide emissions equivalent to 400 automobiles. Since brussels sprouts, like all members of the cabbage family, are rich in sulphur compounds, they were able to quickly and effectively remove most of these hydrocarbons and nitrous oxides from his bloodstream before they made him ill with headaches.

Something Worth Waiting For

Brussels Sprouts-And-Broccoli Frittata

Needed: Vegetable cooking spray; 1/2 cup chopped onion; 1 cup small fresh broccoli florets; 1 cup thinly sliced brussels sprouts; 1/2 cup diced turkey ham; 1 garlic clove, minced; 2 eggs; 4 egg whites; 1/3 cup cooked spaghetti (about 1/2 oz. uncooked); 1/4 cup grated Parmesan cheese, divided; 1/4 tsp. crushed red pepper.

Coat a large nonstick skillet with cooking spray; place over medium heat until hot. Add the onion; sauté 3 minutes. Add the broccoli and next 3 ingredients, and sautee 2 more minutes. Remove from heat, and set aside. Then lightly beat the eggs and egg whites in a bowl with a whisk; stir in the broccoli mixture, pasta, and 2 tbsp. cheese. Recoat the skillet with cooking spray; place over medium heat until hot. Pour egg mixture into skillet. Cover, reduce heat to medium-low, and cook 8 minutes or until almost set. Sprinkle with remaining cheese and pepper. Wrap handle of skillet with foil; broil 3 inches from heat 1 minute. Cut into wedges; serve warm. Yield: 4 servings.

Buckwheat

(See under GRAINS)

Bulghur

(See under GRAINS)

Burdock
(Arctium lappa)

Brief Description

There are basically two kinds of burdock. Common burdock (*A. minus* is the kind more commonly found intercropped with corn and wheat in the Midwest. On the other hand, greater burdock is the one primarily harvested for its root as an important source of food for the Japanese. They use it there as we use carrots here. This variety of burdock has the big, round, brown bristly burrs, hence the common name of cocklebur.

Unsurpassed Blood Purifier

Burdock root is perhaps the most widely used of all blood purifiers, among the best the herbal kingdom has to offer for this, and *the* most important herb for treating chronic skin problems. It's one of the few that can effectively treat eczema, acne, psoriasis, boils, herpes and syphilitic sores, styes, carbuncles, cankers, and the like.

To make an effective tea, bring 1 quart of water to a boil. Reduce heat to simmer, adding 4 tsp. cut, dried root. Cover and let simmer for 7 minutes, then remove from heat and let steep for 2 hours longer. Drink a minimum of 2 cups per day on an empty stomach (more if chronic skin problems persist). A larger quantity can be made and used to wash the skin with often. Or capsules (4 per day) from your local health food store under the Nature's Way label may be taken instead.

Eliminating Stones

A great remedy for getting rid of some stubborn kidney and gallstones is to make a burdock-catnip tea. Bring 4 cups of water to a boil and add 2 tbsp. of chopped, fresh, or cut dried root. Reduce to simmer for about 10 minutes, covered. After which, remove from heat and add 3 tsp. chopped or cut fresh or dried catnip herb.

Let steep for 1 $^1/_2$ hours. Then strain and to each cup add 1 tsp. lemon juice and $^1/_2$ tsp. pure maple syrup or blackstrap molasses to sweeten. Drink slowly. Exactly 10 minutes later, take orally 1 tbsp. of pure virgin olive oil.

Repeat this regimen 3 times each day. The tea soothes irritated tissues and helps to break up or partially dissolve the stones, while the oil acts as a lubricant to remove them from the body more easily. It's *very important* that *no* greasy, fried foods, soft drinks, refined carbohydrates like white flour or white sugar products or red meat and poultry be consumed during this treatment; otherwise, absolute and complete success cannot be fully guaranteed.

After taking the last cup of tea and spoonful of oil at night just before retiring, be sure to sleep on your right side with a pillow underneath your armpit. This resting posture, some claim, seems to expedite the removal of stones from the body more quickly.

A Wild Vegetable Dessert

Who would ever think that vegetables would serve as delicious desserts in place of more standard fare like pie and ice cream? Well, in the case of burdock root, you have such a tummy pleaser fit for a king.

BURDOCK ROOTS, HAWAIIAN STYLE

Needed: 2 tbsp. sweet butter, $^1/_4$ cup packed brown sugar; 1 tsp. lemon juice; 1 cup canned, drained pineapples chunks (save juice); $^1/_2$ cup pineapple syrup drained from chunks; 2 tbsp. cornstarch; 2 cups burdock roots, cut into rounds and precooked until tender.

Melt butter in skillet over low heat, add brown sugar and lemon juice, stir. Mix pineapple syrup with the cornstarch, stir well and add to the butter and sugar mixture. Stir constantly over low heat until the mixture is a thick sauce, about 20 minutes. Add the burdock roots and pineapple chunks to the sauce and heat through. Serve warm.

A Nourishing Herb-Fish Soup

An alternative-care medical doctor from Tokyo, Japan told me what he prescribed to many of his patients recuperating from recent illness or surgery, or who just needed extra vim, vigor, and vitality, at a recent conference in Rhode Island.

First, secure a freshwater fish from your local fish market or supermarket. Carp, salmon, and trout are the best. You will need about a pound of fish, along with 1 $^1/_2$ lb. fresh burdock root; 1 tbsp. sesame seed oil; $^2/_3$ cup uncooked barley pearls; 1 tsp. fresh, grated ginger root; several unused green tea bags; $^1/_2$ cup chopped chives; 2 tbsp. lime juice; some kelp.

Nothing of the fish should be removed; head, fins, scales, and bones must all be kept intact. Chop the entire fish into 1 $^1/_4$″ chunks (about a dozen pieces); then cut the head into several more pieces, removing the eyes. Next cut the burdock root into *exceedingly thin* slices. Sauté this herb root for half an hour in sesame seed oil. After this place the pieces of fish on the bed of sautéed burdock. Cover with just enough water to maintain a nearly 3″ level over the fish.

Next scatter the barley over the fish and roots, along with the chives and kelp. Then place several unused tea bags in opposite corners of the pot *and on top* of the fish. Cover and bring to a boil, then reduce the heat to a lower setting and slowly cook for at least 5 $^1/_2$ hours. At the end of this period of time, uncover the pot and remove the tea bags. Then add the grated ginger and lime juice, cover and simmer again for an additional 15–20 minutes. The entire preparation can be consumed over several days time.

BUTTERNUT

(See under NUTS)

CABBAGE (GREEN, RED, SAVOY)
(Brassica oleracea)

Brief Description

Cabbage is believed to have originated in northern Europe and seems to have spread southward and eastward into the Mediterranean basin. If it had arrived through the Middle East, as some authorities have suggested, it would have been known to the Hebrews, but nowhere is it mentioned in the Bible. The Greeks and Romans ate a lot of cabbage and couldn't seem to get enough of it to satisfy their demands.

Besides being one of the oldest cultivated vegetables around, cabbage is also one of the most versatile. In addition to those plants that the layperson recognizes as cabbage—head cabbage, red cabbage—there are others so different in appearance and taste that it does not occur to most people to call them cabbages, yet they are. Broccoli, brussels sprouts, cauliflower, collards, kale, and kohlrabi are not only all cabbages, they even belong to the same species, *B. oleracea*. These different plants were not developed by hybridization or by mutation, but simply by encouraging the development of one element or another already present in the original plant. Any leaves with curling veins, like those of the cabbage, can be persuaded to form heads, so from what was probably a collardlike plant, the encouragement of this tendency produced head cabbage—Milan, Savoy, bell, and similar varieties. Cabbage contains anthocyanin, which tends to turn plants blue, red, or shades in between; development of this element produced red cabbage. By emphasizing the ability of cabbage to form budding heads at the junction of stem and leaves, brussels sprouts were created. Concentration on the flower resulted in both cauliflower (centripetal inflorescence) and broccoli (centrifugal inflorescence). Swelling pith gave rise to kohlrabi.

There is a big departure from *B. oleracea* with Chinese cabbage, of which there are several principal varieties. One of the most familiar is bok choy, which grows like celery and has a thick white stalk and drooping, dark-green leaves that make it look a lot like kale.

Strong Cancer Preventative

An overwhelming abundance of medical and scientific evidence has been published in the last decade to show that cabbage and its kind can help to prevent cancer if used in the diet properly.

Prestigious journals such as *Federation Proceedings* (May 1976), *Cancer Research* (May 1978), *Journal of the National Cancer Institute* (Sept. 1978), *Mutation Research* (Volume 77, 1980), and *Science News* (April 13, 1985) have reported significant research showing that the sulphur and histidine in broccoli, brussels sprouts, cabbage, cauliflower, kale, kohlrabi, and mustard greens inhibit the growth of tumors, prevent cancer of the colon and rectum, detoxify the system of harmful chemical additives and increase our body's cancer-fighting compounds.

The bottom line to all this, then, is that you should be consuming lots of these vegetables if you ever expect to substantially reduce your risks of getting cancer.

Lowers Serum Cholesterol

Cabbage and its kind, especially varieties like brussels sprouts, can dramatically lower what is often called "bad" cholesterol (low-density lipoproteins). This "bad" cholesterol usually causes hardening of the arteries in the course of time. At least two scientific publications have reported these findings. By including cabbage and its other members in your diet more often, you stand a very good chance of not developing coronary heart disease later on in life.

Increases Elimination

Cabbage promotes increased bowel movements. A 1936 medical journal noted that for every gram of powdered cabbage leaves fed to three healthy male medical students, their respective stool weights increased by 18 grams each. A more recent study showed that when 19 healthy male hospital staff volunteers consumed finely powdered cabbage supplements with their regular diets over a three-week period, their stool weights increased by 20%. This can be attributed to the water-holding capacity of cabbage fiber. About 5 cups of shredded cabbage, raw or cooked, is suggested twice a week in the diet for improved colon function.

Fights Yeast Infection

Some native folk healers in various American ethnic groups, such as Hispanics and African-Americans, have prescribed raw cabbage juice for yeast infections covering the head, skin, hands and feet, as well as for treating premenstrual syndrome (PMS). In giving women relief from PMS, the cabbage juice has been taken internally or else used as a vaginal douche or both in some cases with relatively good success. Luc De Schepper, M.D., a Los Angeles physician experienced in candida research, lists cabbage, brussels sprouts, broccoli, watercress, chives, and spinach as mandatory sulphur foods for suppressing yeast infection.

A Fantastic Ulcer Healer

Different medical journals have separately confirmed that *raw* cabbage and raw cabbage juice is just the ticket for relieving *and healing* any kind of grastrointestinal ulcer, be it duodenal, peptic, or what have you. Half a cup, morning and night, of raw juice is a terrific antacid remedy plus a great ulcer healer.

A relatively new antiulcer product which some Florida practitioners have been prescribing to their patients of late comes from herbs used for stomach problems by the horse-and-buggy Amish of Ohio. It may be obtained by mail order from Old Amish Herbs under the name of Cabbage Compound (see Appendix).

Radiation Protection

Two important medical studies published nearly 20 years apart, showed that rabbits and guinea pigs previously exposed to lethal doses of uranium and X-rays were afforded considerable protection against their harmful effects when cabbage leaves were added to their basic diets. Considering the excessive radiation to which we're constantly exposed today, ranging from home computers and microwave ovens to color TVs and high-tension power lines running through the neighborhood, it may be a good idea to add more cabbage to your weekly diet for extra protection.

Two Kinds of Coleslaw

Cabbage Slaw with Buttermilk Dressing

Needed: 4 cups thinly sliced red cabbage; 4 cups thinly sliced bok choy (Chinese) cabbage; 1 cup thinly sliced fresh brussels sprouts; $^1/_4$ cup chopped red onion; $^1/_4$ cup chopped fresh parsley; 1 tsp. sugar; 1 tsp. celery seeds; 1/2 tsp. salt; $^1/_4$ tsp. pepper; $^1/_3$ cup low-fat sour cream; $^1/_3$ cup nonfat buttermilk; 1 tbsp. tarragon white wine vinegar; 1 tsp. Worcestershire sauce.

Combine the first 5 ingredients in a large bowl; toss well, and set aside. Combine the sugar and next 7 ingredients, and stir well. Pour over the cabbage mixture, and toss gently. Cover and chill 1 hour. Yield: 8 servings.

Real Creamy Coleslaw

Needed: 1 $^1/_2$ cups plain yogurt; 6 $^1/_2$ tbsp. brown sugar; 1 tbsp. pure maple syrup; 3 $^1/_2$ tbsp. apple cider vinegar; $^3/_4$ cup olive oil; $^1/_3$ tsp. each powdered garlic, onion, mustard and celery; dash of kelp; 1 $^1/_2$ tbsp. lemon juice; $^3/_4$ cup half and half; $^1/_2$ tsp. sea salt; 1 head each green and red cabbage, very finely shredded.

Blend together the yogurt, sugar, syrup and oil. Add spice powders, kelp, lemon juice, half-and-half, and salt. Stir until smooth. Pour over the coleslaw in a large bowl and toss until cabbage is well coated. If you wish, add just half of the dressing to a head of cabbage, saving the rest to dress fruit salad or other salads with. Keeps in a closed container in the fridge for nearly a week. Recipe makes about 1 quart of dressing.

CANTALOUPE

(See under MELONS)

CAPSICUM

(See under PEPPERS)

CARAMBOLA

(See under TROPICAL FRUITS)

CAROB
(Ceratonial siliqua)

Brief Description

Carob pods grow on a dome-shaped evergreen tree with dark-green compound leaves consisting of two to five pairs of large, rounded glossy leaflets. The tree can reach a towering height of nearly 50 feet and is native to southwestern Europe and western Asia, but is also widely cultivated in the Mediterranean region.

The pods are the so-called "locusts" consumed by John the Baptist during his wilderness residency, hence the other common name of "St. John's bread." Seeds were used in ancient times as weight units for gold from which the term "carat" is reportedly derived. The Prodigal Son in Jesus' famous parable subsisted on discarded carob pods, as mentioned in Luke 15:16—"And he would fain have filled his belly with the husks that the swine did eat."

Kidney Failure Reversed

In the medical journal *Nouv-Presse-Med.*, a French physician related how a clinical case of chronic kidney failure was successfully reversed with carob gum. Approximately 2 level tsp. of carob powder in cranberry juice or milk, taken 4 to 5 times daily, should be of some use in stimulating inactive kidneys.

Remarkable Antidiarrheal

One of the very best remedies for human infant and adult and livestock diarrhea is carob powder. A back issue of *Western Dairy Journal* advocated the mixing in of carob powder with regular feed to cure and prevent scours or diarrhea in heifer calves.

An even more effective use for halting the same condition in people has been found as the following episode with a Montana sheep rancher residing near Harlowtown shows:

> I am 65 and have had diarrhea for years. I was always listless and worn out. The least work would wear me out. I didn't realize that diarrhea was nearly all of my trouble.

> I tried carob powder and it worked wonders. In my case, it takes a highly heaping soupspoonful to fix me up. I take it each meal . . . I have used carob and it has helped me a lot. I wish I had known about it a few years earlier.
>
> Medical doctors gave me medicine similar to Kaopectate, but I had to take it in enormous quantities to get relief. About 4 fluid ounces of Kaopectate was required per night in many cases, or a little less of the prescribed medicine for the same results. Carob works much better.

When traveling to foreign countries, always carry a can of carob powder with you to use. Carob pods are particularly rich in one class of tannins, which manifest strong antiviral properties. Thus, it would appear that carob powder might work just as well as some kinds of antibiotic drugs usually given to treat bacterial-induced diarrhea.

And for diarrhea in infants—quite a common problem it seems—carob powder rates magnificent in its performance here. A number of related articles in back issues of the *Journal of Pediatrics* conclusively show that carob is able to completely correct this dangerous nuisance in any infants suffering from diarrhea. The powder can be given in a formula drink preparation or if the baby is old enough to eat solids, it can be mixed in apple sauce or some kind of pudding and fed that way with satisfying results.

Chocolate-Cocoa Substitute

There are many advantages to using carob in place of either chocolate or cocoa. First, it has far less calories than either cocoa or sweet chocolate. Second, it's a darned sight cheaper. Third, because it's so naturally sweet, it takes a lot less to make brownies or a shake. Fourth, it doesn't have the addictive substance, caffeine, like the other two have, which can be potentially harmful to children. Finally, it doesn't interfere with calcium assimilation as chocolate or cocoa do. And, it doesn't cause acne either.

Indulging Recipes

Here are some recipes you can really indulge yourself in without fear of getting fat, incurring pimples, or feeling guilty for eating unhealthy things.

Carob Brownies

Needed: $^1/_2$ cup honey; $^1/_2$ cup safflower oil; $^1/_2$ cup carob powder; 2 beaten eggs; 2 cups whole wheat flour; $^1/_4$ tsp. sea salt; $^1/_2$ cup granular lecithin; 1 tsp. almond extract; $^1/_2$ cup ground or finely chopped nuts.

Blend honey, oil and carob powder. Add beaten eggs. Stir in flower and salt. Add lecithin and almond extract, then the nuts, Pour into an oiled 8″ square pan. Bake at 350° F. for 30 minutes. Cool and cut into squares.

Carrot
(Daucus sativus)

Brief Description

Carrots around the world grow in all shapes and colors. Westerners would mistake the Asian types, with their bulbous purplish red roots, for beets. Other colors are pale and deep yellow, red and white. The roots range from spherical to cylindrical. One variety in the Far East grows up to a yard long.

Unusual Mouth Wash

French herbalist, Maurice Mességué advises that carrot tops be made into a tea for an effective mouth wash and gargle, due to their strong antiseptic action. Bring 3 cups of water to a boil, adding 1/2 cup chopped carrot tops. Simmer for 20 minutes, steeping another 30 minutes. Strain and store in refrigerator. Use this mouthwash each morning.

Heals Burns and Scalds

Shauna Wilson of Atlanta, Georgia, wrote to me several years back of a treatment she used for a burn on her 4-year-old daughter's arm, after the girl had accidentally pulled a pan of boiling hot water down on herself from the stove. "I soaked her arm in ice water first, then dressed the injury with some gauze which had been dipped in carrot juice and lightly wrung out," she said. "I repeated this several times more the next day, and by the third day most of the swelling and inflammation was gone."

Lowers Cholesterol and Cancer

A diet supplemented by raw carrots each day helps to lower cholesterol in the body. It's recommended that you eat a carrot salad or munch on a carrot every time you consume greasy food.

Carrots contain high amounts of the anticancer nutrient, beta-carotene. Scientific documentation has shown that this root vegetable really does help to keep the incidence of cancer down in those who consume it regularly versus those who seldom include it in their diets. For instance, the December 1986 issue of *Epidemiology* noted that pancreatic cancer risks are substantially higher in those who consume fried and grilled meat and margarine, but significantly reduced with almost daily consumption of carrots and citrus fruits.

Energy Stimulant

Ever have a glass of carrot juice and feel a surge of energy afterward? It's probably due to the high natural sugar content. Some of the Pennsylvanian Amish folks use a tonic called Carrot Concoction from Old Amish Herbs (see Appendix) to give them more vim and vigor when fresh carrot juice isn't available. Equal parts of carrot and pineapple juice make a nice energy drink for those with hypoglycemia.

Infantile Diarrhea Cured

Some three dozen cases of infantile diarrhea in Innsbruck, Austria were cleared up with nothing but carrot soup and carob. The chronic diarrhea, lasting nearly a week or more, had been caused by *E.*

coli bacteria. But carrot soup containing 2% carob powder was able to block the activity of this virus within the infants' upper small intestinal tract, resulting in no more diarrhea. (See the recipe on page 88 for making carrot soup.)

Vegetable Laxative

Carrots also clear up constipation. Carrots produce some looseness of the stool, while grains and leaves increase bowel gas. Two separate medical studies published 42 years apart cited the laxative properties of raw carrot pulp (*Journal of Nutrition*, 1936, and *Lancet*, 1978). The 1978 report observed quicker actions when cabbage and carrots were combined in the same meal. A 1931 nutrition review cited that for each gram of carrots consumed by three healthy male medical students, their respective stool residues increased in weight by 19 grams or 1/2 oz. (See the recipe on page 89 for a good carrot-cabbage raisin salad that really gets the bowels moving!)

Protects Against Toxic Chemicals

There's nothing quite like carrot fiber to protect the body against certain chemical pollutants. Rats experiencing loss of weight, extensive diarrhea, and an unthrifty appearance when given harmful chemicals in their diets had a complete reversal of all these symptoms when carrot root and cabbage powders were added to their daily regimens. And, interestingly enough, many Soviet doctors recommend carrots in the diets of factory workers exposed to harmful chemicals.

Kick the Smoking Habit

In the 1983 comic strip, "Wizard of Id" by Parker and Hart, two of the king's guards were chatting in the courtyard. One asks the other, "I understand you used to smoke 4 packs a day?" His companion agrees, adding, "Now I eat a carrot whenever I crave a cigarette." "How's it working out?" asks the first guard. "Fine," replies the second, as he hops away like a bunny rabbit.

But there's more truth to this than meets the eye. Over the years in my travels around the world I've heard some pretty good stories relating to the use of carrots for knocking the nicotine habit. But none quite so dramatic or simple in detail as what a lady friend of mine in Indonesia related tome in October 1986. Her name is Josephine Hetarihon, age 35, and she works as an executive secretary in Jakarta. This is her story:

> It was on September 26, 1979 that I finally managed to quit smoking. I had been smoking 2 packs a day of Dunhill before this.
>
> I started smoking when I was in junior high school at the age of 15. A friend suggested that I use carrots to help me quit.
>
> It took me about two weeks on this carrot program until I was able to quit smoking altogether. I would eat about 2–3 carrots a day. I found that the sweet taste of the carrots satisfied me enough so that I didn't crave a cigarette.

Relieves Asthma

Reverend John Wesley, the English theologian who founded the Methodist church, recommended eating boiled carrots and drinking the warm broth thereof as a "seldom fail" remedy for relieving asthma. Lukewarm carrot juice also has a similar effect.

An Odyssey of Health

This is a true story about carrot juice, capitalism, and zoos. But more about the first than the last two. The Hogles are an old and well-established Salt Lake City family dating back to more than a century. They were part of the non-Mormon element who migrated to Utah and did their share in making the Beehive State one of the best in the nation.

James A. Hogle was one in this line of capitalists who had a talent for making money. His wife Mary C. Hogle had an equal gift for spending what he made—not only to car dealers and shoe salesman, but also to various charities and other worthy causes. One of these was the establishment of a zoo on the east bench at the mouth of Emigration Canyon (so called because of the early Mormon pioneers coming through it into the Salt Lake Valley for the first time on July 24, 1847). Over the years Hogle Zoo has become one of the best known and top-rated zoos in the West; each year it is visited by hundreds of thousands of people and is considered to be a major tourist attraction.

In 1916 Mary Hogle suffered an attack of appendicitis and had her appendix removed. By 1921 she had begun to experience a number of health problems which marked the beginning of a decade-long bout with illness. Her chief complaints seemed to center on her gastrointestinal tract. She experimented with a number of different fasting programs in hopes of getting well. But just the opposite occurred, and such extreme measures actually made her condition worse instead of better.

Shortly after her return to Salt Lake in early 1930, Mary rose from her sickbed to answer the doorbell, and there stood Mr. Holderness, a salesman who was peddling peanut brittle, sewing machine oil, and magazine subscriptions. Mary was impressed with the man's healthy appearance. His looks were unquestionably striking; he was tall, erect, almost totally bald, with a light fringe of gray hair. He had sharp features, a straight nose, and high cheekbones. He was hatless and his bald head was darkly tanned. His shirt was open at the front and he wore blue-and-white striped dairyman's overalls. He also had a heavy Australian accent, but what impressed Mary the most was his complexion. Mr. Holderness's skin was a deep golden color—not a tan, exactly; he seemed to glow from within. In her private writings, Mary described what happened next:

> During the ensuing conversation he told me of his own previous affliction and remarkable restoration to health by drinking carrot juice. His symptoms had been so similar to mine that I was glad to listen to him and, of course, I was more or less receptive to the idea because of articles by Dr. Kellogg and others which I had previously read.
>
> Carrot Juice! . . . Drink your vegetables. It was worth trying. Anything is worth trying when you are desperate . . . . Within a few days after commencing to take the carrot juice, I was convinced that my body was actually [recovering] . . . I noted with great satisfaction that my erratic, wobbly, fluttering heart was picking up strength

> and vigor. With a period of several months, all the distressing symptoms, however long withstanding, had disappeared. My recovery is so complete that the present years find me enjoying the best health of my entire adult life.

Mary Hogle's condition improved so dramatically that she had no doubt what had caused the miraculous cure. For many years to come, the "carrot juice man," Mr. Holderness, would be employed by the Hogles as a gardener and caretaker at their home, but his primary responsibility was manufacturing the golden elixir that had restored her to health.

Mary began to deliver quarts of carrot juice to sick friends and relatives. What had worked for her could work for them. She then went into the carrot juice manufacturing business by converting the basement of her mansion into the kind of factory that only Bugs Bunny of cartoon fame might have operated, but with Mr. Holderness at the controls.

> As soon as I was convinced of the remarkable healing action of carrot juice, I wanted other people who were suffering to have it. I had especially constructed electrical machinery; I set up and subsidized the manufacture of it on a rather large scale, ten gallons daily—sending much of it to sick and ailing individuals. I discovered that it rapidly corrected acid conditions and found it valuable in the treatment of infective disorders. It appeared to be the specific treatment for blood disorders, blood poisoning, anemia, and other bad blood conditions. It had the effect of improving the quality and texture of teeth. People afflicted with serious wasting disease could, by taking enough of it, become strong, active, and robust . . . A most gratifying revelation was that several who were afflicted with cancer were helped by it.

The apparatus that Holderness employed for extracting the carrot juice for Mary Hogle consisted of a hand grater, a number of clean flour sacks, and a cider press. The carrots were grated by hand; the shavings were put in the flour sacks and then squeezed under the large wooden screw-type press. The juice ran out the sides of the press and dripped onto a grooved board which collected the drippings and directed them into a wide-mouthed funnel inserted in a bottle. It was an arduous procedure and no doubt part of the benefit of drinking the juice was the exercise one got in going through the collection process. Also, the pulp was unfortunately discarded, thereby wasting valuable vegetable fiber which we now know is a definite health benefit to the body.

Things have come a long way since Mary Hogle's time, however. Today there are different kinds of juicing machines which can make carrot juice in just a matter of minutes with minimal effort. I cover this entire topic of juicing and juicers in one of my other health best-sellers entitled *Heinerman's Encyclopedia of Healing Juices* (Prentice Hall, 1994). One of the most popular juice machines in America is the Vita-Mix Total Nutrition Center. It can make juice out of just about anything and, best of all, *includes* the fiberous pulp *with* the juice! Nothing is wasted and everything is preserved.

Here is a quick and easy way of making carrot juice. Cut 1 1/2 cups unpeeled carrots in 1″ pieces. Put them in the Vita-Mix container along with 1 cup distilled water and 1 cup ice cubes (made from distilled water). Secure the two-part lid in place by locking under the tabs. Move the black speed control lever to HIGH. Lift the black lever to ON position and allow the machine to run for 1 1/2 minutes until smooth. Makes 2 cups of delicious and refreshing juice! For more information, write or call Vita-mix corp., 81615 Usher Road, Cleveland, OH 44138 (1–800–848–2649).

Therapeutic Recipes

Our culinary recipes should also be our household remedies. The two recipes that follow are, therefore, intended not only to taste good, but also to keep you in a fit state of health as well.

MIRACLE CARROT SOUP

Needed: 5 1/2 washed, unpeeled carrots; 8 cups water; 1/2 tsp. kelp; 1/2 tsp. sea salt; 1/2 tsp. honey; 1/2 cup chopped onion; 6 tbsp. melted butter; 6 tbsp. flour; 2 cups hot milk.

Chop carrots fine. Combine them with the water, kelp, sea salt, and honey. Cook for 1 hour. Add the chopped onions and simmer on low heat for 10 more minutes. Melt butter in skillet on medium to low heat. Slowly stir in flour, but don't brown. Simmer for 10 minutes without burning, stirring frequently. Then add hot milk to skillet and whip smooth by hand with a wire whip. Add this to the soup, whipping thoroughly by hand until well mixed. Serves 6 or can be refrigerated and used for medicinal purposes whenever necessary.

PROTECTIVE CARROT-CABBAGE SALAD

Needed: 2 $^1/_4$ finely shredded carrots; 2 $^1/_2$ cups finely shredded or chopped cabbage; $^1/_3$ cup finely chopped green bell pepper; 2 tbsp. raisins; 3 tbsp. canned pineapple chunks; 1 tbsp. pineapple juice from can; 1 cup plain yogurt; 10 chopped dates.

Mix the carrots, cabbage and bell pepper together in a large mixing bowl. Add the raisins, pineapple chunks, juice, and dates. Stir everything together again, mixing well. Finally, turn in the yogurt with a wooden spoon and mix thoroughly until a smooth, somewhat tight consistency to the salad is formed. Serve on Romaine lettuce leaves. Or refrigerate and eat some every day to protect yourself against the harmful chemicals in our environment and food and water supplies. Yield: 6 servings.

CASABA

(See under MELONS)

CAULIFLOWER
(Brassica oleracea botrytis)

Brief Description

American humorist Mark Twain wrote this one-liner in regard to the vegetable discussed here: "Cauliflower is nothing but cabbage with a college education." Starting out as a prospector in Carson City, Nevada, he quickly turned to newspaper reporting in 1862 in Virginia City, Nevada when his unlucky streak for finding gold left him short of cash.

He and fellow humorist Artemus Ward (pseudonynm for Charles Farrar Browne) had lunch together one time. Cauliflower happened to be served as the accompanying vegetable to the meal and both men quickly put their witty talents together to come up with a brief but funny dialogue concerning this member of the cabbage family. it was on this occasion that Ward encouraged Twain in his writing.

The name Mark Twain was taken from a term originally used by leadsmen taking soundings on the mighty Mississippi; as a boy Clemens used to ride the river boats often and delighted to hear them call out "mark twain", meaning that the water ahead was "two fathoms deep."

Twain traveled to various Mediterranean countries, where he encountered a large number of subvarieties of this vegetable: the white cauliflower in Milan, green cauliflower in Rome, purple cauliflower in Catania, and a giant cauliflower in Naples. From there he went to the Holy Land looking for a religious experience of some sort, but only met with disappointment. This prompted him to remark that he'd rather "go to Hell where all of the interesting people are, than to Heaven where only boring people reside." He incorporated material from this trip into a book, *The Innocents Abroad* (1869).

He is best remembered for his children's classics, *The Adventures of Tom Sawyer* (1876) and the *Adventures of Huckleberry Finn* (1885). While cauliflower isn't mentioned in either of them, yet Twain's different biographies show that he enjoyed this "intellectual vegetable" often while writing both novels.

Cauliflower is prized for its broad, curdlike, white or purple flowerheads of closely packed buds. The head is produced on a short stalk, after the plant has made a rosette of leaves with some elongation of the stem. The leaves are long and slender, very much like the Calabrese broccoli, which the plant closely resembles. There are two general types: the white short-season type commonly found in vegetable hors d'oeuvres and the purple long-season type usually referred to as cauliflower broccoli.

Satisfying the Hungries

Creative people are often high-strung, nervous types; therefore, it helps if they have something in hand that they can occasionally munch on, drink up, or hold in their fingers while working. This is why so many writers, artists, composers, and poets are in the habit of chewing gum, eating snack foods, drinking endless cups of coffee or mugs of soda pop, or smoking cigarettes. It just seems to help ease their nervous tension a bit during those periods when their creative juices are flowing the best.

It is said by some of his biographers that Twain was in the habit of munching on raw cauliflower buds or radishes while thinking out the plot and story line of a work of literature. It has been stated on good authority that he consumed a great deal of cauliflower while writing *A Connecticut Yankee at King Arthur's Court* (1889).

Those who are in similar situations or who suffer from nervous tension and frequently have the hungries ought to take a tip from Mark Twain and snack on something healthy like cauliflower buds. They are better for you than traditional snack foods like potato chips, candy, or salted nuts, and best of all, they won't make you fat!

Getting Rid of Headaches

I don't know exactly how this remedy worked for Mark Twain, but, suffice it to say, *it did*! He invested much of his fortune in the publishing house of Charles L. Webster & Co. in 1896; soon, thereafter, the business failed and Twain nearly lost his shirt in the deal. Frustration and bitterness set in, which soon gave him frequent headaches.

He turned to frequently nibbling on cauliflower buds and radishes to alleviate some of the stress and tension. Perhaps it was the constant, methodical chewing that relaxed the blood vessels leading to his forehead thereby increasing circulation, that could have helped. Or it may simply have been some of the sulphur components in the vegetables themselves that accounted for his relief. Twain took to the lecture stage to clear his obligations and resumed his writing thereafter.

A Yummy Recipe

Cream of Cauliflower and Pea Soup

Needed: 2 cups chopped cauliflower; 3 cups shelled peas; 8 cups water; 1 cup sliced onion; 1 tbsp. olive oil; 1 whole bay leaf; 1 tbsp. arrowroot flour; 1 cup cold water; $^1/_2$ tsp. granulated kelp; 4 cups liquid from boiling peas; more kelp to flavor.

Chop up the cauliflower. Then shell the fresh peas and set them aside. Prepare the water. Slice the onion. Heat a Chinese wok, add the olive oil and chopped onion. Sauté until the onion is transparent. Add cauliflower and lightly sauté it. Add some water to cover the vegetables, and the bay leaf. Cover and simmer 12 minutes. Remove the bay leaf and blend the soup until creamy. Add more water if necessary. Dissolve the arrowroot in some cold water; add kelp to soup. Stir the arrowroot into the soup and bring it to a rolling boil. Pour into individual bowls.

Next bring the peas to a rolling boil in 4 cups water. Cook 5 minutes. Blend in a small amount of the cooking water until creamy. Add the rest of the cooking water to measure 4 cups or until desired consistency is reached. Pour into a saucepan and heat until boiling. Slowly pour into one side of the serving bowl already containing the cauliflower mixture. Using a spoon, gently coax the peas in. Yield: 4–6 servings.

CELERY
(Apium graveolens)

Brief Description

Celery is one of the oldest vegetables in recorded history. The ancient Egyptians were known to gather celery from marshy seaside areas for food. it is a plant of many uses and little waste: the leaves and dried seeds make good seasoning; the outer ribs are best cooked and the inner ribs may be consumed raw, as they are good for the heart.

The variety most commonly available is the light-green to medium-green Pascal celery. Stalks are firm and solid with a maximum of green leaves. They usually have a glossy surface and snap easily. As a member of the distinguished parsley family, it enjoys some of the same reputable medical claims often attributed to the former herb. The ancient Greeks on the Isthmus of Corinth around 450 B.C. regularly crowned their winning athletes with crowns of celery stems and leaves.

In 1982, the average American ate 7.8 lbs. of celery, 11% more than 5 years earlier. In 1983, it was a $235 million crop in America, compared with $184.5 million for carrots and only $152 million for broccoli. California supplies the nation with more than 60% of all celery production.

Calm Frayed Nerves

The seeds and the stalks of celery both contain a sedative compound called "phthalide" (the "ph" is silent). In mainland China, celery juice was useful in reducing hypertension in 14 out of 16 patients. The juice was mixed with equal amounts of honey and about 8 tsp. was taken orally three times each day for up to a week. Make your own celery juice at home with a juicer or buy it fresh from a health food store. Mix equal parts of it and carrot juice together and drink an 8 oz. glass once a day to help strengthen frayed nerves and calm you down.

Quick Relief for Hornet Stings

Several years ago while participating as an instructor in a plant identification hike with 20 others in Provo Canyon, near Robert Redford's Sundance Ski Resort, one of our number who was barefoot, accidentally stepped on a black hornet. Intense pain and swelling commenced within minutes.

I asked everyone to look around for some plantain or yarrow, but to no avail. Another member of our small group had some celery sticks in a plastic bag. I asked for one, started chewing it vigorously until I had ground it to a pulp with my back molars. I then applied this wad of celery and saliva directly to the wounds and held it there for about 15 minutes. The throbbing ceased and she felt more comfortable thereafter. A crudely made mud pack with more chewed celery inside brought added relief for some hours later. She managed to hobble along with the rest of us, supported between two of her friends.

Helps Keep Weight Off

Robin W. Yeaton, Ph.D., who worked for the UCLA Program Development Symposia in October 1985 told me on an airplane flight that she lost over 30 lbs., in 2 $1/_2$ months, by nibbling on a lot of celery sticks whenever the urge to snack came over her. The sodium seemed to have a positive effect in helping her to shed additional pounds too.

Using Celery in Recipes

Be sure that the celery you use is always fresh and crisp as possible. Avoid any kinds that are wilted, brown or diseased-looking. And above all, do *not* store celery in your refrigerator for longer than 3 weeks. According to the November–December 1985 *Journal of Agricultural Food Chemistry*, the furocoumarins present in very small amounts in fresh celery can increase 25 times or more after about 3 weeks' storage. Old celery has caused cancer in animals.

For a simple dish on the stir-fry principle, add sliced celery to bite-sized pieces of chicken and allow to simmer covered for 20–30 minutes. Or use lamb with celery in much the same manner.

Creamed celery, another wintertime Sunday dinner favorite of the 1930s and 1940s, can be adapted to today's cooking styles by spicing it up with either cheese or nut meats. Toasted almonds add the crunch to this interesting version.

Creamed Celery with Almonds

Needed: 8–10 celery branches with leafy tops intact; 1 tbsp. diced shallots (an onionlike plant); 3 tbsp. butter; 1/4 tsp. sea salt; 1 tbsp. whole wheat flour; 1/2 cup cream; 1/2 cup chicken broth; 1 cup toasted almonds.

Slice the celery on the diagonal, melt the butter in a heavy pan with a tight-fitting lid. Add the shallots first, then the celery. Cover the pan; cook until celery is tender, about 8 minutes. You should not have to add liquid. Shake the pan every now and then to prevent scorching. When the celery is tender, add the sea salt and sprinkle in the flour. Toss celery with a mixing spoon to distribute flour.

Place the pan over a double boiler; add cream and chicken broth. Cook until the raw flour is gone, and mixture thickens slightly, about 5 minutes. Add 3/4 cup of the toasted almonds and toss. Place the celery mixture in a serving dish; top with the remaining almonds. Sprinkle paprika over the top and serve. Serves 3–4.

Chard

(See under BEETS AND SWISS CHARD)

Cherry
(Prunus avium)

Brief Description

Since the caveman era, wild cherries have existed in the temperate parts of Asia, Europe, and North America. The hundreds of varieties on the market today may be classified in terms of sweetness and color. Bing and Royal Ann cherries are both sweet, but Bings have deeply colored juice, whereas the juice of the other variety is colorless. Sour cherries—the ones most favored for pies, tarts and turnovers—are similarly divided: morellos have colored juice and amarelles colorless liquid. The very popular tart cherry, Montmorency, is light to dark red with red juice. Sweet cherries are available from May through August, while sour cherries go from late June to mid-August.

Great Gout Cure

Nothing works better for gout than either raw sweet cherries (15 per day), cherry juice concentrate (1 tbsp. three times daily); or else a tea made of the stems. To make the tea just bring 2 pints of water to a boil, then throw in half a handful of stems, reduce heat and simmer for 7 minutes, then remove and let steep, covered, for 20 more minutes. Drink at least 2 cups a day to keep the gout under control. All these remedies also work well for arthritis.

Controls Hacking Coughs

The bark of a related species, wild black cherry (*P. serotina*), is frequently used in many cold and cough preparations, such as Smith Brothers' Cough Drops. To make your own terrific cold and cough syrup is relatively easy. Just combine in a stainless steel pot 3 cups of water, $^1/_2$ cup of whiskey (any brand), and $^1/_2$ cup blackstrap molasses and bring to a boil.

Reduce heat to the lowest setting possible, then add 16 tbsp. or 1 cup of cut, dried wild cherry bark purchased from any local health food store. Stir well with a wooden ladle, cover and let simmer for about 25 minutes until the mixture has been reduced a little in volume and is somewhat thick. Stir every so often as necessary. Then let steep for about 15 minutes, after which it should be strained through a coarse strainer into a clean fruit jar or bottle.

Store in a cool, dry place and take 2–3 tbsp. of the syrup several times a day to control hacking cough due to asthma, bronchitis, emphysema, smoking, and the flu. A tea can be made with the same ingredient amounts, except the molasses and whiskey, to relieve sinus congestion and other symptoms of the common cold.

Antidote for Fish Food Poisoning

A tea made from cherry bark and other ingredients is a useful antidote to counteract the effects of bad shellfish and spoiled fish in general. Bring 1 pint of water to a boil. Then add 1 tsp. each of cherry or wild cherry bark, fresh, grated ginger root and finely chopped Bermuda onion. Cover, reduce heat and simmer for 7 minutes. Then remove from stove and steep for an additional 20 minutes or so. Drink both cups when lukewarm.

Dessert Recipe

The following recipe has been adapted from Better Homes & Garden's excellent work, *Eating Healthy Cook Book*. The kindness of the editors is greatly appreciated.

A Delicious Cherry Cobbler

Remember to spoon the batter onto a *hot* cherry filling. That way, the biscuit topper cooks faster and more uniformly.

Needed: 1 cup flour; 1 tbsp. pure maple syrup; 1 tsp. baking powder; $^1/_4$ cup butter; 1 slightly beaten egg; $^1/_4$ cup canned goat milk; 4 cups fresh or frozen, unsweetened, pitted tart red cherries; $^1/_3$ cup brown sugar; $^1/_3$ cup water; 1 tbsp. quick-cooking tapioca.

For biscuit topper, stir together flour, maple syrup and baking powder. Cut in butter until the flour mixture resembles coarse crumbs. In a small mixing bowl stir together the egg and goat's milk. Add milk mixture all at once to the flour mixture, stirring just enough to moisten. Then set aside.

For cherry filling, in a medium stainless steel saucepan combine cherries, brown sugar, water, and tapioca. Let stand for 5 minutes, stirring occasionally. Cook and stir until it begins to bubble like a hot spring or small geyser.

Turn the hot cherry filling into an 8″ × 1 $^1/_2$″ round baking dish or a 1 $^1/_2$ quart casserole. Immediately spoon biscuit topper on top of the cherry filling to form 8 mounds. Bake in a 400° F. oven about 25 minutes or until a wooden toothpick inserted in the center of the topper comes out clean as a hound's tooth. Serve warm. Should make 8 servings.

Fruitcake Frenzy

"Fruitcakes . . . all tend to be the same," humorist Erma Bombeck once wrote. "Each [of them] having an assortment of incompatible fruits and the distinction of weighing more than the stove they're cooked in. They defy all of the culinary rules in the book. No one ever says, 'This fruitcake is so light you don't know you're eating it.' "

A radio station in Houston, Texas even organized a Fruitcake Olympics back in 1991, in which people played shuffleboard with them and whacked at them with bats. A liquor company once put up billboards with a picture of its product and the slogan underneath: "It sure beats fruitcake!"

But in the tiny Texas town of Corsicana, no one makes fun of fruitcake. This is the home to the Collin Street Bakery, which is, by the estimate of its owner, William McNutt, Jr., the largest fruitcake mail order emporium in the nation. Say what you will about fruitcake elsewhere in the country, but here in little Corsicana, fruitcake is king!

I paid a visit to this place in late 1994. I found the noise coming from the assembly line at the Collin Street Bakery to be quite deafening; it can be equated with the clamor of a bowling alley and a Midas muffler shop rolled into one. Pans clanged, conveyor belts rattled, machines thumped and whumped, all punctuated by a steady pneumatic whoosh from somewhere near the huge ovens.

'Twas the season, indeed, and the company was in full fruitcake frenzy, churning out 33,000 of them per day. Collin Street will weigh in with about 2,000 tons for the entire year. That is the approximate weight of 14 blue whales or 33 African elephants.

Collin Street makes 1.5 *million* fruitcakes a year. Scoff all you want, but demand was up at the time of my visit some 11% over the previous year and had risen steadily for decades. At least that's what Bob McNutt, a company vice president and the owner's son, claimed.

Located 50 miles south of Dallas in the depleted oil town of Corsicana, Collin Street is part of a region of bakeries known, as they could only be in Texas, as the Fruitcake Triangle. There are several other nearby bakeries such as Mary of Puddin' Hill, The Original Texas Ya-Hoo Cake Co., and Eilenberger's. Fruitcake is sufficiently serious business in these parts that a 1994 conference of newspaper food writers in Texas featured a panel discussion of it.

I interviewed some of these other competitors while I was there and came away with at least one very valuable piece of culinary information. The *secret* to really good fruitcake is in the *cherries*. Forget about the pecans, papaya, pineapple, honey, and heaven's knows what else goes into making one of these doorstops or paperweights. The general consensus of all those bakeries down in Corsicana churning out fruitcake by the hundreds of thousands is that it's the cherries that count the most out of all the ingredients included.

And where do these various fruitcake bakeries get their best quality cherries from? Why, the Pacific Northwest I was told repeatedly time and again. Collin Street gets theirs from an undisclosed place in Oregon. The Ya-Hoo Cake Co. tried cherries from several states in the intermountain West, but after receiving numerous complaints settled for an unnamed supplier in Washington State. And so it goes.

None of these bakeries was willing to share with me even so much as a smidgen of any of their individual recipes. Corporate secrets, they said. But at least we now know this much if anyone is ambitious enough to make his or her own fruitcake—you'd better get your cherries from the Northwest or expect your gift to be continually passed around, unopened, by well-meaning friends, neighbors, and relatives who won't care much for tasting it.

CHICKPEA

(See under BEANS)

CHERVIL
(Anthricus cerefolium)

Brief Description

Chervil comes from a Greek word meaning "leaf of rejoicing" or "cheer-leaf." The sixteenth-century English herbalist Gerare confirmed this original meaning when he wrote: "It is good for old people—it rejoiceth and comforteth the heart and increaseth their strength."

Chervil is of East European origin and is to be found growing wild in southeast Russia and most of Iran. It found its way to England thanks to the ancient Romans. Today, however, the leaves and stems are principally used in France for seasonings, salads, soups, and as a pot-herb.

This annual plant has a round, finely grooved, branched stem which grows 12 to 26 inches high from a thin, whitish root. The leaves are opposite, light green and bipinnate, the lower leaves petioled, the upper sessile on stem sheaths. The small, white flowers grow in compound umbels from May to July. The elongated, segment seeds ripen in August and September.

Excellent for Eye Disorders

Chervil has an outstanding track record in parts of Europe (especially France) for successfully treating a variety of eye disorders, among them being severe inflammation of the deeper structures of the eye (ophthalmitis), separation of the retina from the choroid (detached retina), and loss of eye lens transparency (cataract). And when used in conjunction with other eye herb remedies such as eyebright, the results are nothing short of simply amazing.

A distinguished oculist in Paris in the last century used chervil locally in ophthalmia. He proposed applying chervil poultices to the affected eye and also washing the eye with a decoction of the same plant. This treatment has been recommended due to the good results obtained by other specialists.

The medicinal virtues of this herb are very much linked with its smell, and this is quickly destroyed by heat. So it's a plant that should not be cooked, not even broiled. "The ancients used it for eye troubles," notes Maurice Mességué, a famous French folk healer, "and I have been able to confirm for myself its value in such cases. I, myself, like to use parsley and chervil against conjunctivitis and other eye inflammations. Steep the chopped leaves in boiling water, cool to body temperature and apply the solution with an eye cup. It soothes the burning sensation and acts as a disinfectant."

One of the most successful formulas for many eye disorders to ever come out of France has been attributed to Professor Leon Binet, a prolific author of medical books and a former dean of the Faculty of Medicine in Paris. His remedy calls for equal parts (or 1 tsp. each) of freshly cut chervil, parsley, Roman chamomile (*Anthemis nobilis, not* German chamomile), and lavender flowers, all to be added to 1 pint of boiling water and permitted to steep away from any heat for about 20–30 minutes. I recommend that an equal amount (1 tbsp.) of fresh or dried eyebright herb also be added to the solution, which is later strained and applied to both eyes with an eye cup three times a day. This is good for cataracts, detached retinas and occasionally glaucoma.

CHICKWEED
(Stellaria media)

Brief Description

This apparently feeble member of the pink group is actually a lusty annual with matted to upright green stems that take over many areas. Commencing its growth in the fall, it vigorously thrives through the sleet and snowstorms of winter, even in the far north, survives most weed killers, beginning to bloom while the snow is often still on the ground, and many times it finishes its seed production in the springtime. It's so abundantly fruitful, however, that it flowers throughout most of the country every month of the year.

Growing to a foot high in matted to upright trailing stems, it has egg-shaped lower and median leaves and stemless and highly variable upper leaves. In the star or great chickweed (*S. pubera*), the characteristics blooms, brightly white and about $1/2$" across, have such deeply notched petals that they 5 appear more like 10, the number of stamens. Usually gathering themselves together at night and on cloudy or foggy days, they unfurl under the brilliant sun.

Antidote for Blood Poisoning/Uremic Poisoning

Chickweed ranks beside herbs such as burdock root as being terrific blood cleansers! Where there exists a threat of blood poisoning or tetanus due to chemical dye or dirt getting into the bloodstream, here's what you should do. First, make a poultice and apply it directly to the affected area in order to draw out as much of the poison as possible. To make the poultice, simply blend together 1 tbsp. each of the powdered ginger root, capsicum and kelp, adding just enough honey/wheat germ oil (equal parts) to form a nice, smooth paste of even consistency. Spread this on clean surgical gauze and apply to the area. Cover and leave for up to 7 hours before changing again, if necessary.

At the same time administer internally capsules of chickweed (6 at a time) or a tea (2 cups at a time) made by adding 1 tbsp. dried herb to 2 cups boiling water and steeped for 20 minutes before straining and drinking. The same steps can also be followed with great success in treating carbuncles, boils, venereal disease, herpes sores, swollen testicles and breasts, and so forth. All the herbs mentioned here can be purchased from your local health food store under the Nature's Way label.

Nice Salve Relief for Itching, Rashes

Chickweed brings great comfort to the miseries of chronic itching and severe rashes. Just make a salve using fresh chickweed, if possible; otherwise, the dried powder will have to be used instead.

Needed: 1 1/2 cups coarsely cut fresh chickweed (or 1/2 cup liquid chlorophyll with 1 cup powdered chickweed); 2 cups pure virgin olive oil; 6 tbsp. beeswax. Warm up the oil and beeswax in a pan on top of the stove on medium heat. Then combine all the ingredients in a heavy cast iron skillet or small heavy roast pan and place in the oven for about two hours on just the "warm" setting. Then strain through a fine wire strainer while mixture is still hot, pour into small clean jars and seal tightly.

Herbal Weight-Reducing Program

Most of the herbal literature, past and present, recommends using chickweed in treating obesity. My friend, Mike Tierra, a licensed, practicing herbalist in Santa Cruz, California, mentioned in his *Way of Herbs* that "chickweed is particularly useful for reducing excess fat, having both mild diuretic and laxative properties."

Mike then gives his own weight-reducing program which has helped many of his heavier patients shed unwanted pounds. *Needed:* The following powdered herbs—kelp (5 parts), cascara sagrada (1 part), senna leaf (1 part), cinnamon (1 part), and licorice root (1 part). (I've omitted Mike's 1 part of poke root because I don't consider it that safe to use.) Fill some "00" gelatin capsules purchased from any drugstore with the above herb mixture and take 1-2 capsules three times daily, *before* meals, with a cup of herbal tea.

To make the herbal tea, combine equal parts of these cut, dried herbs—chickweed, cleavers or bedstraw and fennel seed—or approximately 2 1/2 tbsp. of the same in 1 pint of boiling water. Let steep for half an hour before drinking. Both methods should have the scales soon dropping in your favor and pleasing you very much.

Naturalist's Recipes

The late great nature lover and herb forager, Euell Gibbons, devised several recipes using fresh chickweed.

GIBBONS'S CHICKWEED AND GREENS

Needed: "Chickweed is so tender that it cooks almost instantly," he wrote, "and it should always be short-cooked to preserve the maximum amount of its health-giving nutrients. To make Chickweed and Greens, I always use about 2 parts chickweed and 1 part stronger greens. The stronger greens are put on first, covered with boiling water, and cooked about 10 minutes; then the chickweed is added, and after the water has regained a boil it's cooked about 2 minutes more. Drain, but don't throw away that cooking water. Chop the greens right in the cooking pot, using kitchen shears; season with salt, butter, a little pepper and some finely chopped raw onion. Sprinkle each serving with some crumbled crisp bacon. This makes a hearty and palatable dish that requires no apologies."

I've made the above omitting the fried bacon because I so intensely dislike pork, believing it's bad for your health, and have substituted kelp for the salt and capsicum for the black pepper. Also, I've squirted the juice of halves of lemon and lime over the cooked green and stirred just before serving. They give the dish an extra lip-smacking goodness. He suggests that the chickweed/other green cooking water be drunk for obesity problems.

CHICKWEED AND CHEESE SOUFFLÉ

Needed: 2 tbsp. butter; 2 tbsp. whole wheat flour; 1 cup milk; $^1/_2$ cup chickweed, washed and finely chopped (tightly packed); $^1/_2$ cup cheddar cheese, grated; $^1/_2$ tsp. salt; $^1/_4$ tsp. white pepper; 3 egg yolks; 5 egg whites.

Preheat the over to 350° F. In a 2-3 quart saucepan over low heat, melt the butter, and when the foam subsides, add the flour. Stirring constantly, cook for a minute or two. Gradually add the milk, stirring constantly, and allow just to begin to boil. Remove from the heat, and add the chickweed, cheese, salt, pepper, and egg yolks. Mix thoroughly.

Immediately before baking beat the egg whites into the chickweed mixture, and then gently fold in the rest of the whites. Pour the soufflé mixture into a 1 quart soufflé dish or equivalent-sized casserole. Bake in the middle of the oven for 30 minutes or until the soufflé is risen, browned, and set. Serves 2-4.

Chicory, Endive, and Escarole

(Cichorium intybus, C. endiva-latifolia)

Brief Description

Chicory is a scruffy, weedy perennial that is usually cultivated and also found wild in the United States and Europe. The plant has many 2 or 3 foot, sticklike stems; open, widely spaced foliage and milky sap. The striking thing about chicory, however, is its bright, almost iridescent blue flowers that bloom incongruously on the stems as if stapled to the wrong plant. The rootstock is light yellow outside, white inside and contains a bitter milky juice too.

In the United States the name endive usually refers to the small, pale, cigar-shaped plant, while escarole refers to the broad, bushy head with waxy leaves. Endives have a slightly bitter taste. All three salad plants in this family—endive, chicory, and escarole—were believed to have been some of the bitter herbs consumed by the Children of Israel during the Passover before their hasty exodus from Egypt. Chicory root is frequently used in natural coffee substitutes and added to regular coffee to give it a richer flavor and reduce its caffeine content somewhat.

Chicory Coffee for Male Birth Control

There is some clinical evidence that chicory root might be helpful in rendering male sperm temporarily infertile. Scientists in Ahmedadbad, Gujarat, India, administered brewed water extracts of dried powdered chicory roots to 30 male adult Swiss mice, while a corresponding group received no chicory at all, only water to drink instead. In a week and a half, the mice were autopsied. Those that had been on the chicory root brew registered considerable infertile sperm counts as well as decreased weight in their testicles. This information may prove helpful for men not desiring to use condoms.

Good-quality roasted chicory root, either cut or powdered, may be obtained from better health food stores or through mail order from Indiana Botanic Gardens in Hammond. Brewing the roasted root with the drip method of coffee making works best, to give you a very flavorful and rich blend. Making the chicory coffee extra strong and drinking up to 6 cups per day should be enough to render the average adult male's sperm infertile for at least a week without having to resort to other means of birth control.

Effective Liver Cleanser

The same above brewing method makes an excellent drink for cleansing the liver and spleen as well as treating jaundice. An average of 2 cups per day for these purposes is sufficient it seems.

Combats Fat in the System

Lab rodents that were deliberately fed a very-high-fat diet containing chicory roots, experienced a remarkable decrease in their blood cholesterol levels later on. This suggests that whenever deep-fried foods or fatty meats are to be consumed, a cup or two of chicory root brew be consumed with the meal for protection against eventual hardening of the arteries.

Lowers Rapid Heartbeat

Over a dozen years ago a group of Egyptian scientists investigated the potential use of chicory root in treating tachycardia or rapid heartbeat. Their study showed the presence of a digitalislike principle in both the dried and roasted root which actually decreased the rate and volume of heartbeat. Its effects were demonstrated in the toad heart, for instance.

While further research obviously still remains to be done before determining its full impact on human health, it would seem that a cup or two of the root brew by the drip method might just help alleviate this condition somewhat, whenever it occurs.

Neutralizes Acid Indigestion

A cup of the *cold* root brew is excellent for settling an upset stomach or correcting acid indigestion and heartburn.

Dissolves Gallstones

Chicory root and endive tea is very good for getting rid of gallstones. To 1 quart of boiling water, add 3 tbsp. cut root. Reduce heat and simmer for 20 minutes, then remove from the heat and add half a cup of finely cut, raw endive, cover and steep for 45 minutes. Drink several cups at a time twice a day in between meals, but especially so about 2 hours before retiring for the night.

A Complete Chicory Snack

To enjoy something different and slightly off the beaten path that's both healthy and exhilarating, try a snack using all three kinds of chicory species. Accompany the following "wild" salad idea with a cup of instant Country Beverage containing chicory from Old Amish Herbs (see the Appendix).

WILTED CHICORY GREENS SALAD

Needed: 1 medium-sized onion, sliced and separated into rings; 1 cup sliced, fresh mushrooms; 1 minced clove garlic; 2 tsp. butter; $^1/_2$ tsp. dried, crushed basil; $^1/_2$ cup loosely packed raisins and 2 cups each of endive and escarole snipped into pieces with kitchen shears or scissors.

In a large saucepan cook the onion, mushrooms and garlic in butter on low heat until tender, but don't brown. Stir in the dry basil and a dash of kelp if you like. Then add both kinds of chicory greens and 2 tbsp. of apple cider vinegar. Cook and toss mixture occasionally for 2 $^1/_2$ minutes or until the greens begin to turn limp or wilty-looking. Just before removing from the pan, add the raisins and give everything a final stir.

Transfer right away to a serving dish. Should be eaten relatively soon while still warm. Two pieces of lightly buttered pumpernickel toast also make a great accompaniment to this salad and warm chicory coffee.

CHIVES

(See under GARLIC AND ONIONS)

CHOCOLATE (FROM CACAO BEANS)

(Theobrama cacao)

Brief Description

Chocolate is obtained from the ground, roasted beans of the cacao tree: This evergreen with leathery, oblong leaves reaches a height of nearly 30 feet and a trunk width of about half a foot. The leaves are typically evergreen, with whitish or yellowish flowers slightly tinted with orange and pink. The berries are borne directly on the trunk and the branches may be red, yellow, purple, or brown in color. Inside the thick, ridged and furrowed fruit rind, are a white or pinkish acid pulp enclosing 25–60 brown or purple, bitter and somewhat oily seeds.

It is these seeds or cacao beans which are of prime economic importance and yield cocoa powder, cocoa butter, and chocolate upon curing by fermentation and drying, followed next with roasting, and finally by grinding while still very hot. Chocolate was first introduced to Cortez in 1519 by Montezuma. Currently, Americans now spend an average of some $3 billion to satisfy a craving for 1.8 billion pounds of the stuff.

Bitter Baker's Chocolate Stops Diarrhea

On Thursday, January 5, 1995, my father Jacob, my brother Joseph, and I went to the Lion House in downtown Salt Lake City to celebrate my father's 81st birthday.

The rather unappetizing subject of diarrhea was brought up during our lunch by my brother, who complained of having suffered with the problem for the previous four days. He said it had left him extremely weak and fatigued. He asked what remedy might be good for correcting this. Before I could answer him, my 81-year-old father spoke up and mentioned a remedy I had never heard before.

My dad said that his father Jacob Heinerman, Sr., would always resort to dark, bitter baker's chocolate whenever he contracted diarrhea. He would bite a piece off and thoroughly chew it before swallowing. Dad said he claimed that it always worked and promptly stopped his diarrhea.

I then gave my brother one of my own remedies which I've relied on a lot whenever I've traveled to foreign countries. I recommended that he take 1 tsp. of cornstarch and stir it in some milk and drink that. It also is just as effective as the bitter chocolate that my grandfather used.

Removes Skin Wrinkles

Cocoa or cacao butter, from which chocolate is made, can be used to help remove wrinkles on the neck in a skin condition known as "turkey neck," removes wrinkles around the corners of the eyes ("crow's feet"), and also those at the corners of the mouth. Put a little cocoa butter on the ends of your middle three fingers and gently massage in a rotating motion into the wrinkled skin every morning and evening. Perform this procedure each time for about 10 minutes. Within a couple of weeks, most of the smaller wrinkles should be fainter in appearance and the remaining deeper ones not so apparent and noticeable as before.

Facial Mask for Dry Skin

Cocoa powder isn't just for drinking or making cakes with. It's a very good facial mask for softening up old, dry skin that has a slightly weathered look to it. Add enough dairy cream and a little olive oil to about 2 cups or less of cocoa powder to make a dough that's easily pliable and not runny or too stiff. It should be of a thick enough consistency so that you can apply it all over your face much as you would do with a mud pack or green clay poultice. The addition of the olive oil helps to give it greater elasticity and prevents it from drying out too soon or becoming somewhat cakey or chalky-like.

Now 2 cups of plain, dry cocoa powder yield slightly over 2 tbsp. of pure fat, of which about 10–12% is linoleic acid. Add to this approximately 3 tbsp. of olive oil, which gives an additional 12% or so of linoleic acid, and you have a facial mask containing at least 25% linoleic acid not to mention what the dairy cream is apt to provide as well. With the cream, the total linoleic acid content can be expected to be near 30% for that matter.

Why all the fuss over linoleic acid? Simply because it's one of the ingredients to healthier, more youthful-looking and lovelier skin. Dermatitis, a skin disease characterized by scaling, flaking, thickening, and color change, can be treated with linoleic acid. And the main nutrient which gives baby's skin such a nice smooth, subtle, almost velvety softness, is none other than linoleic acid.

This then is why I've recommended this particular mask for dry, rough skin. In a couple of weeks or less, you should begin to notice a definite change for the better in regaining some of that youthful complexion back.

Helps Reduce Hypertension

Dry cocoa powder is used by some native practitioners in the Philippines for treating high blood pressure. They attribute this to the theobromine present, which enlarges the constricted blood vessels common in hypertensive victims.

The high potassium content shouldn't be overlooked either. Now the National Academy of Sciences has set a minimum level of daily potassium intake at 1,875 mg. Putting 2 1/4 tbsp. of cocoa powder in one cup of goat's milk and drinking the same twice a day is one way of illustrating that you'd be getting close to 80% of this minimum amount set.

The research work of Dr. Louis Tobian with the University of Minnesota School of Medicine in Minneapolis and Dr. Elizabeth B. Connor at the University of California at San Diego has proven beyond a doubt that a diet high in potassium is one of the real secrets to lowering blood pressure levels and holding them in check. Therefore, cocoa powder may be of some dietary value in helping to reduce hypertension, but neither cocoa nor chocolate are recommended for migraine headaches.

The Ultimate Mousetrap Bait

Want to get rid of mice, but don't have a cat? Or tried various kinds of bait, and nothing seems to work? Well, just bait new traps with fresh chocolate. And voilá! Your mouse troubles are over. They'll practically kill themselves to get it.

Giving Your Love Life a Boost

It's not so far-fetched as you may think to say that chocolate might be a reasonable aphrodisiac to give your love life a boost of sorts. Chocolate, you see, is loaded with the same kind of stuff called phenylethylamine (PEA) that the brain cranks out in big quantities when stimulated by the passions of love. In his fascinating book, *The Chemistry of Love*, New York City psychiatrist Michael R. Liebowitz, M.D. says that romantically depressed people tend to crave chocolates. He speculates that their PEA levels are low and that chocolate gives them a big PEA boost.

Chocolate Remedies from the Ancient Maya

To understand what the Maya ate and drank in the way of food and used as medicinal agents as part of their daily lives, we must turn to that famous observer of things Maya, Diego de Landa.

Seldom has there been a more two-faced figure. A Franciscan who was named bishop of the Yucatan Peninsula in 1572, late in his life, he burned well over 30 Maya books containing valuable historical, literary, and religious information about their fascinating culture. Yet he recorded an "alphabet" which provided a vital clue for the recent decipherment of the Maya writings which had escaped his holocaust. His portrait still hangs in the vast religious complex which he built on top of an ancient Maya pyramid in Izamal, Yucatan, and depicts him with downcast eyes and a hypocritical, cunning smile of feigned meekness and presumed innocence. However, he clearly overstepped the limits of his clerical authority to such an extent that he was deported back to Spain to be tried for breaking the ecclesiastical rules of his superiors. He subjected his Maya flock to hideous tortures, yet left an exhaustive account of their daily life. It is to him that we must now turn to find out what the sixteenth-century Maya ate and used in the way of remedies.

The bishop mentioned that chocolate was of great ceremonial and social importance to the classic Maya. This has since been verified archaeologically with elegantly painted cylindrical clay chocolate vessels being found in some of their tombs. One of these vessels has painted on it a scene of chocolate preparation, with a buxom young lady pouring the contents of one cylindrical chocolate vessel into another such vessel standing on the ground. The Aztecs employed the same method for raising a fine head on liquid chocolate almost a thousand years later.

Landa claimed that chocolate was added to *pinolli*, a toasted maize (corn) powder carried by travelers, as an alternative to chile. He said that this mixture helped them to endure great distances of travel in hot, humid weather without becoming so easily exhausted. I've tried this myself a few times when in the Yucatan doing field research on folk medicine. I've usually mixed equal parts of chocolate in with toasted corn powder and then stirred the mixture into some hot water. It gives a tired body dissipated by heat renewed energy that seems to last for hours!

Landa also spoke of chocolate drunk with honey, chile peppers, or just plain with water. The first was for helping to settle an upset stomach, the second chocolate-with-chile peppers was intended to stimulate the bowels in the event of constipation, and the last was for helping to ease mild headaches due to slight fevers.

Finally, Landa related a rather curious use for chocolate that may surprise most folks. The classic Maya made a red paste called *achiote* from the outside coating of arnatto (*Bixa orellana*) seeds, growing on a handsome shrub with decorative flowers that is still sometimes planted in tropical courtyards throughout much of Central America. The red coloring matter would then be combined with chocolate which would make the mixture a dark crimson color. The Maya would then slather this stuff all over their bodies to keep themselves *cool* from the intense rays of the hot sun. Believe it or not, from their version of "suntan lotion" came the original term of "*redskins*" which the Spaniards applied to them after seeing their skins covered with this *achiote*-chocolate paste.

The same thing is currently employed by descendants of the ancient Maya in the Guatemalan Highlands for treating small wounds, skin eruptions, and burns to assist healing without scarring.

Wickedly Delightful Recipe

If you're trying desperately hard to lose weight or curb that sweet tooth of yours, then I *strongly* advise you to quickly flip the page and go on to the next entry. *Don't even bother reading this!* Because if you do I'm afraid you might be tempted to try making this next recipe, which Queen Elizabeth of England and Queen Juliana of the Netherlands have indulged themselves in, in times past. It's a dessert treat fit for royalty with a divine taste that is almost out of this world. It's to the point of being sinfully delicious if you're counting calories these days.

A Cake Fit for a Queen

Needed: 6 oz. (approximately 31 1/2 tbsp.) semisweet or bittersweet baking chocolate; 6 oz. (approx. 31 1/2 tbsp.) creamery sweet butter; 4 large eggs; 1/4 cup brown sugar; 1/4 cup blackstrap molasses; 1/2 cup ground, toasted hazelnuts; 4 tbsp. sifted flour; 1/4 tsp. almond extract; 1/4 cup dark honey; 1/8 tsp. cream of tarter; pinch sea salt.

Preheat over to 375° F. Grease an 8″ × 3″ round cake pan with lecithin (purchased from your health food store), then flour the bottom and sides of it. Melt chocolate and butter over low heat in a small saucepan placed in a larger pan partly filled with water. Stir occasionally until melted and smooth, then remove from heat. Meanwhile separate eggs, placing the whites in a clean, dry mixing bowl with the salt and cream of tartar. In another bowl whisk the yolks with 1/4 cup sugar, 1/4 cup molasses and the almond extract, until the entire mixtures forms a nice ribbon when the beater is raised up. Then stir in the warm chocolate mixture, nuts, and flour. Set aside for the time being.

Next beat the egg whites, sea salt and cream of tartar until soft peaks form. Slowly pour in the honey until the whites are stiff but not dry. (Note: 1/8 cup honey and 1/8 cup brown sugar may be necessary to achieve desired stiffness.) Fold about one-third of the whites thoroughly into the chocolate batter to lighten it, then quickly fold in the remaining whites. Turn mixture into the prepared pan and bake for 45–50 minutes. A toothpick inserted into the center of the cake should show moist crumbs—not too dry, not too runny—just right! Cool the cake in the pan, then glaze with the chocolate cognac glaze.

Chocolate Cognac Glaze

Needed: 4 oz. (approximately 21 tbsp.) sweet butter cut into small morsels; 6 oz. (approx. 31 1/2 tbsp.) semisweet or bittersweet baking chocolate cut into tidbits; 1 tbsp. blackstrap molasses; 1 tsp. pure maple syrup; 2–3 tsp. cognac.

Place chocolate, butter, molasses, and maple syrup in a small saucepan and warm gently in a water bath over low heat. Stir frequently until the glaze is silky smooth and completely melted. Be careful, though, that you do not get it too hot. Remove from heat immediately, stir in cognac and set aside until nearly thickened or set up. Refrigerate if you are in a rush.

After the glaze is cool, until almost set but still spreadable, you are then ready to apply it to the cake. Run a knife around the edges of the completely cooled cake to release it from the sides of the pan. Cooled cake will have settled in the center leaving a high rim around the sides. Press this rim firmly with your fingers so it's level with the center. Now reverse the cake onto a cardboard circle but exactly to fit the cake itself. Place on a decorating turntable or on a work surface covered with wax paper.

The bottom of the cake has now become the top instead. Spread the sides and top with just enough cooled glaze to smooth out any imperfections, crevices or rough places. This is termed the "crumb coat," being an undercoating to prepare for a smooth final glaze. Gently reheat the remaining glaze over just barely warm water until it's smooth and pours easily, with a consistency of heavy dairy cream.

It should be just lukewarm by now. Strain the glaze through a very fine strainer to remove any air bubbles or crumbs that might be present. Pour all the glaze onto the center of your cake top. Use a metal spatula to coax the glaze over the edges, coating all sides. Use as few strokes as possible. When the cake is coated, lift it off the wax paper or decorating turntable and let it dry on a rack before moving it to a serving platter.

Now this cake may be presented as it is or decorated with chopped, toasted hazelnuts if you wish, which have been pressed around the sides of the cake just before the glaze hardens. Or melted chocolate can be piped through a paper cone for a more elaborate decoration.

To toast your hazelnuts or filberts, just put them on a cookie sheet in a 375° F. oven for about 20 minutes. Let the nuts cool, then rub off most of their skins between your hands. After this, pulverize them, a handful at a time, in your blender or food processor, using an on-off quick action to prevent making nut butter out of them.

These recipes have been adapted and somewhat modified by me from those provided by Jinx Morgan and her husband, Jeff, who've been innkeepers on the beautiful island of Tortola in the British Virgin Islands.

CHRYSANTHEMUM

(See under ORNAMENTAL FLOWERS)

COCONUT

(See under NUTS)

COFFEE
(Coffea arabica, C. canephora)

Brief Description

Different types of coffee are preferred in various parts of the world. Arabica coffee is produced mostly in South and Central America, particularly Brazil, Columbia, Mexico, and Guatemala, while robusta coffee is produced mainly by African countries such as the Ivory Coast, Uganda, Angola and so forth. In the United States, Colombian and Central American coffees are preferred over Brazilian and African coffees. The March 1981 *National Geographic* concluded that the world's annual bean production could make 3,644,000,000 cubic feet of liquid coffee, a volume equal to the Mississippi's outflow for an hour and a half. And although internal consumption can possibly cause a variety of health problems, ranging from pancreatic cancer and genetic birth defects to elevated serum cholesterol levels and hypoglycemia, it does have several therapeutic benefits.

Grounds Make a Brisk Body Rub

Some health authorities have recommended rubbing the skin with a dry luffa brush in order to enliven the skin more. But in Japan, people are buried up to their necks in roasted coffee grounds and rub the grounds all over their bodies to shed old dead skin and stimulate circulation. You may try the same thing on a more limited scale with the warm ground rubbed on your face and neck in a rotating fashion. You'll find your skin will feel like new in a short time!

Coffee Enema for Really Good Clean-out

Robert Downs, D.C. an Albuquerque chiropractor, claims that an occasional enema several times a year is good for getting rid of hidden toxins that might lurk somewhere in the colon. And the *Journal of the American Medical Association* for October 3, 1980 mentioned that coffee enemas, in particular, have become very popular throughout the country for treating chronic, degenerative diseases.

Fill a hot water bottle two-third full of lukewarm, *freshly brewed* coffee. The coffee should be as strong as possible. Next add one-third lukewarm water to which has been added 2 tbsp. of olive oil. Spread some newspapers on the bathroom floor. Affix the hot water bottle somewhere toward the top half of the door, making sure you have attached the long hose and closed the end of the syringe beforehand so water doesn't run out.

Then lay down on your back, bending both legs at the knees and spreading them apart some distance. Lubricate the syringe with a little petroleum jelly or some saliva from spitting on it to make insertion easier and less painful. Gently work the syringe into your rectum with one hand, while at the same time using the other hand to pull one cheek of your gluteus maximus aside.

Once the syringe is all the way in, you may then bring your upraised knees and legs together a little and release the control stem just above the syringe on the hose. Water will commence to flow into your bowels, but you should keep your fingers on top of the control stem in case the water needs to be quickly shut off for some reason. It's a good idea to permit the water to enter in short spurts, rather than in longer moments, by pressing down and then releasing the control stem every 10 seconds or so.

This way more water can safely enter the colon without causing undue discomfort. *Take in only that amount of water which your bowels can adequately handle with minimum pain!* To attempt more than what can be contained is only asking for trouble—not only in the mess created on the floor, but also the potential damage that could be done internally as well if one doesn't use good judgment.

Neither the coffee nor the olive oil is going to hurt you. In fact, they are probably the best combination of enema ingredients I know of to really tackle impacted fecal material in a quick and direct way! Unfortunately, some overzealous health enthusiasts and cancer quacks working in substandard Mexican border town clinics tend to overdo a good thing and administer several such coffee enemas in a single 24-hour period, day after day and sometimes week after week. Common sense says this is excessive and harmful, to say the least. Soap suds enemas ought to be avoided, according to *Postgraduate Medicine* (Volume 83), because they may be harmful to the delicate membrane linings in the colon and rectum.

COLLARDS

(See under KALE AND COLLARDS)

CORNSTARCH AND CORNMEAL

(Zea mays) (See also Grains)

Brief Description

Depending on where you happen to be in the world, corn can have a number of different meanings. If you are Native American and live in the United States, then it would be called maize. But if you live anywhere in the British Isles, it can have two different interpretations: In London, it is wheat, and in Edinburgh or Belfast, it is oats. In northern Germany, corn is rye, in southern Germany, wheat; were *Volkornbrot* (literally "whole corn bread") means black bread made from the whole kernel of rye, this indicates that northern influence is dominant; where it means simply bread made from any coarsely ground unpolished grain, usually wheat, here you'll find southern influence is dominant. In South Africa, Bantu corn is millet. Actually, all that corn basically means is "grain"; each locality in the world interprets it as standing for its own most familiar cereal.

Corn is not simply food to Native Americans. To many tribes it is the basis of religion and the symbol of fertility and beneficence. "Seed of seeds," "Sacred Mother," "Blessed Daughter," and "Giver of Life" are all appellations by which the sacred corn food is addressed. Indian tribes of the American Southwest are well noted for their many different varieties and colors of corn. In Zuñi Indian ceremonies yellow corn is used to represent the north, blue for the west, red for the south, white for the east, all colors for the zenith, and black for the nadir. The Tewas of New Mexico distinguish seven principal varieties of corn—six colors and a dwarf variety. The Hopi of northern Arizona have as many as twenty varieties of corn. Much ceremony was connected with the planting of corn. Among the Papagos there were songs for every stage of its progress: leaf appearing, stalking growing high, ear forming, and tassel forming. Corn was roasted on coals or in adobe ovens and dried, or it was dried uncooked and then stored on the cob. The monumental task that remained was the grinding of the corn every time meal was needed for mush or bread; this was always left up to the squaws and young maidens to attend to.

An excellent book on the subject is *The Story of Corn* by Betty Fussell (New York: Alfred A. Knopf, 1992). It explores the myths and history, the culture and agriculture, and the art and science of America's most quintessential crop.

Stop Diarrhea Fast

A handy remedy for sudden and unexplained bouts with diarrhea is plain old cornstarch. In 1 cup of milk or water thoroughly stir one heaping tablespoonful of cornstarch. Drink quickly before it settles. Repeat as needed.

Facial Masks for Skin Problems

A good face mask once or twice a week is an excellent way of sealing moisture in your skin, drawing out impurities and tightening pores. The tightening power of masks that dry on your skin helps draw the blood to the surface—an excellent aid to circulation, causing a flushing, restoring process. This is an important function, particularly as you get older and may be less active. Here are some suggestions you can use alone or in combination—just remember to use *equal portions* of each ingredient:

FOR DRY SKIN: Cornstarch with any of these: egg yolks, mashed bananas, peaches, avocados, olive oil, yogurt, powdered milk, cucumbers, sour cream, honey, lemon, grated carrot.

FOR OILY SKIN: Cornmeal plus any of these: papaya, strawberries, tomatoes, lemon juice, cucumber, apple juice, whipped egg white.

FOR NORMAL SKIN: Strawberries, bananas, egg yolk, yogurt, mayonnaise.

CRANBERRY

(See under BERRIES)

CUCUMBER
(Cucumis sativus)

Brief Description

This ancient plant is a native of southwestern Asia, where cultivated seeds almost 12,000 years old have been discovered. The Egyptian pharoahs fed them to their Hebrew slaves, which caused these people later on to grumble against Moses in the Wilderness of Sin for not having any more "cucumbers and melons which we did eat in Egypt freely of."

The cucumber is related to melons and, like them, has a high water content, which keeps its interior flesh cool in the hottest weather; hence, the expression "cool as a cucumber." Cukes are divided into three classes: the standard field-grown slicing kind; the smaller pickling kind, also field grown, and the newer greenhouse varieties, some of which are seedless. A warm tea or cool vegetable drink made of peeled cukes is wonderful for eliminating excess fluid accumulations in body tissues, especially in chronic cases of gout and edema.

Beauty Secrets from Women in Bingham

On the extreme west side of the Salt Lake Valley lays the little mining town of Bingham, Utah. Many years ago a huge mountain towered behind it. But this was gradually replaced over the decades by the world's largest open-pit copper mine, owned and operated by Kennecott Copper Corporation. A book on the town and its inhabitants was written some years ago by two of its inhabitants.

In *Upstairs to a Mine* (Logan: Utah State University, 1976, pp. 69, 71), authors Violet Boyce and Mabel Harmer mentioned some of the cosmetic applications for cucumber by a few of Bingham's older women.

"Millie's Swedish mother guarded her fair Scandinavian complexion," they wrote, "with a mixture composed of half a cucumber, four strawberries, half a peach, and a cup of sour cream. She stirred it well and allowed Millie and me to try it out. After it had caked for a while she wiped it off with a soft cloth and witch hazel. It was wonderfully soothing and relaxing, regardless of what it did for our faces."

One of them mentioned what her cousin Fern used cucumber for. "She was studying to be a stenographer and wanted nice, soft hands. She always had glycerine and rose water on her dresser which kept her hands soft and sweet smelling. Fern also used cucumber to white her hands."

From these couple of examples, it is apparent that cucumber has amazing cosmetic applications to help women stay beautiful. A little bit later in this section I'll be sharing with you a wonderful wrinkle remover from ancient Egypt.

Relieves Tired, Inflamed Eyes

Putting a slice of cucumber over each eye after a long day does wonders to soothe tired, inflamed eyes. This also works well for bloodshot eyes if a mashed poultice of cucumbers in cheesecloth is applied directly on the lids for half an hour or so. It works also for itching and inflammation due to hayfever and related allergies during the summer time.

Perfect Remedy for Wasp Stings

The following was related to me several years ago on an airplane flight by Joy Adkins, then with the Department of Criminal Justice at Marshall University in Huntington, West Virginia:

> When my boy was seven, we went 'seng (ginseng) hunting together. I turned over a branch and disturbed a nest of yellow jackets. They came charging out at us like World War II dive-bombers. My son headed for the creek, swatting and swinging wildly at them every step of the way. I got stung hitting them, too.
>
> Well, we managed to get back to our house but in very painful condition. I went out and got some cukes from the garden and cut them into thin slices and put them right on top of all our stings. And you know what? So help me, you could actually feel the drawing power of these cukes as they drew the sting poison right through the skin. In fact, it kind of hurt a little, like when a person pulls the hairs on your arm. Well, the swelling went down in just a matter of minutes it seemed.

Soothes Aching Feet

Want to give your feet the ultimate treat after so many hours crammed into socks and shoes with little or no air? Cut up several large, unpeeled cucumbers and add to your blender. Whip them up into a thick mush. Refrigerate until cold. Get two pans large enough to accommodate your feet and pour equal parts of this cold cucumber mush in each. Then put your bare feet in them, squishing the mush around between your toes and so forth. And just lay back in an easy chair or chaise lounge and enjoy what some have called "sheer delight" for sore feet.

Nature's Own "Chap Stick"

Have you ever felt the outside of a cucumber? Notice the slightly greasy or oily feeling to the skin. Cucumbers have moderate amounts of fat in their skins. When your lips are chapped and dry, take an unpeeled cucumber, wipe it off good, pucker up your lips as if to put on lipstick, then just run the skin surface slowly across them in a back-and-forth motion. This does a fairly nice job of lubricating them, I've found.

Cleopatra's Wrinkle Remover Secret

Back in the late summer of 1980 on my return home from mainland China, I made a brief stop in Cairo, Egypt. While there, I was introduced to an Egyptologist by the name of Mostafa Abdel El-Selim, who was an expert in deciphering ancient hieroglyphic writings. He took me into an adjacent room of his small museum in the city of Nazlet El-Simman and proceeded to show me some very old-looking and highly brittle fragments of papyri estimated to be almost 2,000 years old and written in the time of Cleopatra, one of the great queens of Egypt.

Knowing of my deep interest in herbs and folk remedies, he handed me an English transliteration of several columns of hieroglyphics which he had recently deciphered after considerable effort (part of the manuscript fragment was missing). But what I was able to read thoroughly excited me. He told me that I was holding in my hands one of the oldest beauty secrets, which according to legend, kept Cleopatra virtually wrinkle-free most of her life.

Herewith for the very first time in book form has this been reproduced with the kind consent of my friends in Egypt. I've made a few slight changes to accommodate our own twentieth-century technology and conveniences, such as substituting the protein from fresh oxen blood to protein from fresh dairy cream and replacing smooth stones for pounding the cucumbers with a blender to whip them up. Otherwise everything else pretty much remains the same.

Slice 2 unpeeled cucumbers lengthwise and widthwise. Put them into your blender with just enough dairy whipping cream to make a nice, thick, smooth mixture that "flows as evenly as the Nile" or, to my way of thinking, has the consistency of cooked Cream-'o-Cheat cereal. Then add 1 tbsp. of olive oil and blend again. Follow this with 1 tbsp. of honey and repeat the blending process again. Finally, add a little mud (I substituted cornstarch instead) and blend for a few more seconds. Set in the refrigerator for half an hour to cool more.

Scrub your face, forehead, neck, and throat with several lime or lemon halves, but don't wipe dry. Immediately lay in a reclining position and slowly apply this cucumber cream mixture as you would shaving cream to the face or legs. Leave this make on for about 1 1/4 hours. Afterward, when the face is thoroughly clean, dip your fingers into a cup half full of some more dairy whipping cream (*un*whipped, however). Rub in a well in a rotating fashion. Let dry in the air. *Under no circumstances apply anything else*. Within a short time, you should begin to experience fewer wrinkles, and if you do this often enough, who's to say that you might not be as beautiful as Cleopatra was? I should also point out that the juice obtained from liquefying a *peeled* cucumber in a food blender is most excellent for teenagers to use for reducing their acne problems caused by a steady diet of hamburgers, French fries, milkshakes, and soft drinks.

A Wild Salad Dressing

Would you like to try something different on your salads for a change, and turn those "vegetable blahs" into exciting "green hurrahs"? Well then just top 'em with this wild idea of a dressing that you're going to find is unbelievably tasty when you and your guests try it.

Cucumber-Yogurt-Goat Milk Dressing

Needed: One 8-oz. carton of plain yogurt; 1/4 cup crumbled gorgonzola cheese (specialty shop or deli for this); 1/4 cup crumbled limburger cheese; 1 tbsp. apple cider vinegar; dash of kelp (health food store for this); 3 tbsp. pure maple syrup; $^3/_4$ of a large unpeeled and shredded cucumber; 1 tbsp. canned goat's milk (goat cheese may be substituted for the above cheeses if you like a milder dressing).

In a bowl combine yogurt, two cheeses, maple syrup, cider vinegar and kelp. Stir in the cucumber. Cover and chill. Then add goat milk to thin a little. Should make 1 $^2/_3$ cups or about 10 good servings.

Cucumbers have always been credited with being cool and refreshing, which is not surprising since they are 96% water. Once seen as only as a salad ingredient, the cuke is now often used as a vegetable in its own right. Delicious salads of cucumber, yogurt, and garlic (but without the goat milk), such *cacik* from Turkey and *raita* from India, are borrowed from Oriental cooking.

Cucumber can also be served hot: diced and cooked in butter it makes a good companion to chicken and is a wonderful revelation with fresh or saltwater fish.

Nice Salad

Thai Cucumber Salad

Needed: $^1/_3$ cup minced shallot; $^1/_3$ cup sliced green onions; 4 medium cucumbers (about 2 $^1/_2$ lb.), peeled, halved lengthwise, seeded, and thinly sliced; 2-4 small red hot chiles, halved lengthwise, seeded, and thinly sliced (1-2 tbsp.); $^1/_2$ cup rice vinegar; 2 tbsp. brown sugar; $^1/_2$ tsp. salt; $^1/_4$ cup chopped fresh cilantro.

Combine the first 4 ingredients in a large bowl. Next combine the vinegar, sugar, and salt, and stir well. Add to the cucumber mixture, tossing to coat it thoroughly. Stir in the cilantro. Yield: Serves 10.

Currants

(See under BERRIES)

D

Daffodil and Daisy

(See under ORNAMENTAL FLOWERS)

Dandelion

(Taraxacum officinale)

Brief Description

The name dandelion is sometimes loosely applied to other milky-sapped weeds with fluffy yellow flowers. But true dandelion is that ubiquitous weed growing prolifically in millions of lawns, backyards and pastures throughout America. This perennial herb has deeply cut leaves forming a basal rosette in the spring and flower heads born on long stalks. All leaves and the hollow flower stems grow directly from the rootstock. The creator of the comic strip "Marvin" once had his adorable diapered hero surveying a clump of dandelions and then thinking to himself, "Dandelions are Nature's way of giving dignity to weeds!"

Grandpa Walton's Wart and Liver Spot Remover

I recall almost a decade ago of being part of a studio audience on a late-night television talk show that featured the late screen actor

Will Geer, who portrayed Grandpa Walton on "The Waltons." Geer was discussing the practical uses for the milky sap contained in the stems of dandelions.

"You just take some of them, break them open, and rub that juice on any wart you have," he told his host, while at the same time illustrating it by a circular motion of his fingers on the back of his band. "You just do that two or three times a day, and I'll guarantee that you won't be plagued with warts anymore."

He also confirmed that this same milky sap was excellent for reducing dark "liver spots" which generally appear on the backs of the hands of elderly people. "I just do the same thing with them that I'd do with warts," he said, "only I use more of the juice and rub it in more thoroughly." He then held up both his hands in front of the camera for a close-up view. From the TV monitors located in the studio, we the audience were able to clearly see just how well this remedy had worked for him. Most of his liver spots had become so faded that one almost had to strain his or her eyes in order to detect any faint signs of them that might have still been barely visible.

Good for Hypertension

In the spring dandelion leaves and roots produce mannitol, a substance used in the treatment of hypertension and weak heart throughout Europe. A tea made of the roots and leaves is good to take during this period, from about mid-March to mid-May. Bring 1 quart of water to a boil, reduce heat, and add 2 tbsp. cleaned and chopped fresh roots. Simmer for 1 minute, covered, then remove from heat and add 2 tbsp. chopped, freshly picked leaves. Steep for 40 minutes. Strain and drink 2 cups per day.

Wonderful Liver Medicine

The late naturopathic physician John Lust stated in his *Herb Book* that dandelion root is good for all kinds of liver problems, including hepatitis, cirrhosis, jaundice, and toxicity in general, as well as getting rid of gallstones. Bring 1 quart of water to a boil, reduce heat to low and add about 20 tbsp. of fresh dandelion leaves, stems, and clean, chopped root. Simmer as long as it takes for the liquid to be reduced to just a pint, then strain. Take 3 tbsp. six times daily, Dr. Lust recommended.

For those desiring something more convenient in capsule form, there is the nice AKN Formula from Nature's Way, which contains considerable dandelion root and other cleansing herbs. It can be obtained from any local health food store.

Remedy for Diabetes

Dr. David Potterton, a licensed, practicing medical herbalist in Great Britain, once wrote that the high insulin content of the root may be regarded as "a sugar substitute to prescribe for people with diabetes mellitus." Three capsules of the dried root each day is recommended for this. The Nature's Way brand from your local health food store is often purchased with this in mind.

Flowers Improve Night Blindness

For those with a problem of being able to see clearly in the dark, the substance called helenin found in dandelion flowers may be just the ticket. According to the *Journal of the American Medical Association* for June 23, 1951, which carried this report, the blossoms also contain vitamins A and B-2 (riboflavin). Steep a handful of freshly picked flowers in a pint of hot water for about 20 minutes. Drink 1 cup twice a day.

Reduces Fever of Childhood Infections

If your child or grandchild comes down with measles, mumps or chickenpox, three common infectious diseases of childhood years, then dandelion tea is the thing to give him. Bring 1 quart of water to a boil. Reduce the heat and add 2 1/2 tbsp. dried, cut root and simmer, covered, for 12 minutes. Remove from heat and add 3 tsp. dried, cut leaves. Steep for half an hour. Strain, sweeten with 1 tsp. pure maple syrup or 1 tsp. blackstrap molasses per cup of tea, and give lukewarm to a child every 5 hours or so until fever breaks and lung congestion clears up. This tea is also excellent for all types of upper respiratory infections, ranging from pneumonia to chronic bronchitis.

Homemade Wine

Delicious dandelion wine is fun and easy to make at home.

DANDELION WINE

Needed: 2 quarts dandelion flowers (make sure they're not sprayed); 1/2 gallon water; 1 orange; 1/2 lemon; 1 1/4 lb. brown sugar; 1/2 cake yeast. Carefully remove all traces of stems. Place flowers in some kind of crockware. Add the sugar, sliced orange, and lemon, then pour boiling water over everything. Let set 2 days, stirring occasionally. On the third day, strain into another crock and add yeast. Let ferment 2 weeks in a warm place.

Delicious Salad

This is essentially Crab Louis, a classic dish.

DANDELION, PURSLANE, AND CRABMEAT SALAD

Needed: 3 eggs; 2 cups (about 3 oz.) purslane, washed, thoroughly dried with paper towel or in a vegetable spinner and chopped medium; 1 cup dandelion, washed, thoroughly dried with paper towel or in a vegetable spinner and chopped medium fine (about 1 1/2 oz.); 1 tomato cut into 8 wedges; 3 oz. cooked king crab meat, broken into bits with all the cartilage removed; 3 tbsp. mayonnaise; 3 tsp. red wine vinegar; 1 tbsp. olive oil; 3 tbsp. medium picante sauce.

Hard boil the eggs, peel, and slice them. Combine the purslane and dandelion. Place them in the center of a large plate. Surround with the tomato and sliced eggs. Sprinkle the crab meat on top. In a small bowl thoroughly mix the mayonnaise, vinegar, oil, and picante sauce. Pour over the greens and crab, and serve. Serves 4.

DATES
(Phoenix dactylifera)

Brief Description

David the Psalmist is quoted in the Old Testament as saying that "the righteous shall flourish like the (date) palm tree." Modern Arabs claim that there are as many uses for dates as there are days in the year.

These sugary fruits are a boon to desert dwellers, growing in hot, dry regions where most food plants cannot—yet they have their own rather temperamental requirements. Date palms must have a source of underground water, but any moisture in the air will keep the fruit from setting, and temperatures below 70° F. will keep it from ripening. The trees themselves can survive in cooler, wetter areas, but their nutritious fruits cannot.

Sun-ripened dates are plump and shiny, with lighter, smoother skins than the dried ones. The latter may contain added sweeteners and preservatives. Fresh dates are often described as either "soft," "semidry" or "dry," depending on the softness of the ripe fruit. Covered and refrigerated, they usually keep indefinitely.

Folk Remedy for Some Cancers

My friend and colleague Jim Duke thinks dates hold some limited merits in the treatment of various kinds of cancer. Jim is with the United States Department of Agriculture Germplasm Resource Laboratory in Beltsville, Maryland, and knows a lot about herbs.

In his recent work, *Medicinal Plants of the Bible*, he devotes a couple of inches of space to their potential with cancer. For instance, he says that a poultice made from crushed date pits and date meat may help testicular tumors. And the fresh fruit, "prepared in various manners," he asserts, may "remedy cancer of the stomach and uterus, abdominal tumors, hardness of the liver and spleen and ulcerated and nonulcerated cancers." A drink made of the fresh, pitted fruit in any kind of juice (orange, carrot or pineapple) would be the most logical way to take it internally, while a handful of pitted dates made into a purée could be spread on external eruptions with apparently satisfying results. Of course, in anything this serious, competent medical treatment should also be sought, besides relying on useful folk remedies like these.

Dynamic Laxative

Ben Harris, a popular health writer, once proposed that six dates be boiled in a pint of hot water for several minutes, and then the resultant liquid drunk warm, morning and night; or six dates eaten raw followed with a glass of warm water twice a day, in order to promote active frequent bowel movements.

Good Digestive Aid

Close to 20% of the total amino acid content of dates is the nonessential glutamic acid. The sourness of its properties is helpful in diluting excess gastric acid in the gut and relieving heartburn. So whenever your stomach feels upset, just eat a few dates or soak several in a cup of hot water for a couple of minutes and then drink the liquid.

Ingenious Ways to Use Dates

A nourishing drink for a perfect midafternoon pickup can be made using dates, milk, and powdered coconut. Take about half a dozen pitted dates, 1/2 cup canned goat's milk, 1/2 cup unsweetened pineapple juice, 1 tsp. powdered coconut, and an even 1/4 tsp. of pure maple syrup, and blend together for a tasty, refreshing drink when your get-up-and-go got up and went!

A spicy, lip-smacking fruit spread could be made by blending together 1 cup pitted dates, 1/2 cup chopped nuts, 1/2 cup plain yogurt, 1/2 cup powdered milk, and 1/2 tsp. ground cinnamon. Refrigerate until ready for use.

Incredible Date/Fig–Nut Bread

Needed: 1/4 cup warm water; 1 tbsp. granular yeast; 1 cup warm goat's milk (canned); 1 tbsp. blackstrap molasses; 1 tsp. sea salt; 1/2 tsp. ground cinnamon; 1/2 tsp. ground cardamom; 2 tbsp.

olive oil; $^1/_2$ cup whole wheat flour; 2 1/2 cups white flour; $^1/_4$ cup pitted, chopped dates; $^1/_4$ cup chopped figs; $^1/_2$ cup coarsely chopped pine nuts.

Sprinkle yeast over water; let stand 2–3 minutes and stir until dissolved. Add goat's milk, molasses, sea salt, cinnamon, cardamom, and olive oil. Stir in whole wheat flour and 1 cup of white flour. Beat well. Add dates, figs and pine nuts and enough additional white flour to make a dough that will clean the sides of the bowl and can be gathered into a ball. Turn out onto a lightly floured board and knead 10 minutes. Cover dough with a cloth and let rest 20 minutes; punch dough down and divide in half. Form into 2 loaves and place in greased 7 $^3/_8 \times 3\ ^5/_8 \times 2\ ^1/_4$ loaf pans (or pans close to these dimensions). Cover with a cloth and let rise in warm place until double in bulk or until dough reaches top of pan. Bake in a preheated, 375° F. oven for about 30 minutes or until bread sounds hollow when tapped. Brush with remaining oil and remove to a rack to cool. Makes 2 loaves. (This recipe has been considerably modified from Eileen Gaten's *Biblical Garden Cookery*, courtesy of the publisher.)

DAY LILY

(See under ORNAMENTAL FLOWERS)

DEWBERRY

(See under BERRIES)

DURIAN
(Durio zibethinus)

Brief Description

The durian, whose stench has been linked to a backed-up sewer, is so offensive that it is banned from Singapore's subways, hotels, and airplanes. But when this spiky pea-soup-green fruit is in season in Southeast Asia, the atmosphere becomes festive. That's because the durian's taste is nothing like its corpse-rotting smell—in fact, it's heavenly.

The flavor of durian fruit has been described as a delightful mingling of garlic, strawberries, custard, onions, pears, and other unlikely combinations. Usually the hard shell is split open and the rich, creamy pulp is eaten raw, but there are other treats such as durian ice cream. Animals don't seem to mind the odor. Durians are devoured by ordinarily carnivorous tigers, and elephants swallow them whole.

Durians grow on trees that can be more than 100 feet tall; the ripened fruits plummet to the ground. The farmers scooping them up must protect themselves from more falling fruits, which are nearly the size of soccer balls. Because they blossom at night, durian trees are pollinated only by bats and fireflies. Even the flowers of these trees have stinky perfume, giving off the scent of clabbered or sour milk.

Durian-growing countries include Myanmar (Burma), Thailand, Vietnam, the Philippines, Malaysia, and Indonesia.

Driving Away Malaria

During the time I spent in trekking all over Indonesia (from 1984 to 1986), I discovered a rather remarkable use for this fruit with the highly offensive odor. Different folk healers with whom I closely worked beat the ripe durian flesh to a juicy pulp and had their patients afflicted with malaria drink and eat this stuff in adequate quantities. I never knew them to fail curing cases of malaria, even when regular prescription drugs failed to eradicate the problem.

Getting Rid of Worms

Intestinal parasites are another problem plaguing many of the countries in the South China Sea. Again, durian fruit is employed for the specific purpose of eliminating them from the G.I. tract. When I asked one of the local folk healers how the fruit achieved this, he simply laughed and replied, "Because they [meaning the worms] can't stand it!"

I presumed the same thing held true for malaria. Here is probably the world's smelliest fruit, that only skunks and grave-diggers might enjoy, successfully curing two pervasive and difficult health problems in that part of the world.

E

Eggplant

(Solanum melongena)

Eggplant's bad reputation for bitter taste goes way back into history. An ancient joke circulated in the Near East many centuries ago concerned a Bedouin in ninth-century Baghdad who was asked to give his opinion of eggplant. His response was that "its color resembles a scorpion's belly and its flavor that of the scorpion's sting." This is still the perception in which eggplant is held today—as something nasty, almost with a bitter sting to it.

But in Southeast Asia, where this vegetable was first domesticated, just the opposite is true. People there tend to view bitter flavors in a much different light than they do in the West. Bitter flavors are greatly valued; consequently, eggplant is held in high esteem. For example, in Thailand a tiny eggplant relative called pea eggplant, which is the size of a marble, gives added zest to soup, curries, and chile sauce on account of its intense bitterness and tartness.

Technically, eggplant is a fruit and not a vegetable, though almost always it's eaten as the latter. When eggplant left the Orient and reached India, something unfortunate happened to it: it became associated with madness. It still had this association when brought to the Near East in the early Middle Ages. Even today, if somebody contradicts himself in Egypt, his listener is likely to say "*Adi zaman il-*

bitin-gan"—"Uh-oh, it's eggplant season." The Tuaregs of the Sahara have informed anthropologists studying their interesting culture that if someone ate eggplant twice a day for 40 days, that person was apt to go crazy. They swore that somebody tried this once and at the end of 40 days he walked around naked, mumbling to himself, "I've eaten eggplant twice a day for 40 days, and now just look at me—I'm drooling and out of my mind!" (Either they really believed this or they enjoyed putting one over on these anthropologists.)

But ultimately the eggplant triumphed. Today it is widely consumed not just in the Middle East and Asia, but also in Africa, southern Europe, and Latin America. The common eggplant in America and in much of Europe, remains the familiar big purple-skinned variety, which ranges in shape from almost round to an elongated globe. But in other parts of the world, there are almost endless variations on the thing: the same shapes occur; however, they may be green or white or multicolored.

Relieves Abdominal Pains

The April 1982 issue of *Tropical Doctor* reported an effective native folk remedy used by Nigerian healers to stop abdominal pains. Equal parts of chopped, dried eggplant, and ripe tamarind fruits were simmered in 1 quart of hot water for half an hour, strained, and then drunk, 2 cups at a time, to relieve any kind of abdominal pains which women might be experiencing.

Calix Tea for Alcoholism

The calix part of the eggplant that is attached to the stem thereof makes a dandy infusion to mellow the effects of alcohol in the bloodstream to neutralize the side effects of certain berry and mushroom poisonings and to stop smokers' hacking coughs. Simmer 2 1/2 tbsp. of chopped eggplant calix in 2 cups of water for 20 minutes. Drink when warm.

Lowers Serum Cholesterol

Eggplant, along with onions, apples, and yogurt, to name just a few, are foods that can really help lower excessive cholesterol in your blood. Eggplant, in fact, contains substances that actually bind up cholesterol in the intestines and carry it out of the body so it can't be absorbed. Adding more eggplant to your diet seems to be a good way of preventing fatty buildup in your heart.

Powdered Eggplant for Dental Health

A special dental preparation can be made from the calix or top part by which the eggplant is attached to its stem. The top parts are spread out on a cookie sheet and *slowly* roasted at 175° F. for 24–30 hours or until they can be easily reduced to a powder by either grinding them up or crushing them with a wooden mallet or hammer.

Next 1/4 cup of sea salt and 2 tbsp. of powdered kelp should be mixed together thoroughly and then evenly spread out on another clean cookie sheet, and also slowly roasted at the same temperature for about half the time. When sufficiently cooled, the sea salt and kelp should be mixed in with the powdered eggplant calixes.

Toothaches can be relieved by rubbing some of this mixture into the gums near the painful tooth for a couple of minutes. The purulent gum disease, pyorrhea, which afflicts several million Americans can be successfully treated by lightly brushing the teeth and gums with some of this eggplant-salt-kelp mixture before retiring for the night. After brushing rinse the mouth with cold water, and, then, with a moistened forefinger dipped into some of this mixture, rub some more of it on the outside of the gums. Leave it on for awhile before rinsing away.

Stops Nosebleeds and Bleeding Ulcers

This same eggplant-salt-kelp powdered blend is very useful for stopping bleeding. In case of a nosebleed, wet the corner of a clean cloth, handkerchief or paper towel with water, squeeze out the excess, then dip it into the mixture and insert the same in the nose. And for internal bleeding, just mix 2/3 tsp. of the blend with 1/2 cup of water and drink at once. Repeat this, as needed, to stop ulcers from bleeding. It should be emphasized here that those with hypertension should not take this internally!

What a Way to Go

This is probably one of the few, if not the only place, in this book that you'll find a pizza recipe. But this one beats 'em all for tasty excitement while still remaining a healthy dish that's good nourishment for your body.

EGGPLANT PIZZA

The preparation time on this is 20 minutes and baking time 10–15 minutes. *Needed:* 1/4 cup olive oil; 2/3 cup warm water; 2 cups multigrain biscuit mix (from any health food store); 2 tbsp. stoneground cornmeal (health food store); 1 cup natural spaghetti sauce; 1 tsp. oregano; 1 1/2 tsp. dried basil; 1 finely minced garlic clove; 1/4 cup shredded mozzarella cheese; 1/2 cup shredded goat cheese (from a local deli); 1/4 cup grated Parmesan cheese; 1 cup thinly sliced eggplant.

Preheat oven to 500° F. Place oil and water in large mixing bowl and stir in biscuit mix. Scrape dough onto a pastry board or other flat surface that has been sprinkled with additional biscuit mix. Knead for about 2 minutes, gradually adding the cornmeal until the dough is smooth and elastic. If the dough seems dry, add a few drops of olive oil.

If using a cookie sheet, rub it with olive oil before rolling out the dough into a circle or any shape you desire. Prior to adding the topping, spread thinly sliced eggplant on another cookie sheet, lightly brush them with olive oil and broil in the oven for just a couple of minutes. Then remove and place them immediately on top of the pizza dough. Next spread on the spaghetti sauce and sprinkle with oregano, basil, and garlic. Distribute the cheeses evenly over the entire surface. Bake for 10–15 minutes until the sides and bottom of the crust are golden brown and the cheese has pretty well melted.

If you think the pizza recipe was different, here is an even more daring idea with plenty of snappy flavor in it.

EGGPLANT CAKE

Needed: 3 medium eggplants; 5 tbsp. olive oil; 1 yellow onion, chopped; 1 clove garlic, chopped; 1 tbsp. butter; 1/2 cup seasoned bread crumbs; 2 eggs; 1 tsp. ground cardamom; 3 drops Tabasco sauce; 3/4 cup heavy cream; sea salt and granulated kelp to taste (from any health food store); and 1 roasted red pepper, peeled and chopped (roast in oven at 375° F. for about 25 minutes).

Preheat oven to 375° F. Trim and halve eggplants lengthwise. Sprinkle with 4 tbsp. olive oil and bake 30 minutes. Then cool.

Cook onion and garlic in remaining oil over medium heat until tender. Scrape the flesh out of eggplants and coarsely chop. Transfer to a strainer to drain the bitter juices. Cut the skin into lengths 1″ wide.

Butter an 8″ springform pan. Line the bottom and sides with skin strips, being sure to keep the shiny side down. Then sprinkle bread crumbs into the pan and shake it well so that the entire interior surface is adequately coated.

Beat together the eggs, cardamom, Tabasco sauce, and cream. Season to taste with sea salt and granulated kelp. Fold in eggplant, onion, garlic, and red pepper. Pour into the prepared pan and then bake for 35 minutes. Remove from the oven. Cool for 15 minutes, then unmold onto a serving platter. Allow to cool for an hour. Serve with chopped fresh tomatoes mixed with olive oil and granulated kelp. Serves 4–6.

Fig
(Ficus carica)

Brief Description

The genus to which figs belong is extremely large, containing some 800 different species of widely varied tropical vines, shrubs, and trees. *Ficus* differs from other genera of the Mulberry family in that the hundreds of tiny female flowers are borne on the inside of a fleshy fruitlike receptacle (called a synocium), with a small opening at the apex.

The common fig is a native of the Mediterranean area. It has been bred and cultivated from early times for its commercially valuable fruit and has been naturalized in other parts of the world that have a mild, semiarid climate. King Nebuchadnezzar grew figs in his famous Hanging Gardens of ancient Babylon. Roman gladiators consumed ample quantities of them prior to combat in the amphitheaters to give them extra strength and stamina, and an advantage over their opponents. Jesus enjoyed eating both figs and dates as the four New Testament gospels verify.

In the United States, figs, are grown in California, Texas, Utah, Oregon, and Washington. Jesuit priests planted figs at the first Catholic mission in San Diego; this so-called black mission fig is still an important variety in that state, which currently grows about 98% of the entire U.S. fig crop.

Some edible varieties can be pollinated only by the fig wasp; the Smyrna fig is representative of this. The insect passes its larval stage inside the inedible fruit of a wild variety called the caprifig. To produce mature fruit, the cultivated variety is subjected to a process called caprification; flowering branches of caprifig are hung in the tree so that the emerging wasps will transfer caprifig pollen to the edible fig. After entering the receptacle and laying its eggs, the wasp dies, and its body and eggs are absorbed by the developing fruit; only the eggs laid inside the caprifig survive. Other edible varieties like the Adriatic or mission fig, bear larger fruits when caprificated. The ripe fruit contains masses of tiny seeds and is soft and pear-shaped; it may be greenish, yellow to orange, or purple in color.

The name fig is also applied to various unrelated plants that either resemble the fig tree or bear figlike fruits.

Relieves Sore Throat and Lungs

Bring 2 cups of water to a boil, adding 5 1/4 tbsp. chopped figs. Simmer on low heat for 5 minutes. Cover and steep until cool. Sip half-cupfuls every 4 hours or so to relieve sore throat and lungs.

Nice Fruit Laxative

In 4 cups of boiling water, put 10 1/2 tbsp. each of figs, raisins, and uncooked barley. Simmer on low heat for 15 minutes, then add 2

$^1/_4$ tbsp. cut, dried licorice root and remove from heat, permitting to steep for 30 minutes or so. When cold, stir and strain. Take 1 cup at night and again in the morning as a laxative. In Egypt a few raw figs are consumed to relieve digestive problems caused by eating too much red meat, fish, eggs, cheese, or milk.

Poultice for Sores and Boils

Put three to four figs in a pie tin with enough milk to cover them. Cover with another inverted pie tin and place in an oven set on a very low temperature for an hour. By that time the figs should have absorbed all the milk. Cut the figs open and lay them directly on the sore or boil. They soon draw out all purulent infection.

Some country folks used powdered figs in a paste to apply to old wounds and sores so they heal much faster. Old Amish Herbs makes such a Fig Paste for livestock and human use, both internally as well as externally (see the Appendix).

Helps Clean Teeth

In parts of Africa and Central America, ripe figs cut in half are used to clean the teeth by rubbing the cut side against the enamel for several minutes.

Figs for Cancer

Scientist Jonathan L. Hartwell listed figs as a useful treatment for different kinds of cancer in his five-year survey, "Plants Used Against Cancer," which was published in the scientific journal *Lloydia* from 1967–1971. In the July 1978 issue of *Agricultural & Biological Chemistry*, a team of Japanese scientists identified the anticancer component in a steam distillation of figs which reduced tumors by 39% as being benzaldehyde. A subsequent follow-up study with 57 cancer patients showed a 50% regression of tumors with the administration of benzaldehyde from figs, according to *Cancer Treatment Reports* for January 1980. Benzaldehyde also occurs in large amounts in edible mushrooms such as *Agaricus bisporus* and Japanese shiitake mushrooms, as well as in sweet almond oil.

This is not to suggest that figs in and of themselves are going to cure cancer. But when used with regular medical care and alternative therapies, it seems to be a very useful food in the treatment of cancer, based upon the data just presented.

Removes Arthritic Pain

Figs, like pineapple (bromelain) and papaya (papain), have an important sulphur compound called ficin, which is valuable in the treatment of chronic joint inflammation and swelling of soft tissues common to rheumatoid arthritis and traumatic injuries, such as a twisted ankle or pulled muscle ligaments.

Soak about 6 figs in 2 $^{1}/_{2}$ cups of boiling water for a few minutes to soften them up a bit; then mash into a poultice and apply directly to any area of stiffness and soreness on the body. Cover with a heavy towel or warm flannel cloth and keep on for half an hour or so. Or apply a thin cloth over them and then a heating pad, set on low heat. It will really bring incredible relief, even to lower backaches—a type of pain that is sometimes hard to get rid of.

Hot Fruit Appetizer

Here's something to really please your palate as a pre-meal warmup item.

Broiled Fresh Figs and Dates

Needed: 9 slices bacon; 12 fresh figs; 12 unpitted dates; some Roquefort cheese, cream cheese, and slices of ham. Cut bacon into pieces long enough to wrap one time around the figs and dates. Remove hard stem end of figs and make a gash on sides with a sharp paring knife. Mix together equal quantities of roquefort and cream cheeses and fill the figs; wrap in bacon, securing with a wooden pick. Pit dates and fill with equal quantities of cream cheese and ham; wrap in bacon, too. Thread the figs and dates on a long skewer and broil, turning several times, until the bacon is crisp. Serve hot from the skewer or keep hot in a small covered dish. This recipe comes from Eileen Gaden's *Biblical Garden Cookery* with the kindness of the publisher.

Healthful Snack

This wonderful recipe not only tastes great and is good for you, but also delivers an amazing amount of energy to the body.

FIG DELIGHTS

Needed: 2 cups freshly-milled brown rice flour; 1 1/2 cups freshly-ground sunflower seeds; 1/2 cup tofu; 1 1/2 cups dried figs; 1 cup apple juice; 1 tsp. ground cloves.

Simmer the figs in apple juice until they become soft. Then blend this fig-apple mixture in a Vita-Mix total food machine or push it through a wire sieve. The quantity should make one cup of fig purée. Combine this purée with the remaining ingredients.

Then add just enough water to form a stiff dough. If the dough is too sloppy, add some more rice flour. If it's too dry, add more liquid. The important thing here is to have a soft, moist dough that forms easily into a log shape.

Wrap wax paper around the log of dough and put it in the freezer for 20 minutes or in the refrigerator for 1 1/2 hours. It can even stay in there overnight, if necessary.

Preheat oven to 350°F. With a thin-bladed sharp knife, slice the log into thin biscuits (about 1 1/4" thick) and lay them out on an ungreased stainless steel biscuit tray. Place one biscuit tray in the oven for 25-35 minutes, depending on the flour type and dough temperature. Try a biscuit after 25 minutes to see if they are done. If required, bake longer.

Remove these biscuits from the tray to prevent further cooking. Cover them with a towel while they cool. When cold, store biscuits in an airtight container.

G

Garlic and Onions

Chives
(Allium schoenoprasum)

Garlic
(Allium sativum)

Green Onion (A Scallion)
(Allium fistulosum)

Leek (A Scallion)
(Allium porrum)

Onion
(Allium cepa)

SHALLOT (A SCALLION)
(Allium ascalonicum)

Brief Description

Chives (usually referred to in the plural) are the smallest, though one of the finest-flavored member of the onion tribe. It belongs to the same *Allium* species that garlic and onions do. Seldom found in the wild anymore, this hardy perennial is cultivated all over the world, from Corsica, Greece, and Sweden to Siberia and throughout North America. The bulbs grow quite close together in dense clusters and are of an elongated form, with white, rather firm sheaths. Chives are used to season omelets, cottage cheese, baked or mashed potatoes, sour cream sauces or dips, salad greens, and salad dressings.

Garlic is a close kin to the onion and was widely used throughout antiquity as an aphrodisiac of sorts, a plague repellent, an antidote to ward off demons and vampires with an embalming agent, not to mention being a popular culinary spice as well. The leaves are long, narrow, and flat like lawn grass. The bulb is of a compound nature, consisting of numerous bulblets or cloves, grouped together between the membranous scales and enclosed within a whitish skin, which holds them as in a sac.

Believe it or not, the elegant and beautiful Easter lily and the dry, old smelly onion are close cousins, both coming from the lily family. This is the largest branch of the onion family, consisting of a mind-boggling number of varieties ranging from mild to sweet. Their skins may be pearly white, bronze, or red. The interiors can be either yellow or white. The globe-shaped onions or American variety are the most pungent and may be white, yellow, or red in color. Bermudas or Spanish onions are much milder in flavor, kind of large and flat and either white or yellow in appearance. A curious custom in ancient Egypt was for people about to take an oath to raise one hand and put the other on an onion, much as we do on a Bible in courts of law today.

The last three members of the onion family are also thought of as scallions. Green onions have bright green tops and very small bulbs, and can be mild or slightly sweet in flavor. Leeks are to French cuisine what beer and hot dogs are to American ball games. These big onions have flat ribbonlike leaves and resemble their diminutive cousins, green onions, in appearance, but not flavor. This was one of several vegetables that the Hebrews moaned and groaned to Moses about missing when they left Egypt for the Promised Land (see Numbers 11). Finally, shallots are definitely for the true gourmet, who prefers a delicate and distinctive flavor that's sweet yet bitey. The small bulbs resemble garlic in appearance.

A closing genealogical tidbit for those who like unusual facts is that asparagus is also a close cousin to all of these onion relatives, but without their noxious odors.

Relieves Headaches Due to Colds

A simple remedy for relieving headaches due to the sinus congestion caused by colds and flu is a tea made from chives and ginger. In 1 cup of boiling water put 1 $^1/_2$ tbsps. finely chopped chives and $^1/_2$ tsp. finely shredded ginger root. Cover with a flat plate and steep for half an hour. Strain, then drink lukewarm. Headache usually goes away in 20 minutes or less. Repeat as often as needed.

Garlic Acts as a Natural Antibiotic

The medical world is gradually accepting the long-held view of many traditional folk healers from around the world that garlic is Nature's own antibiotic penicillin of sorts. The journal *Medical Hypotheses* (Vol. 12, pp. 227–237) noted in 1983 that "there appears to be sufficient data to indicate that garlic is indeed a natural antibiotic" and speculated that "garlic may play a role in preventive medicine and holds a promising position as a broad-spectrum therapeutic agent."

As an antibiotic, garlic has helped cure 82% of spinal meningitis cases compared with only 15% for the drug, Amphetericin (*BEPHA Bulletin*, July 1986) or, along with leeks, has reduced the incidence of poliomyelitis by better than 30% compared with an untreated control group (*Antibiotics Annual*, 1958–59). But it's in the area of cancer that garlic seems to have racked up the greatest medical successes achieved thus far. The following four selected references are but a mere handful of the many published reports implicating garlic extract in the reduction of tumors and other types of malignancies:

- *American Journal of Chinese Medicine,* Vol. 11, pp. 69–73.
- *Science*, Vol. 126, pp. 1112–1114.
- *Journal of Urology,* Vol. 136, pp. 701–705; Vol. 137, pp. 359–362.

Now one of the foremost leaders in current cancer research employing garlic therapy is Benjamin Lau, M.D., Ph.D., a professor with the Department of Microbiology at the Loma Linda University School of Medicine in Loma Linda, California. In one study which appeared in *Current Microbiology* (Vol. 13, pp. 73–76) for 1986. Dr. Lau reported that an extract of garlic completely inhibited the activity and further progress of a parasitic fungus (*Coccidioides immitis*) which has recently become associated with some AIDS victims and is known to produce fever, pneumonialike symptoms, and inflamed skin leisons.

Some of Dr. Lau's more dramatic work has been with the effect of odorless garlic extract by itself and in conjunction with a certain killed vaccine (*Corynebacterium parvum*) in controlling transitional cell carcinoma and bladder cancer. Dr. Lau observed, "*Allium sativum* was shown to elicit macrophages (large scavenger cells) and lymphocytes (white blood cells) leading to cytotoxic destruction of tumor cells."

In a personal letter written to me in the early part of July 1987, Dr. Lau said, "We think most highly of garlic as a natural remedy for various ailments. We further believe garlic to be a valuable supplement to one's diet in terms of overall good health. I am currently working on a project in which we will attempt to isolate and identify anti-tumor and immune-stimulating components of garlic." There are many odorless garlic products available, but only one which has been repeatedly proven to have medical and scientific merit to it. This is Kyolic aged garlic extract, the same one which Dr. Lau used in her experiments and available under the Wakunaga label.

Good Salt Substitute

For those on a sodium-restricted diet due to their hypertension, a green onion or two with your meals may be just the ticket to satisfy your taste buds for salty things. For storage, trim and wash a bunch of green onions, then wrap loosely in a damp paper towel to retain moisture. Refrigerate for up to a week, changing moist paper towel every other day to retain as much freshness as possible. A green onion goes especially well with a chunk of good crusty bread.

Therapy for Burns, Insect Stings

A favorite French recipe is also a favorite French remedy as well. Vichyssoise, the cold leek and potato soup, created by a famous French chef has also been used by several renowned French herbalists as a near perfect remedy for serious burns and bee, wasp and hornet stings, and red ant and centipede bites.

To make a good vichyssoise for healing rather than eating purposes, only the following items in their given amounts are necessary: $^2/_3$ cup each of thinly sliced leeks (green onion may be substituted if true leeks are unavailable) and thinly sliced white onion; $^1/_2$ tbsp. olive oil; $^2/_3$ cup very thinly sliced, peeled potatoes; $^1/_2$ cup each of water and half-and-half.

First, slowly brown the leeks and onions in the oil on medium heat without burning or smoking until they are a light straw color. Add thinly sliced potatoes and $^1/_2$ cup water. Cover and simmer for about $^1/_2$ hour. While still warm, force contents through a coarse sieve so that the potatoes acquire a purée consistency. After this add the other $^1/_2$ cup of half-and-half, stir well; then cool first before refrigerating.

The consistency of the vichyssoise when used externally should be somewhat thick without being too runny. There is sufficient quantity to cover a burn area approximately 1 $^1/_2$ foot in length and about 1 foot in width. Light strips of gauze may be applied afterward and taped down to hold this vichyssoise poultice in place. Not only is this one of the most cooling remedies I know of for burns, but also one of the most effective to speed the healing process. For insect stings and bites, a small amount may be put on the afflicted site and held in place with a little gauze and tape.

Various on this theme include adding the juice of pressed onion along with the half-and-half or wrapping one slice of peeled onion around an insect sting or bite and leaving it on for 3 hours or so before removing. When this raw onion poultice is removed, the stinger will usually accompany it, having been drawn out in the meantime. Or instead of going to all the effort to make the vichyssoise poultice (which is the best for burns), a shorter alternative may be resorted to by mixing $^1/_2$ cup pressed onion juice with 1 cup of plain yogurt and applying it to the burn instead.

Also some white onions run through a grinder and applied directly to a twisted ankle, banged-up knee, dislocated shoulder, fractured arm, or similar injuries often encountered in sports or hard labor will soon remove the pain and swelling.

Say "Good-Bye" to Earache and Toothache

A Queensland, Australia, grandmother (now deceased) by the name of Edith Evans, who invented a world-famous herbal hair restorative, shared a tried-and-true earache remedy with the Canadian press during an October 1981 visit there. "Bake an onion in the oven," she said, "then cut it into slightly thick slices afterwards. Lay one of these while still quite warm on the outside of your sore ear and cover with warm flannel. Keep the other slices in the oven on a "warm" setting until needed. When the first slice becomes cold, discard and replace with another warm one. Do this until the ear-ache disappears completely. The hot onion actually draws out pain."

Or make a garlic-onion oil by soaking 1/2 peeled, chopped onion and 3 peeled, minced garlic cloves in 1 cup olive oil for 10 days, then with an eye-dropper put between 5–7 drops of warm oil in the ear. Another alternative is to use a commercial herb oil made from green onions, white onions and garlic available from Great American (see the Appendix).

Several drops placed in a hot tablespoon that's been previously heated over a flame, will warm the oil just enough to gently pour in an aching ear. After that the head should be put in a reclining position for awhile and the ear covered with a flannel or soft knap hand towel for a time. The pain should cease very shortly.

And for relieving an excruciating toothache, soak a small wad of cotton thoroughly with some of this special oil before placing next to or on top of the bad tooth. If this oil isn't readily available, just peel and crush a clove of garlic and place it on the tooth instead. Pretty soon the hurt and pain will go away.

Super Hair Conditioner

If you want to try a super conditioner that will leave your hair incredibly soft and further enhance its present shade, then you should try making this solution of plain onion skins, of all things!

The part you want to use is the outer clean, dry, brown or golden onion skin, and not the moist, inner one. Store them in a brown paper bag each time you use an onion for culinary purposes. When you have about 2 1/2 cups of lightly packed onion skins, put them into a pan and add 1 quart of boiling water. Cover and steep them for 50 minutes; then strain through a sieve.

After shampooing your hair, towel-dry it briefly. Then rinse several times with the onion skin rinse, before finally rinsing again with clear water. This rinse not only conditions your hair by giving it a much softer texture, but actually provides a lovely color to your own present shade. In fact, it has even softened some of the gray in older people's hair with weekly use.

Getting Rid of Cold Shivers

An assistant professor of medical chemistry from the University of Puerto Rico in San Juan related an interesting little remedy he had discovered for hypothermia, when we met each other at a scientific conference during July 1987 at the University of Rhode Island in Kingston. He found that by grating a couple of garlic cloves, then mixing them with a pinch of cayenne pepper before wrapping the material in a layer of cheesecloth, and applying it to the base of each heel for awhile, was effective in reducing cold shivers or the sensation of coldness in elderly people. The plaster is to be removed when the heels feel hot.

Stops Hacking Coughs

An old Basque remedy common to the Pyrenées mountain range which runs along the French-Spanish border, quiets any kind of cough, from little nagging ones to the more serious hacking kinds such as whooping, smoker's, and asthmatic coughs.

Take two large Spanish onions, peel them, and then slice very thin. Place in a wooden bowl and nearly cover with 2 cups of dark honey. Next put a flat dish or board over the bowl and let set overnight. In the late morning, strain off the syrup and add to it a jigger of brandy. Bottle and refrigerate, taking 1 tsp. of the syrup every 2–3 hours or as needed to stop tickling of the throat and lungs as well as coughing.

Breaking Up Chest Congestion

Smokers, asthmatics, and those allergic to pollen or who may be suffering from a cold or flu should find ample benefits from this simple salve. Peel and finely mince about 7 garlic cloves. Put them into a wide-mouthed, pint-size fruit jar and add just enough melted Crisco shortening to cover. Then stand the open jar in a pan of boiling water for about 3 hours. Permit it to cool after stirring well; then pour into smaller baby food jars, screw on lids, and store for use. Do not strain.

Rub some of this ointment on the throat, chest, abdomen, and upper portion of the back between the shoulder blades and cover with a large, heavy bath towel for awhile. This treatment is very effective for breaking loose accumulated phlegm and allowing greater ease in breathing.

An optional ingredient might be 1/8 tsp. of eucalyptus oil, added to the melted shortening, which will enhance the effectiveness of the ointment, but isn't really necessary.

"Miracle" Medicines for a Multitude of Problems

Garlic and onion are very good for a wide range of other health problems, often duplicating their benefits for the same illnesses. The following table lists uses for both and explains how either may be applied for the conditions cited. In all instances, they have been proven quite effective, either through clinical verification or folk usage.

Uses	*Methods of Application*
Prevents blood clots	Both onions and garlic prevent proteins from massing to form harmful clots. They should always accompany meal intakes of greasy or fatty foods.
Reduces hypertension Controls insomnia	Prostaglandin A is the antihypertensive factor in onions. Two to three capsules of Kyolic garlic or 5 drops of Great American's green onion-garlic oil (see Appendix) per day are helpful for controlling high blood pressure. Eating cooked onions can relax you. And strangely enough, putting a cut raw onion beneath your pillow should correct insomnia.
Increases longevity	A prominent sociologist from the National Institute of Aging surveyed over 8,500 centenarians a decade ago and found that two preferred foods stood out in a majority of their diets—garlic and onion.
Eliminates worms	An old Amish remedy from Lancaster County, Pennsylvania calls for using garlic

and onion to help expel intestinal parasites from man and beast alike. Slices of these raw herbs or prepared oils of the same are generally consumed.

Helpful in diabetes

The medical journal, *Lancet*, for Sept. 11, 1976, noted that both garlic and onion are very hypoglycemic. Meaning they are quite useful for lowering blood sugar levels in diabetics. Up to 4 capsules of Kyolic garlic or about 7 drops of Great American's green onion-garlic oil (see the Appendix) is suggested for diabetics on a daily basis; but those with low blood sugar levels already should avoid these two herbs as much as possible.

Lowers serum cholesterol

Garlic and onion raise blood levels of "good cholesterol," which in turn clean out the arteries of "bad cholesterol" that can clog them up, choking off the flow of blood through the heart. Be sure to eat lots of garlic and onion when consuming fatty meats and greasy foods. Two capsules of garlic or 5 drops of green onion-garlic oil per day is adequate for protection against atherosclerosis.

Plague preventative
Infection Fighter
Resists bacteria-induced diarrhea
Relieves earache

The U.S. Surgeon General has declared AIDS to be "a national epidemic of plague-like proportions." Some doctors think 4 capsules of garlic or 10 drops of onion-garlic oil per day affords reasonable protection against this virus. A vaginal douche made of 4 chopped garlic cloves steeped in 3 cups of hot water for 30 minutes is good for treating yeast infection. Travelers in foreign countries experiencing diarrhea can be helped by chewing some raw garlic or take up to 10 garlic oil capsules daily to kill bacteria. Warm garlic oil also relieves an excruciating earache. At the First World Congress on Garlic held in Washington, D.C. on August 28–30 1991, a number of international scientists presented a variety of papers to show that Kyolic aged garlic extract was superior to other commercial garlic preparations in fighting infection. Benjamin Lau, M.D., Ph.D., an immunologist from Loma Linda School of Medicine in southern California, presented evidence to show Kyolic's amazing anticancer capabilities. (See Appendix under Wakunaga of America if Kyolic isn't available at your health food store.)

Insect repellant

Rubbing your arms, legs, hands, face and neck with garlic oil will keep gnats, mosquitoes, and other bugs away from you during the summer and fall. Peel and chop 10 garlic cloves, adding them to 1 pint of olive oil. Let set for 10 days before using. Also works well on household pets to help them get rid of fleas and ticks.

Garlic makes a terrific insecticide as well as a repellant. A senior citizen horticulture class in Reedley, California, experimented with a garlic oil spray concocted as follows: Lots of finely minced garlic was soaked in mineral oil for at least 24 hours. About 2 tsps. of the oil were added to 1 pint of water in which 2 1/4 tbsp. of Palmolive soap had been dissolved. This was thoroughly stirred, then strained into a glass container for storage. When used as a spray, 1–2 tbsp. of oil mix was blended into 1 pint of water.

The results reported by the Reedley class were astonishing! Cabbage moths, cabbage loopers, earwigs, potato bugs, grasshoppers, mosquitoes (including larvae), whiteflies, and some aphids were killed *on contact!* Houseflies, June bugs, and squash bugs died within minutes after being sprayed. Cockroaches, lygus bugs, slugs, and hornworms died more slowly.

Garlic Preparations For Staying Well

The following data is taken from one of my other books entitled *From Pharoahs to Pharmacists: The Healing Benefits of Garlic* (Copyright © 1994 by John Heinerman, Ph.D. Published by Keats Publishing, Inc., New Canaan, CT. Reprinted with permission).

When working with fresh garlic for medicinal purposes, it is important to keep in mind that the form in which this spice appears has a lot to do with its ultimate success or failure in treating a particular malady. You would certainly not pour garlic tea into the ear canal to treat an earache; a few drops of warm garlic oil would be much better. Nor would you smoke garlic in a pipe for a toothache; a natural packing of crushed garlic and peanut butter would work much better in reducing the pain and swelling until you get an appointment with your dentist.

In this lengthy discussion, 30 different preparations are carefully described, as are the particular problems best suited for each application.

Some of these preparations are good only for immediate use (such as warm garlic enema, used for treating a colicky or constipated infant), while others, which are alcohol based, can last indefinitely under the right storage conditions in a cool, dry, dark place.

Baths

The Romans were the first civilization to make full use of the mineral baths and hot springs available to them in the scattered parts of their vast empire. Here in the Western Hemisphere, the ancient Mayans developed an elaborate system of baths for similar purposes. Many centuries later, the Plains Indians and other North American tribes adapted the Mayan practice into the familiar sweat-lodge ritual, complete with ceremonial accoutrements.

Full or partial garlic baths come in all sizes and shapes, from the bathtub to the eye cup. Basically, they are baths to which garlic teas have been added. The temperature often determines whether such a garlic bath will be calming or stimulating to the mind and body, whether it will open or close the pores of the skin, and whether it will relieve inflammation, pain, or itching.

When making a decoction (boiled tea) for adding to a full bath, anywhere from a few ounces to several pounds of garlic cloves may be tied or sewed into a linen or other cloth bag and then simmered in a quart to a gallon of water. For partial baths, the only difference is that smaller amounts are used, usually about a third as much as for a full bath. When taking the bath, you can put the bag into the water to extract more of the properties, and you can use it as an herbal "washcloth" to give yourself an invigorating rubdown.

Warm garlic baths should be around 95° F. (35° C.). They can be very calming and soothing to the nerves when combined with an equal amount of dried peppermint leaves. Warm garlic baths are especially helpful for bladder and urinary problems, colds, flus, and fevers. Both hot (100° F. or 43° C. to 113° F. or 45° C.) and cold (55° F. or 13° C. to 65° F. or 18° C.) garlic baths tend to shock the system in a positive way, causing increased heart action (in a cold garlic bath the heart slows down after the initial shock). The hot bath followed by bundling up in wool blankets will invariably induce profuse sweating and can be helpful for treating colds, flus, and fevers, not to mention eliminating body wastes retained because of improper kidney function. By adding other medicinal herbs to the full garlic bath, you can create a bath for just about any purpose you could imagine: to soften, moisturize, or scent the skin, to keep flying insects away in the summer, to remove excess oil, to relieve itching, to stimulate or relax, to tighten or tone the skin, to ease muscular aches, and many more. Don't be afraid to experiment with different herbs in conjunction with the garlic bath to find those that best suit your needs and purposes.

The half garlic bath is halfway between a full garlic bath and the garlic sitzbath. You can sit in water up to the navel with the legs and feet under water, but the upper portion of your body remains out of the water. A cold half garlic bath of no more than a minute and just once a day can be useful for migraines, insomnia, nervousness, overactive thyroid, intestinal gas, and constipation. The warm half garlic bath can be enjoyed somewhat longer, for 10 minutes twice a day if necessary, once in the morning and again at night. It should be about 95° F. (35° C.) and can be used for lower back pain, low blood pressure, and menopausal difficulties. The warm half garlic bath usually includes a vigorous brushing of the skin with a natural bristle brush or luffa sponge and may be concluded with a brief spray of cold water on the back.

The garlic sitzbath involves sitting in a small amount of garlic water. To take a sitzbath, put enough warm or hot garlic bath water (actually garlic soup when you get right down to it) in the tub so that it reaches your navel. Prop your feet up on a hassock or chair beside the tub, then wrap yourself with large towels or blankets so you are completely covered from the neck down. If you are using a bathtub, put in about 4″ of garlic water, keep your knees up, and splash the water onto your abdomen. Remain in the tub for half an hour, then rinse with a short cold bath or shower. Garlic sitzbaths are beneficial for the sexual reproductive organs and the urinary tract, the lower abdominal area, and the rectum. They are also of considerable value for remedying inflammations, pelvic congestion, cramps, hemorrhoids, menstrual problems, and kidney and intestinal pains.

For a garlic footbath, simply place your feet and calves into a deep pot or tub filled with plain garlic water. For coldness in the lower extremities, a hot garlic footbath for half an hour makes a good treatment. It has also been recommended by skilled herbalists for bladder, kidney, throat, and ear inflammations.

Sometimes the exact opposite may work wonders. Consider the case of the man who made headlines some time ago in Altoona, Pennsylvania. He claimed he cured all of his colds and flus simply by immersing his two big toes in ice cold garlic water for an average of 3 minutes daily. He made a strong, clear garlic soup, then placed it in the freezer until it was sufficiently chilled without being frozen. Some have claimed that cold (but not frigid) garlic footbaths are good for tired feet, constipation, headache, and nosebleed. Some of my students and I have tried this for the above conditions and found it to have worked for all of them except constipation.

By alternating between hot and cold garlic footbaths, circulation in the lower extremities is definitely improved, varicose veins are helped, and even weak menstrual flow is increased. Insomnia, migraine, hypertension, and persistently cold feet are assisted as well. First, soak the feet in the hot garlic bath for two minutes, then place them in the cold garlic bath for no more than 30 seconds, then start over again. Alternate between the two baths for 20 minutes, always making sure you end with the cold.

Maurice Mességué, one of Europe's greatest folk healers and herbalists, often treated kings, queens, popes, dictators, artists, musicians, and other dignitaries with nothing more than foot and hand baths used together. His lengthy practice enabled him to confirm the virtues of such garlic baths for successfully treating allergies, asthma, acne (associated with stomach and intestinal problems), and hypertension.

Mességué gave instructions for preparing the full, half, foot, and hand garlic baths in his popular French bestseller, *Health Secrets of Plants and Herbs* (New York: William A. Morrow, 1979). He attributed the recipe for these baths to his grandmother Sophie, who passed it on to his father. Take 30 cloves of garlic and slightly bruise them. Pour over them about 3 gallons of boiling water. Cover with a lid and let stand in a big plastic or metal bucket for half a day. Reheat this entire amount in several large pots until sufficiently hot again, then strain through a colander or large wire sieve into the bathtub and fill the tub with more hot tap water to the desired depth. For half baths, simply cut the above amounts of garlic and water by 50%; for sitzbaths, reduce them to 30% and for hand and foot baths, cut to just 15%. Mességué mentioned that it is not necessary to dilute sitz, hand, or foot baths with extra tap water.

There is also the vapor bath, which is particularly suited for inhalation. The simplest form of the vapor bath, which is also called aromatherapy now, is to hang a sachet filled with cloves of garlic about the neck during episodes of colds, influenza, tuberculosis, or other respiratory diseases until the problem is resolved. For an inhalant garlic vapor bath, you need a chair, a pot containing a steaming garlic tea, something to set the pot on, and enough blankets to enclose you and the entire works completely.

To make the necessary garlic tea, first bruise several cloves of garlic and place them in a small saucepan with just enough water to cover. Simmer gently with the lid on over low heat for about 20 minutes. Leave covered and allow to cool down only enough so that the resulting steam won't damage the mucus linings of the nose and sinus cavities. Add one teaspoon lemon or lime juice to this solution and stir. It is important to keep the pot covered at all times so that valuable steam doesn't escape and cool down too quickly.

Next, arrange the chair and the pot so that you can hold your head over the pot to inhale the garlic vapors. Have someone else drape blankets all around so that you and the pot are entirely enclosed. An easier version is to have the pot resting on a low table or counter top with your head directly above it by several inches and a blanket or large towel draped over everything.

With your head above the pot by several inches, breathe the garlic vapor for 15 minutes. This can sometimes be accompanied by a cold garlic sitz or half bath lasting just a few minutes, followed by a few hours in bed, warmly wrapped in blankets like a mummy. This garlic vapor bath is ideal for colds, flus, sinus and respiratory problems, and inner ear inflammations.

The sauna really provides the most ideal garlic vapor bath, but not everyone is so fortunate as to have one. Since necessity is the mother of invention, one can improvise by building his or her own. A cane chair (one with holes in the seat) is required, along with two pots of steaming garlic tea, a wooden grate, and enough blankets to enclose everything, including you, from the neck or waist down. Place one pot beneath the chair and the other directly in front of it so that you can comfortably rest your feet on the wooden grate when it is placed on top of this second pot. Sit on the chair, put your feet on the grate, and have someone else enclose you and the pots completely. You need to be enclosed just from the waist down, but it may be easier to make a good seal at the neck as well. This particular garlic vapor bath lasts about half an hour, and should be followed by a cold garlic half bath of no more than a minute. The final part of the treatment is bed, as with the inhalant vapor bath. The vapor bath is good for kidney and intestinal pains and for prostate problems. If you have cystitis or prostatitis, omit the cold garlic half bath.

To make a sufficiently hot garlic tea for this homemade sauna treatment, follow the preceding instructions but double the ingredients and the time for cooking.

Be advised that garlic should *never* come into direct contact with the eyes because of its highly irritating sulphur fumes. If an allergic skin reaction occurs with any of these baths, discontinue at once and consult a dermatologist or allergist.

For those not ambitious enough to peel the raw cloves necessary for making a full or half bath, one half to one cup of garlic powder or more can be used in a half tub of hot water. Buy your garlic powder in bulk at any health food or herb store for this purpose.

Bolus

The chief difference between a garlic bolus and a suppository is that the former is for insertion into the vagina for treating vaginitis and similar infections and the latter is for insertion into the anus to help with hemorrhoids.

Powdered garlic is made into a thick claylike consistency using melted cocoa butter, water, or honey and then placed in the refrigerator just long enough to harden, after which it is removed and allowed to warm to room temperature before use. It is rolled into strips about 3/4" thick and cut into segments about an inch in length. When water is used in place of cocoa butter, place the formed boluses on a cookie sheet in the oven at a low temperature of 120° F. (49° C.). When they are dry and hard you can store them in an airtight jar.

The best time to insert the bolus is at night just before retiring. The cocoa butter will melt due to the body heat, releasing the garlic into the system. Garlic seems to be most effective at night. When inserting a water-based bolus into the vagina, use a little lotion or petroleum jelly as a lubricant for easy insertion. When honey is used as the moistening agent, just mix a small amount of it with the garlic powder to make a very stiff clay-like consistency. Form it and then stiffen by storing in the refrigerator. Although a honey bolus can work very well, it will not be as firm as the other two types, though it is usually firm enough for vaginal use.

Capsule

See Gelatin Capsule.

Cold Compress

Soak a cotton cloth or terry hand towel in cool garlic tea, wring out the excess liquid, and apply to any part of the body that is inflammed due to sunburn, windburn, or sunstroke. Try to keep the cold garlic compress away from the eyes, if applying it to the forehead. Leave on until it is warmed by body heat, which takes about 20 minutes. Repeat the application with a fresh cool garlic compress and continue until relief is obtained. This also works well with sciatica, neuralgia, and toothache when alternated, hot and cold. It can also be of great help in reducing the pain and swelling of a sprained or twisted ankle, pulled muscle ligament, injured kneecap, sore wrist or elbow, and so forth.

Cold Extract

A garlic preparation made with ordinary cold water will help to preserve the herb's volatile and essential oils, water-soluble vitamins, and sulphur salts. Cover several bruised garlic cloves with cold water, being sure to use an enamelled pan, wooden bowl, or plastic container instead of a metallic pot. Let the mixture stand a good 16 hours in a cool, dark area, then strain, and the drink is ready. Alternatively, combine 6–8 garlic cloves with 2 cups cold water in a blender. Blend at high speed for a few seconds, let mixture stand overnight, strain, and use.

Decoctions

A decoctin is a simmered tea. When you wish to extract primarily the sulphur and other valuable mineral salts from garlic (rather than the vitamins and volatile ingredients), decoction is the best way to go. For every 1/4 tsp. of coarsely chopped garlic clove, use about 1/4 cups of water in an enamelled pot. Bring the water to a rolling boil first, then add the garlic. Reduce the heat and simmer uncovered for no more than 5 minutes. Then remove from the heat, cover with a lid, and steep for 45 minutes. Strain and drink in half-cup amounts.

Both garlic decoctions and infusions are handy for a variety of respiratory ailments and bacterial and viral infections. Because they can be so potent, however, they should be taken in small amounts spread out over many hours each day. (See also Infusion.)

Douche

A garlic douche is the best treatment for vaginal infections or cleansing the body. However, garlic douches shouldn't be overused as they can upset the delicate balance of natural flora within the vagina itself. A garlic douche is made by preparing a strong decoction. After the decoction has steeped for awhile, add 1 tbsp. of apple cider vinegar or garlic vinegar to help promote acid balance within the vagina. The douche is best applied in an empty bathtub or on the toilet, but never have the douche bag more than two feet above the hips. The garlic douche is slowly and gently inserted while still warm (at body temperature) and retained for up to half an hour, if possible. If the liquid is forced in under too much pressure, it may push the infection upward toward the uterus. Don't douche if you are pregnant.

Electuary

An electuary is an old-fashioned way of giving garlic to sick children who may require a strong but natural antibiotic of some kind. A small amount of finely minced garlic is mixed with dark honey, blackstrap molasses, pure maple syrup, organic peanut butter, or some other tasty medium until a soft pasty mass is formed. The sick child is then given this to eat in small amounts. The sweet flavor will encourage a willingness to consume the garlic without much fuss.

Enema

A lukewarm garlic enema can be administered in cases of stomach flu, intestinal parasites, or constipation. A garlic infusion is usually prepared for this, but in some cases something stronger, such as a garlic decoction, may be necessary.

A garlic enema is taken in while first lying on the right side, then on hands and knees, and finally lying on the left side. This will help the solution to fill the lower intestines better. The garlic fluid should be retained for as long as possible and the procedure repeated every few hours or until two quarts of garlic decoction have been taken in and retained for several minutes at a time. The bathroom is the best place for doing this. Lay some newspaper down on the floor and fill a hot water bottle two-thirds full of warm garlic decoction. Make sure a hose with a syringe attachment is put on tight. Hang the apparatus up on a clothes hook or run the hole at the top through 3 or 4 wire hangers suspended from the shower curtain rod. Apply some olive oil or petroleum jelly to the end of the syringe and insert into the rectum while in a flat position on the floor. Don't take in more than what your body is capable of retaining.

Essence

Dissolve 1 tbsp. of homemade garlic oil in a pint of vodka or brandy. This is a good way to preserve garlic's more volatile compounds, some of which aren't soluble in water. It's also handy for rubbing across the forehead for migraines or on the sides of the face and throat for neuralgia and sore throat. Be careful to keep it out of your eyes. An essence is for external application.

Fluid Extract

A fluid extract should never be confused with a tincture. A commercial fluid extract is made by techniques that utilize multiple solvent extraction and can take up to a month or more to complete. This results in a very concentrated product that is often up to 10 times as potent as a tincture and, therefore, only taken in small quantities, such as 6–8 drops at a time. The recommended amount of tincture is often double or triple these figures. A fluid extract is so concentrated that it should always be diluted with water or juice before taking.

Special equipment and training are needed to make a commercial fluid extract of garlic. The process requires several gallons of an appropriate solvent such as vodka, several hundred cloves of raw garlic, and an elevated container large enough to accommodate these materials. The cloves have to be macerated and the entire material cold-percolated with as little heat as possible to retain most of the active principles. A very slow drip method through connecting hoses or pipes into a secondary container of smaller size is employed to get the final fluid extract, after which the menstruum or original extraction materials are discarded.

Sometimes a fluid extract of garlic that has reached this stage may be subjected to further purification processes to eliminate unnecessary fats or oily resins with additional solvents. The term "purified garlic extract" is used to identify this secondary extraction.

The benefits of a commercially prepared fluid extract of garlic derive from its strength. It is strong enough to combat even the most stubborn of the infectious viruses, that have become resistant to synthetic antibiotics. German and French research chemists have shown that otherwise hardy viruses can be easily vanquished by potent commercial fluid extracts of garlic. Health food stores usually carry the milder and more common garlic tinctures, which are sometimes mislabeled "fluid extracts." Most European herb shops and botanical pharmacies carry various brands of the more powerful fluid extract of garlic. Those traveling abroad would do well to visit stores in the United Kingdom, France, Austria, Belgium, Germany, and the Netherlands to purchase liquid garlic extracts. Some naturopathic doctors in the United States and Canada may also carry these potent garlic extracts in their arsenal of *materia medica*.

An old work from 1908, *A Practical Treatise on Materia Medicia and Therapeutics* (Philadelphia: F. A. Davis, 1908) by John V. Shoemaker, M.D., gave as the standard rule of strength for such fluid extracts this formula: "1,000 cubic centimetres of the fluid extract represents the active principle of 1,000 grams of the crude drug."

Fomentation

A garlic fomentation is an external wet application to treat skin infections, open sores, wounds, swellings, and respiratory ailments in the chest. It is similar in some ways to a garlic poultice but is usually weaker and therefore less dramatic in its actions. Some practicing herbalists equate a compress with a fomentation.

To make a fomentation, make a garlic tea in advance. Take a very moisture-absorbent cotton cloth or terry towel, dip it into the hot garlic tea, wring out the excess liquid, and apply immediately to the area of the body desired. The wet material should be covered by a dry flannel cloth and a heating pad or hot water bottle. This keeps the heat in for a longer period of time. A plastic covering is used to protect bedding if it is to be applied overnight.

Friction

Maurice Mességué, a famous French herbalist and popular European health writer, recommended that his older male patients who suffered from impotence rub some garlic on the tailbone near the base of the spine. Several cloves of garlic are peeled and slightly flattened with a heavy object; they are then firmly grasped between the fingers and rubbed in a circular motion in this area for about 10 mintues every day. Mességué claimed a 35–45% success rate for impotence with this method.

Gelatin Capsule

Most manufactured single-herb or formula products sold in the health food industry today are in gelatin capsules. They are inexpensive to produce, convenient for consumers to take, and more easily digested than tablets. Capsules should be stored at all times in a cool, dry place and kept away from heat and humidity. Their average shelf life is about three years.

Both garlic oil and garlic powder can be put into capsules, then placed in heavy plastic or dark glass bottles which are light and moisture resistant. Manfacturers apply tamperproof seals before airtight lids are screwed on.

Though intended for internal consumption, such capsules are versatile and can also be opened and used for external purposes. Between 6 and 10 capsules may be required to equal a level teaspoon of garlic oil or powder.

Garlic capsules should be taken at night just before retiring to obtain maximum benefits. To eliminate the noticeable garlic odor on the breath, take the capsules with a meal, several parsley tablets, fresh parsley and peppermint, or a mixed green drink (dark leafy greens and carrot juice). With the odorless forms of garlic now available, odor is no longer a problem, although some consumers may have to try more than one brand to find a product that is truly odorless for them. Individual reactions vary.

Garlic capsules offer nearly the same antibiotic advantages that synthetic pharmaceuticals do for many kinds of infections. For disease prevention, an average of two capsules daily is recommended. For treating serious illnesses like hepatitis or cancer, triple this amount twice daily may be necessary.

Garlic capsules are also good for bringing down elevated serum cholesterol and triglyceride levels. Usually about 4 a day is adequate. In cases of early arteriosclerosis, though, double this amount may be advisable.

Because garlic is strongly hypoglycemic, it is of proven benefit to those suffering from insulin-dependent diabetes. A few as 6 and as many as 10 capsules per day may be taken with good results. However, those with the opposite problem, low blood sugar or hypoglycemia, should experiment first with just 1 or 2 capsules to make sure there are no unpleasant reactions to the garlic, such as fatigue, mood swings, nervousness, depression, anxiety, forgetfulness, or insomnia. In this case, the garlic should be discontinued and other herbs tried instead.

In 1990 I asked Benjamin Lau, M.D., Ph.D., a professor of microbiology and internal medicine at Loma Linda Medical School, if he had ever found problems with garlic in his hypoglycemic patients. His response was that there were occasional problems with raw garlic, but never when they took aged garlic extract.

Infusion

When you want a milder garlic broth or tea, an infusion is the way to go. Unlike a decoction, which requires boiling, an infusion is simply steeped. Generally, between one and three macerated cloves are steeped in a pint of boiling water in a tightly covered enamel, porcelain, or glass pot away from the stove. Steeping in this manner usually occupies no more than 20 minutes, but it can take longer if desired.

A novel variation calls for putting several peeled and macerated cloves of raw garlic into a glass pint jar, filling it two-thirds full with spring or distilled water, screwing the lid on tightly, and setting it in the sun for half a day. This makes a wonderful infusion.

Making an infusion rather than a decoction preserves more heat-sensitive nutrients such as vitamin C. Also, an infusion is better for very young children and the recuperating elderly. An infusion is preferred to a decoction when administering an enema to an infant or child.

Juice

Maurice Mességué sometimes used full-strength garlic juice for very stubborn infections or fungal growths which had become drug resistant. He pressed several garlic cloves into a pulp and strained the juice through a fine cloth. Then he added the same amount of 90-proof alcohol and 10 times the amount of distilled water. This is a strong antiseptic and keeps indefinitely.

Liniment

A garlic liniment that includes some eucalyptus oil quickly penetrates the pores of the skin and is ideal for treating strained muscles and ligaments. It can also be used for the relief of rheumatoid arthritis, psoriasis, lupus erthyematosus, and similar inflammations. This combination brings a relaxing warmth to tense muscles and expands the blood vessels to increase the flow of circulation.

Peel and slightly bruise 4 cloves of fresh garlic. Put them into a quart jar. Add 2 1/2 cups of apple cider vinegar and 1/2 cup of gin. Seal with a lid and set aside in a cool, dry place to extract. Shake the contents of the jar three times daily, morning, noon, and night. This process will take about two weeks to complete. Strain and discard the cloves, then add 15 to 20 drops of eucalyptus oil. Shake well and reseal until needed.

The use of the vinegar and alcohol causes the liniment to feel cool and evaporate quickly, leaving no messy residue behind. The aromatic eucalyptus oil guarantees fast penetration and rapid relief.

Liqueur

Maurice Mességué, the renowned French folk healer, made the following liqueur from garlic and gave it to some of his older patients as an occasional tonic for increasing their vitality. In a quart fruit jar put 5 crushed cloves of fresh garlic and 3 cups red burgundy wine. Seal with a lid and set on the window sill so it is exposed to plenty of sunlight for 14 days. Shake every so often. Strain through fine cheesecloth and rebottle in a dark glass bottle with a tight-fitting lid. Take 2 drops the first day, 4 the second, 6 the third, and so on until you reach 24; then reverse the procedure until you arrive back at 2 again.

Oil

Homemade garlic oil is one of the finest household remedies to have around. It is especially valuable for infants and young children for the treatment of earaches, inner ear infections, teething, thrush (oral candida), diaper rash, athlete's foot, genital itch, bed sores, and minor burns. Garlic oil will keep its intended potency for up to three months if properly stored in a dark glass bottle with a tight-fitting lid inside the refrigerator. To avoid rancidity, be sure to refrigerate and add a few drops of eucalyptus oil or glycerine as a preservative. For convenience, use a dropper bottle. One teaspoon of garlic oil is equivalent to one clove garlic.

Finely mince enough fresh garlic cloves to equal three quarters of a cup of garlic. Put into a large wooden or plastic mixing bowl and add 3/4 of a cup of pure virgin olive oil. Add the oil slowly 1/4 cup at a time while stirring. Put in a covered glass jar and set on the window sill where it can get lots of sunshine. Let it stand for 1 1/2 weeks. Shake the jar gently 2 or 3 times a day. On the eleventh day, strain the contents through several layers of cheesecloth and store the oil in the refrigerator. To avoid rancidity, be sure to refrigerate and add a few drops of eucalyptus oil or glycerine as a preservative.

To warm a little of the oil for dropping into the ear to relieve an earache or treat an inner ear infection in a young child, put about 6 to 8 drops on a clean teaspoon. Hold a candle under it or hold the spoon over a gas burner for no more than a minute, until the oil becomes lukewarm and pleasant when dropped on the back of the hand. Place the oil in an empty dropper. With the child's head tilted sideways, place the dropper close to the opening of the ear and gently squeeze 3 to 4 drops into the ear. More can be added later on, but do not put too much (more than 8 drops) into the child's ear at any one time.

Ointment

An old-fashioned garlic ointment made in Hong Kong some years ago by an aging Chinese herbalist is useful for any external skin infections or oral problems such as cold sores. One whole garlic bulb and its cloves are peeled and coarsely chopped, then placed in 1 pint of hot water. It is boiled, uncovered on low heat long enough to permit half of the water to evaporate. The remaining liquid is then strained into a hot cup of olive oil, vegetable shortening, or petroleum jelly. Some beeswax is slowly added afterward until a firm consistency is obtained. A little gum benzoin or tincture of benzoin added just before the mixture becomes stiff helps preserve the ointment longer.

Packing

A unique garlic packing devised by herbalist Lalitha Thomas has proven very helpful in treating open sores, major wounds, bad cuts, skin fissures, insect bites and stings, animal bites and scratches, and similar injuries.

Her remedy calls for equal parts of powdered slippery elm, garlic, and myrrh, mixed together dry in a large bowl. Then enough raw, uncooked honey, aloe vera gel, glycerin, or blackstrap molasses is added to make a smooth, even paste. This material is packed into or upon any part of the skin in need of soothing and healing. It works well for sunburn, other minor burns, and general inflammation.

Pill

Garlic pills are used in much the same way as garlic capsules, but they have the added advantage of not being affected by heat so they last longer. Peel 5 cloves of garlic and finely mince them. Add a small amount of powdered slippery elm bark and powdered marshmallow root (equal parts), not to exceed 10% of the mixture. Slowly add distilled or spring water and mix it with the garlic and powdered herbs until a doughy consistency has been achieved, or use a little gum arabic dissolved in boiling water as an adhesive. Roll the dough into little balls about the size of a pea. The pills may be taken immediately, but to preserve them for later use, dry them in the warm air or in an oven on a cookie sheet at low heat for 30 minutes or longer. These pea-sized pills contain about one quarter the dose of a gelatin capsule, so you would need to take approximately four of them to equal a capsule. Such pills are taken for all of the same uses listed under gelatin capsules.

Poultice and Plaster

A garlic poultice is a warm, moist mass of either powdered or well-macerated cloves that is applied directly to the skin for treating blood poisoning, venomous bites, disease eruptions (measles, mumps, chickenpox), insect stings, rose bush or cat scratches, slivers, genital herpes, venereal disease, Kaposi sarcoma (common with AIDS victims), and so forth.

It is a good idea to add a little finely grated fresh ginger root or a pinch of powdered cayenne pepper to the garlic to promote better circulation. It is also necessary to add a small amount of a mucilaginous herb such as comfrey root, marshmallow root, or slippery elm powder for adhesion.

The garlic and powdered herbs can be moistened with hot black or green tea or a tincture of witch hazel, available from any pharmacy. Spread this moistened paste or pulp on a wet, hot wash cloth, apply with the plant material against the skin, and then cover with another dry cloth to keep the moisture and heat inside as long as possible. The first cloth can be moistened from time to time with additional teaspoons of hot water before being covered up again.

A simpler version of this calls for the raw cloves of garlic to be peeled and macerated, then applied directly to the skin with nothing over them. Because raw garlic may cause blistering in some cases, test a small area first.

A plaster is similar to a poultice, but the pounded garlic mass is either placed between two thin pieces of linen or combined in a thick base with powdered herbs before being applied to the skin.

An old-fashioned garlic-mustard plaster, though somewhat irritating to the skin, is ideal for aches, sprains, pulled ligaments, spasms, lower back pain, sciatica, and coldness in the extremities.

It is made by combining equal parts of powdered yellow mustard and garlic powder with just enough tincture of witch hazel to make a thick paste. The paste is spread on a cotton cloth. Another thin cloth is placed on the skin and the mustard cloth put over that. The plaster should remain on until the skin begins to redden and a burning sensation is felt. The plaster is removed and the residue is washed away with cold water.

This plaster shouldn't be used on tender, sensitive areas of the body, such as around the eyes or the groin area. If the mustard powder seems too strong, it can be cut with a little rye or whole wheat flour. After removing this plaster, the skin should be powdered with rice flour, whole wheat flour, or powdered slippery elm bark, then wrapped with dry cotton.

Another excellent way to cool the skin following the application of a garlic-mustard plaster is to use tofu. Put some tofu in a cheesecloth bag or fine-mesh wire strainer and squeeze all of the excess water out of it. Then mash the tofu together with 20 percent pastry flour and 5 percent grated fresh ginger root. This is applied directly to the skin for cooling off the area.

Powder

There are advantages and disadvantages to using commercially prepared garlic powder for remedial means. On the plus side, a powder is much easier to work with than are macerated, wet, pulpy cloves. The powder can be mixed with any type of liquid and formed into paste, poultice, plaster, pills, and tablets. The down side is that if high heat was used to create the powder, valuable vitamins, minerals, enzymes, and fats may have been lost, thereby reducing some of the garlic's healing powers.

To make your own garlic powder, peel and coarsely chop fresh cloves of garlic, spread them on a cookie sheet, and place in an oven at a low heat for several hours, until thoroughly dried. To preserve as many nutrients as possible, you can avoid heat altogether and dry garlic at room temperature in dry weather by spreading chopped garlic on cheesecloth spread over wire racks near a fan or strong breeze. Grind the dried garlic in a coffee grinder or spice mill. The resulting garlic powder should be kept in an airtight glass or plastic container and stored in a dry, cool place.

Salve

A salve is synonymous with ointment.

Smoking

Of all the ways in which garlic can be used, this is perhaps the strangest. But when combined with other herbs and smoked, garlic is efficacious in three specific areas: relieving coughs, clearing up bronchial congestion, and relieving the addiction to tobacco.

The best way to smoke garlic with other herbs is in a pipe or in self-rolled cigarette paper. For bronchical congestion, smoking garlic in combination with coltsfoot and mullein leaves will bring great relief. This is especially true for asthma sufferers. For the relief of a hacking cough, smoking garlic with peppermint leaves and rosemary herb will yield satisfying results.

To aid in quitting the tobacco habit, lobelia (also known as Indian tobacco) is smoked with garlic. It contains lobeline, which is similar to nicotine but doesn't have the same effects. Lobeline and some of the aromatic sulphur components in garlic will reduce the sensation of need for nicotine, but they don't themselves lead to addictive smoking.

Mugwort and catnip have been smoked with garlic in India, China, and Great Britain because of their calming effects in treating insomnia and restlessness. They also tend to reduce some of the stress and nervousness which prompts people to smoke in the first place.

To make garlic suitable for smoking, here are some simple directions. First, peel and coarsely chop the cloves in two bulbs of raw, whole garlic. Spread these evenly on a cookie sheet and set in the oven on low heat for several hours until they are thoroughly dried. Allow them to stand on a counter or stove top overnight until all moisture is gone.

Spread half of them on a clean white cloth or dish towel and cover with the remaining portion. With a rolling pin, go back and forth several times over these dried garlic pieces to reduce them in size a little more. Be careful not to roll too many times, because you don't want powder but rather particles about the size of BB shot.

At this stage, combine the garlic with some of the dried herbs previously mentioned. The entire mixture is then ready for smoking. Some of it can be tapped into a pipe (I'm told that corncob pipes seem to work best for this), or a sheet of tobacco paper can be used.

Smoking herbs in this manner is strictly for medicinal purposes and not intended for pleasure or smoking satisfaction. Smoking is a bad habit and should be avoided whenever possible. But the use of garlic in conjunction with other herbs in this manner is medically relevant when other treatments may have failed.

Suppository

A garlic suppository is for insertion into the anus, for treating hemorrhoids or anal infections. AIDS victims often suffer from severe anal infections and garlic suppositories will be of great value to them. To make a garlic suppository, consult the section under *Bolus* in this chapter and follow the instructions given there, using either cocoa butter or water. Shape the suppository for easy insertion into the anus.

Syrup

Garlic syrup is a great way to administer garlic to young children and the elderly in case of coughing, laryngitis, sore throat, or tonsillitis. There are two basic garlic syrups from which to choose.

Syrup #1 calls for half a cup of peeled, finely minced raw cloves of garlic placed into a small stainless steel, glass or enamelled cooking pot. Add enough raw dark honey or blackstrap molasses to cover the garlic and very slowly simmer on low heat until the garlic seems to have disappeared into the syrup, about 20 minutes. Cover the pot to lessen the evaporation of essential nutrients, but stir frequently with a wooden spoon to avoid burning. While simmering, the syrup can be diluted with distilled or spring water, if you wish. Strain (if you don't like the garlic bits left in) and store in the refrigerator. This syrup can be used as frequently as you want. It is commonly taken as needed or every hour in teaspoon doses for small children and tablespoon doses for teenagers and adults.

Prepare syrup #2 the same as you would #1. Depending upon your need, add any or all of the following to the garlic and honey or molasses:

- 1 tsp. cloves (whole or powdered) for pain relief.
- 1 1/2 tbsp. slippery elm bark or marshmallow root (powdered) for additional decongesting and healing of injured throat and lung tissues.
- 1 1/2 tbsp. grated fresh ginger root or 1 tsp. dried ginger root powder for increasing overall circulation, warmth, and effectiveness of the syrup.

Both of these syrups were originally devised by my friend Lalitha Thomas of Prescott, Arizona, with onion as the chief ingredient. I adapted them to garlic and have used the syrups successfully in a number of cases.

Tablet

Some herbal manufacturers still make garlic tablets. Tablets are handy because they do not deteriorate in heat as gelatin capsules do, so in hot and humid parts of the world, such as the jungles of Central America or Indonesia, where I've done my share of trekking, they are preferable to capsules.

Unfortunately, compression and heat are required to mold and bake these tablets, which destroys some of the garlic's valuable enzymes and vitamins. You need to take about three garlic tablets for every garlic capsule to get the same strength.

Tea

(See DECOCTION and INFUSION)

Tincture

A garlic tincture is similar to a fluid extract of garlic, only not as potent. It is used for exactly the same conditions that are treated by a fluid extract of garlic. (See Fluid Extract).

To make a good garlic tincture, peel and coarsely chop 4 to 5 cloves of raw garlic. Place them in one pint of vermouth, vodka, gin, brandy, or rum. Shake twice daily, morning and evening, to agitate the contents and expedite the extraction process. Do this for a maximum of 21 days, then filter the fluid through a strain or cheesecloth.

Michael Tierra, a practicing herbalist in Santa Cruz, California, recommends that all herbal tinctures be made "on the new moon and strained off on the full moon so that the drawing power of the waxing moon will help extract the herbal properties."

Had my own Hungarian grandmother, Barbara Lieberhardt Heinerman, not done this herself years ago, I would have suspected this piece of advice. But having experimented with making garlic tincture by different lunar phases, I can attest to the stronger properties of the new moon garlic tincture. I attribute this, in part, to the earths electromagnetic forces as well as the moon's gravitational pull.

Average intake for a garlic tincture should be between 10 and 15 drops once or twice daily, as needed. Because it's weaker than a fluid extract of garlic, this tincture doesn't need to be diluted in water or juice but can be taken straight underneath the tongue.

Vinegar

Maurice Mességué, the great French folk healer, often used garlic vinegar for disinfecting bed sores, diabetic leg ulcers, wounds, skin conditions such as scabies and ringworm, and the successful removal of corns, warts (general, genital, and anal), and callouses.

To make garlic vinegar, he would peel and macerate 3 cloves of grated garlic in 3 1/2 cups of wine vinegar (apple cider vinegar may be substituted) for two weeks. To treat the sores, ulcers, and wounds, he applied a cold compress of this garlic vinegar and left it on the affected part until it became warm from body heat, then replaced it with another fresh cold one. For scabies and ringworm, he would wash the skin with the vinegar, rubbing it with a coarse cloth to help it soak in faster. He would usually add 2 drops of camphorated oil to the garlic vinegar for this purpose, to open the pores of the skin.

The best way to use this remedy for getting rid of corns, warts, and callouses is to soak a small cotton ball with garlic vinegar, gently squeeze out the excess, then tape it directly to the growth, leave it on overnight, and change it again the next morning.

Wine

There are two types of garlic wine, both of which are useful in cases of fever, intestinal parasites, and general physical weakness. The first is of French origin and was used extensively by Mességué in his practice. He peeled and slightly bruised a whole bulb of garlic and put it into a stone crock or glass jar with 10 pinches of fresh and finely cut wormwood leaves. To this he added three and a half cups of hot red or white wine and let it set for five days. The dose was two wine glassfuls (a total of 4 fluid onces) a day.

A favorite household remedy in some American and Canadian Chinatowns and also in Hong Kong is a garlic wine kept handy for the cold and flu season. The wine is prepared by soaking 3 peeled garlic bulbs in 1 cup of uncooked rice wine for at least 1 month. Whoever catches a cold or flu takes a tablespoon of this mixture every two hours and again just before retiring. To minimize the undesirable flavor, some Chinese dissolve a little white sugar in boiling water and add this to the garlic wine. But, since sugar isn't good for you, particularly when you are sick, this practice is discouraged.

Dr. Heinerman's book, *From Pharoahs to Pharmacists: The Healing Benefits of Garlic* may be ordered by writing to:

Anthropological Research Center
P. O. Box 11471
Salt Lake City, UT 84147

A "garlic hotline" (212-746-1616) is now available to dispense the latest medical and nutritional information available concerning garlic. It is free to health care professionals, the public, and the media. This new component of the Nutrition Information Center of New York Hospital-Cornell University Medical Center in New York City was started in January 1995 with a generous educational grant from Wakunaga of America—makers of the world's premier selling aged garlic extract known as Kyolic.

An "Onion Family" Delight

A real gourmet treat you're not likely to forget is found in the two recipes that follow, which together constitute a complete meal.

Lusty Spanish Rice

Needed: 2 large Spanish onions, sliced very thin; 2 cloves minced garlic; 1 finely diced leek; 1 finely minced shallot bulb; 4 tbsp. olive oil; 2 cups brown rice; 1 cup chopped, shelled, unsalted walnuts; 4 $^{1}/_{2}$ cups boiling water; 2 sweet green bell peppers, sliced with center seed cores finely diced; 2 tsp. turmeric; 3 tbsp. freshly chopped parsley.

Sauté the onions, garlic, leek and shallot in olive oil until nice and brown. Then add the rice and chopped walnuts. Stir well and cook until all of the oil is absorbed. Next add the water, bringing to a boil. Cover and reduce heat to medium, cooking until all the liquid has been absorbed. In the meantime, sauté peppers and their diced centers. Remove rice from pan, adding turmeric, parsley and peppers. Serve while still hot. This makes a very tasty dish for 6 when topped with the sauce below.

Zesty Leek-Chives Sauce

Needed: 4 cups peeled and diced Pontiac (red) potatoes; 3 $^{1}/_{2}$ cups thinly sliced leeks; $^{1}/_{2}$ cup each of finely chopped chives and green onion; 3 $^{1}/_{4}$ cups water; 1 tbsp. kelp; 1/2 cup half-and-half; some sour cream; a little chopped parsley.

Simmer potatoes, leeks, chives, green onion, water and kelp in a large heavy saucepan for 45 minutes, or until contents are tender. Mash the vegetables with a fork or potato masher and then purée in a blender. Return to pan and reheat a bit. Then remove from stove and stir in half-and-half, sour cream, and parsley. Should be of the consistency of gravy. Use as a sauce to pour over helpings of Spanish rice.

Geranium

(See under ORNAMENTAL FLOWERS)

GINGER
(Zingiber officianle)

Brief Description

Ginger is an erect perennial herb with an aromatic, knotty rootstock that is thick, fibrous, and whitish or buff-colored in appearance. The plant reaches a height of 3–4 feet, the leaving growing 6–12″ long. It is extensively cultivated in the tropics (e.g., India, China, Haiti, and Nigeria), especially in Jamaica.

Antinausea Remedy

A fellow colleague I've known for some years, Dr. Daniel B. Mowry of the Department of Psychology at Bringham Young University in Provo, Utah, conducted an amazing experiment to show that powdered ginger root is the best thing for nausea and vomiting, surpassing even Dramamine, the medication usually recommended for motion sickness.

Thirty-six undergraduate students were asked to take either 100 mg. of Dramamine, 2 capsules of powdered ginger root, or 2 capsules of a placebo (powdered chickweed). Then each one was individually blindfolded and led to a special tilted chair that rotated when the motor was turned on.

Slightly less than half an hour was allowed to elapse after each volunteer swallowed one of the above substances before motion sickness was induced by the rotating chair. None of those who had taken either the Dramamine or placebo were able to last the full 6 minutes in the chair, whereas 50% of those who swallowed the ginger capsules remained in the chair for the full time.

The ginger root group experienced no vomiting, which suggests that the herb is good to take when traveling on an airplane, train, or ship. A product made by Great American Natural Products called Ginger-Up (see Appendix) has become very popular of late for travelers and pregnant mothers experiencing morning sickness, to use, often 2 capsules at a time.

Natural Blood Thinner

People frequently subject to blood clots are generally prescribed oral anticoagulants to help keep their blood relatively thin. One of the most commonly used drugs for this is warfarin sodium (better known as coumadin). Unfortunately, it's also used as a potent rat poison and can lead to serious internal hemorrhaging over an extended period of time. Ginger root is an ideal replacement for such synthetic blood thinners. An average of 2 capsules twice daily in between meals appears to have helped a small number of those with such problems.

Incredible Relief for Aches and Pains

Nothing seems to work quite like a hot ginger compress on muscular aches and pains, joint stiffness, abdominal cramps, kidney stone attacks, stiff neck, neuralgia, toothache, bladder inflammation, prostatitis, and extreme body tension. But keep in mind that as wonderful a remedy as it is, time, considerable effort, patience and a certain change in life-style are all required in order to make it totally successful.

Dr. Koji Yamoda, an M.D. from Tokyo, shares this cure with me.

Bring a gallon of distilled or spring water to a boil in a large enamel pot with a lid on top. Meanwhile wash 1 $^{1}/_{2}$ fresh ginger roots, but *don't* peel them. Then proceed to grate these roots by hand, using a rotating, clockwise motion instead of the usual back-and-forth movements. This keeps those tough fibers from building up on the grater, Dr. Yamoda said.

Next put this grated ginger root in the center of a clean muslin cloth that has been cut to form an 8″ square and slightly moistened. Then draw the corners together to form a nice, little bag and tie the top with string, thread or fish line. Be sure to leave plenty of room inside the bag for air and water to circulate.

Before putting this ginger bag in the hot water, make absolutely certain that the heat has been turned down and that the water is no longer boiling. Now uncover the pot and gently squeeze the juice from the bag into the water, before dropping it into the pot. Cover and permit the contents to simmer an additional 7 minutes.

Dr. Yamoda informed me that the resulting liquid would acquire the hue of gold and yield a distinctive ginger aroma. The bag may be pressed against the sides of the pot with a wooden ladle to turn the water yellow, if the process seems to be a little slow in happening. Remove the pot from the stove when ready and set aside.

To be effective, ginger compresses must be applied relatively hot, he insisted, but not so much as to seriously scald the skin of the patient. Besides being used for compresses this ginger broth can also be added to bath water to soak an aching back or sore muscles in or for soaking tired, aching feet in as well. The patient should be laying flat on the floor to receive the full benefits of these compresses front or back.

A terry cloth hand towel is dipped into the pot, while holding both ends. The towel is lifted out and excess water is gently squeezed back into the pot. The steaming towel is then refolded to the desired width and applied directly to the site of pain. A second such compress can be placed immediately over or next to the first one, after which a large and fluffy, dry bath towel is placed over both compresses in order to retain as much heat as possible for the greatest length of time. The bath towel should be folded in half at least once before covering the compresses. Under these conditions, the compresses should remain fairly warm for up to 15–20 minutes. Dr. Yamoda recommended that another set of compresses be applied after this for a total treatment time of 45 minutes or so, and repeated again about 4–6 hours later or as needed.

He explained that in all of his years of clinical practice, nothing seems to have relieved most kinds of physical aches and pains as well as this remedy has. He has even used such ginger compresses on the chests of patients suffering from extreme asthma and bronchitis, with their mucus congestions breaking up in no time at all. Smaller washcloth-sized compresses can be applied against the side of the neck, throat or jaw to relieve neuralgia, stiffness, swollen glands, and toothache.

Relief for Hypertensive Headaches

Mix enough powdered ginger and cold water together in a small bowl to make a thin, smooth paste. Then apply to the forehead and temples with the back of a large tablespoon and lay down for awhile. This will help to relieve the excruciating pressure building up inside and take away that "exploding" sensation.

Breaks Fever, Eliminates Phlegm

One of the best ways to help break a high fever and get rid of mucus buildup in the sinuses, throat and lungs, is to drink some warm ginger tea. Grate enough fresh ginger root to equal about 2 level tbsp., then add them to 2 cups of boiling water and cover, steeping for 30 minutes. Drink 1 cup while still warm every 2 1/2 hours.

GOURD

(See under PUMPKIN AND SQUASH)

GRAINS
(See also BRANS, BREAD AND PASTA)

Brief Description

BARLEY (*Hordeum vulgare*). Originated in western Asia, where it was one of the first grains to be cultivated. As human food, the larger, white-seeded variety of barley is pearled or ground in a revolving drum until the hull and germ are removed. This reduces the grains to small, starchy balls which are then used to thicken soups. Pot, or hulled, barley is ground enough to remove only the husk.

BUCKWHEAT (*Fagopyrum vulgare*). Native to central Asia. Thought of as a cereal grain, but is really in a family of its own. Mainly used for making flour for pancakes. Groats, or kaska, are kernels with the hulls removed. They are eaten as breakfast food or as thickeners for soup, gravy, and dressing.

BULGHUR (Parboiled cracked wheat. See under WHEAT.)

CORN (*Zea mays*). Native to the Americas. Commercial varieties are either yellow or white in color. Popcorn is distinguished by its small, hard kernels with tough outer covers, while flour corns have soft, starchy kernels. Other recognized types include dent, flint, sweed, pod, and waxy corn. The basic products of refined or processed corn are starch, oil, syrup, hominy grits, cornmeal, and flour.

Indian corn is noted for its unusual variety of colors. For instance, the Hopi of northern Arizona have at least 20 varieties, with multiple legends and religious beliefs being attached to each particular color. Considerable ceremony often attends the planting of corn by some southwestern Native American tribes.

MILLET (*Panicum milliaceum*). Native grain of the East Indies. Is a common name applied to a variety of cultivated grasses with small white or golden kernels. Among the most popular are foxtail and pearl millet. Since it lacks gluten, it's good for people who must avoid this protein.

OATS (*Avena sativa*). Developed from the wild grasses of eastern Europe and Asia. Today there are three general classes and nearly 100 varieties grown. They are equal to corn as a tissue builder. Steel cut oats are simply cracked oat kernels, while rolled oats have been flattened or rolled into thin flakes.

RICE (*Oryza sativa*). An ancient grain cultivated for over 4,000 years. Originated in southeast Asia. Each whole grain has an outside hull, a brownish-colored covering called bran and a finer, lighter-colored layer called polish which surrounds the kernel. Commercial varieties are classified on the basis of size and shape of the kernel—short, medium, and long grain. All have the same food value, but long grain costs more since more kernels break during milling. Brown rice is simply rice that has not had the bran and polish removed.

RYE (*Secale cereale*). Developed from a wild variety still growing in the mountains of eastern Mediterranean countries. It's deficient in the glutinous proteins that give wheat dough the elasticity necessary for good leavening, so pure rye bread is heavy and compact by comparison.

TRITICALE (*Triticum secale*). A hybrid grain produced by cross-breeding wheat and rye. First bred in the 1930s by Swedish agronomists and has since become very popular in the United States and Canada.

WHEAT (*Triticum aestivum*). One of history's first cultivated grains in western Asia. Now covers more of the earth's surface than any other grain crop. For marketing purposes, five classes of wheat based on usuage and habit of growth were established—hard red spring, soft red winter, hard red winter, durum, and white. Generally, the harder translucent varieties are valued for the production of flours while durum is prized for the manufacture of macaroni, spaghetti, and noodles.

WILD RICE (*Zizania aquatica*). Native to the Great Lakes region of the United States and Canada. Although used in the same way ordinary rice is, this really isn't true "rice" as such. Wild rice has a gutsier, chewier, and somewhat smokier taste than conventional rice.

Grain Roto-Rooters for Clogged Arteries

Certain grains such as barley and oats really help to clean out the arteries and valves around the heart that have been plugged up with layers of old fat buildup. And to a somewhat lesser extent, so do rye and wheat as well. All these grain fibers scrub away backlog deposits of grease that have accumulated over a lengthy period of time.

Cereal Grasses for Arthritis, Cancer, and Ulcers

The green juice from young barley shoots possesses strong anti-inflammatory activity. A paper read at the 101st Annual Meeting of the Japan Pharmaceutical Society in April 1981, reported that the powdered juice significantly reduced arthritis and gastric ulcers in lab rodents. And research conducted by Dr. Chiu-Nan Lai at the M.D. Anderson Hospital & Tumor Institute in Houston, Texas, shows that extracts of wheat sprouts can modify, even decrease, the formation of cancer of the esophagus, stomach, liver, breast, and colon, if regularly used in the diet.

One of the best sources for this is Kyo-Green made by Wakunaga of America. Two level tablespoons in 8 oz. water or juice makes a delicious and lovely emerald-green drink. One glass each day will help relieve pain, promote healing, and shrink tumors, as well as to prevent the onset on these diseases. (See the Appendix for more information.)

Grow Your Own Grain Sprouts

You can make your own grain sprouts by following these simple directions. Thoroughly wash 1/3 cup wheat berries, rye berries, or brown rice. Place the grain kernels in a bowl and cover with enough water (about an inch) for grain to swell; then cover with lid. Let stand overnight in a cool place. Drain and rinse them.

Wash three 1 quart jars; place about 1/4 cup of the soaked grain kernels in each jar. Cover tops of jars with two layers of cheesecloth or nylon netting. Fasten the cheesecloth on each jar with two rubber bands or a screw-top canning-jar lid band.

Place the jars on their sides in a warm, dark palce (68–75° F.). Once a day rinse the sprouts by pouring lukewarm water into the jars. Swirl to moisten all the grain kernels, then pour off the water. In 3–4 days, the grains should sprout, with the exception of brown rice which may take 5–6 days instead. Once grains have sprouted, keep them refrigerated until serving time. They should keep for up to a week, this way. Use in salads, sandwiches, soups, or breads.

While most sprouts require a dark place, the cereal grains should be given a few hours of either artificial light or indirect sunlight after their initial sprouting to let them develop chlorophyll. A sprout length of 1–2″ is generally good for this stage of growth. Green cereal sprouts make delicious, healthy drinks when juiced or blended and other vegetable juices like carrot or tomato, for instance.

Barley and Wheat for Body Building

Those in their twenties and thirties, who wish to develop finer physical physiques through strenuous exercise, should consider adding barley and wheat to their diets. Both grains contain growth promoting factors in their young shoots or sprouts. Interestingly enough, the ancient Roman gladiators were called *hordearii* or *barley men* because they consumed so much of this grain just before entering the amphitheaters to battle their opponents. Barley and wheat may be used in cooked breakfast cereals, breads, pancakes, soups, salads and delicious drinks to give increased strength and muscle expansion to those who work out regularly with weights.

Buckwheat for Appetite Control

If you're desperately trying to lose weight, but having a hard time doing it because you're unable to cope with the deadly "munchies," then may I strongly encourage you to start eating more buckwheat pancakes for breakfast. Just two medium-sized pancakes in the morning with a couple of pats of low-fat margarine and some pure maple syrup poured over them, will not only fill you up for the next 4–6 hours, but also prompt *lesser* food intake during your next meal.

In an informal study conducted by our Anthropological Research Center here in Salt Lake City several years ago, 11 overweight people were put on a 2-week program consisting of buckwheat pancakes, along with several other cereal grains. The buckwheat pancakes (2–3 medium-sized) were consumed every other morning and every other evening, with cooked oatmeal being eaten at alternate breakfasts and dinners. Subjects were permitted to eat whatever they wanted for lunch, no matter how sweet or fattening it was. For late night snacks, they were allowed as much shredded wheat and milk they desired, topped off with 2 tbsps. of a commercial brand of granola.

In this period of time, 7 of the 11 volunteers lost an average of 15 3/4 lb., with 4 of the 7 losing in excess of 22 lb. each. All 11 of them, however, noted *a substantial decrease* in the frequency of their snacks as well as in the volume of food actually consumed.

Bulghur Helps Diabetes

Clinical evidence shows that bulghur is one type of food diabetics can safely rely on to help lower their blood sugar levels.

Now there are two ways to make bulghur. One method calls for a cup of whole wheat grain to be boiled in a heavy, covered saucepan with 1 cup of water, after which the heat is reduced and the contents simmered for an hour. The second method calls for the same amount of wheat and water to be put into a small pot of some kind. This pot is then set on a rack in a larger pot which has water in it also. The water should come up almost to the level of the rack. Cover the larger pot and put on high heat for 15 minutes. Reduce heat and steam until wheat absorbs the water (about 45 minutes longer). It can then be consumed either as a breakfast cereal or snack with goat's milk or else used wherever rice would ordinarily be called for.

All Parts of Corn Are Therapeutic

Perhaps no other grain, it can be said, has as many parts of it with therapeutic value as corn does. Consider this: the kernels, the cob, the cornsilk and the meal and starch made from the kernels all have medical significance to them.

When corn is frequently consumed in the diet, either fresh, canned or popped, cholesterol levels go down and bowel movements increase. Soft, boiled corn grits are good to eat every day in case of kidney problems, especially where kidney dysfunction is the cause of swollen legs resulting from lack of urination. Cornmeal made from the ground kernels makes a great facial, opening all of the pores and freeing them from dirt and oils. Just wash the face twice daily with cornmeal instead of soap. You'll find it doesn't even leave the skin as dry and flaky either.

And cornstarch makes a great dusting agent for relieving diaper rash and poison ivy itch, not to mention reducing insect bites and stings, as well as adding a large handful to a tub of lukewarm water to bathe chicken pox, measles, mumps, and hives in. Besides this, taking a box along with you when traveling to a foreign country where the cleanliness of the food and water is in doubt, quickly stops diarrhea. Just add 1–2 level tsp. to a glass of cool water that's been previously boiled. This usually corrects the problem in a couple of hours.

Corncob tea is excellent for treating abdominal swelling, edema in the ankles and wrists and gout in general. Cover 2–3 fresh cobs from which the corn has already been removed or eaten with enough water to cover by 2″. Cook on low heat for about an hour, then strain and cool. Keep in the refrigerator, drinking 2–3 cups a day until problems subside, then reduce intake to only 1 cup per day.

Cornsilk tea is one of the finest remedies for weak or poorly functioning kidneys and kidneystones. It may be used fresh or dried. Steep 2 tsp. of cornsilk in 1 cup of boiling water for 20 minutes, strain, sweeten with honey, and then drink $^1/_2$ cup lukewarm every 3–4 hours. It is ideal to curb bedwetting habits when used with equal parts of catnip leaf and leaves, and valerian root.

Corn Tortillas Ideal Snack in Cool Weather

The holiday period between and after Christmas and Near Year's is the scene for a number of college football bowl games played all across the country. One of the most watched of these was the Cotton Bowl played in the stadium on the state fairgrounds in Dallas, Texas. The opposing teams were the Red Raiders of Texas Tech in Lubbock and the Trojans of USC in Los Angeles.

By the third quarter, USC was 48 and Texas Tech 0; it looked like a virtual shutout for the latter team. That is, until African-American quarterback Zebby Lethridge ran in the only touchdown for Texas Tech, at least putting them on the gameboard with 7 points.

Thousands of Texas Tech fans reacted with wild excitement by hurling hundreds of corn tortillas on to the playing field in celebration of this. The NBC television sportscaster Jim Lamprey, who was calling the game, reported that "many fans actually nibble or eat these tortillas to help stay warm." The other sportscaster kidded him by asking if the fans wrapped their hands up in them. But Lamprey insisted, "No, I mean many claim it gives them enough carbohydrate fuel to keep their bodies warm." (The temperature in Dallas on that Monday afternoon of January 2, 1995, was a cool 46° F.

There may actually be something to this and is well worth checking out the next time you're sitting at a chilly football game and want to stay warm. Try nibbling on some raw tortillas. By the way, USC went on to win with a score of 55 to 7; this was the largest margin of victory ever in the 59-year history of the Cotton Bowl.

Millet Useful for Celiac Disease

Celiac disease is an intestinal disorder caused by the intolerance of some individuals to gluten, a protein found in some cereal grains such as barley, rye, and wheat. Symptoms of celiac disease are weight loss, diarrhea, gas, abdominal pain, and anemia. Malnutrition usually accompanies this disorder because of the greatly reduced absorption of nutrients.

Besides rice and corn, millet is the only other cereal grain that can be safely consumed without a sufferer experiencing further health problems. The addition of cardamom and mace to these other grains makes their gluten more tolerable in those with this disorder.

Oats Lower Bad, Raise Good Cholesterol

In the old Western flicks the good guys usually wore the white hats, while the bad dudes were clothed in conventional black. With cholesterol it's high-density lipoproteins (HDL, the good guys) versus low-density lipoproteins (LDL, the bad guys). HDL protects the heart from fatty deposits, while LDL on the other hand plugs arteries like crazy and contributes to general obesity.

Jim Anderson, M.D., a noted Kentucky researcher working out of the V.A. Hospital with the University of Kentucky College of Medicine in Lexington, has had considerable interest in plant fiber and its effects on cholesterol. His studies, along with recent 1987 research at Syracuse University in New York, both show that high-fiber, low-fat diets alone do not reduce cholesterol as much as when oat bran is added to both diets. Oats also raises HDL, thereby preventing heart attacks and hypertension.

Health food stores around the country carry oat bran in a ready-to-eat breakfast cereal marketed by Health Valley of Montebello, California. Or you can do as I've done for most of my 40+ years—eat a good helping of cooked Quaker Oats every morning. This is my typical breakfast, oatmeal and milk, with little or nothing else accompanying it. Seldom do I have sliced fruit on my cereal, fruit juice, or even toast. And I stopped consuming eggs a long time ago.

Now what effect has this had on my serum cholesterol levels? Well consider this: For someone who seldom exercises, for someone who is about 18 lb., overweight, and for someone who is a heavy meat eater (mostly beef), my most recent medical checkup showed a mere 172 mg. of cholesterol per deciliter of blood and a treadmill test that indicated a good, strong heart and relatively squeaky clean arteries. Anything under 200 mg./dl. is considered by the American Heart Association to be good, with a reading in my range as "very good to excellent." Of course, the additional late hour snack of shredded wheat, granola and milk every night before retiring helps keep my cholesterol and triglyceride levels down too.

Oats Reduce Insulin Dependency

Dr. Anderson's work with many diabetic patients being placed on oats and other cereal grains has amply demonstrated a significant drop in the amount of insulin required each day. Patients thus placed on an oatmeal, Grape Nuts, and All Bran diet, have gone from an average of 26 units of insulin a day to only needing a mere 7.1 units a day. Truly, it can be said that oats and other grains are almost a "miracle food" for diabetics these days.

Kiss Skin Problems "Good-Bye"

With oatmeal as part of your regular skin care routine, you can just about kiss many common skin problems "good-bye." For instance, leading dermatologists now recommend Aveeno oatmeal baths for relieving psoriasis and contact dermatitis. Or bring 6 cups of cold water to a rolling boil. Then scatter in 10 $^1/_2$ tbsp. of old-fashioned rolled oats, reducing heat to a low setting and simmering for half an hour. Strain and use the water to bathe the face with morning and night. A good idea is to dip the edge of a wash cloth in this oatmeal water and scrub the skin in a rotating motion.

Beauty salon expert Paul Neinast of Dallas recommends an oatmeal facial pack to get rid of cysts, blackheads, and other unsightly blemishes, as well as excessively oily skin. Five to 6 tbsp. of freshly cooked oatmeal mixed into a paste with some honey and stiffly beaten egg whites and then applied all over the face, forehead, and throat and left on the skin for half an hour before being washed off, will take away pimples and old, dead skin, leaving the surface much smoother and the complexion more rosy.

Oatmeal is also a real treat for hot, tired, aching feet, especially where corns and calluses are present. Just cook up a large pot of oatmeal, being sure to add only enough to give it kind of a soupy consistency. When still hot enough to tolerate, just stick both of your feet in a pan large enough to accommodate them and the runny cereal solution, and soak for up to an hour. This same oatmeal can be reused several more times before finally discarding, by just heating it up again on the stove.

Can Oats Help You Stop Smoking?

A number of methods and products have been tried, including hypnosis and nicotine chewing gum, to help smokers kick the habit. But success has always remained somewhat marginal to say the least. Now comes a centuries-old Ayurvedic remedy from India that could well prove to be a more workable solution to those looking for new ways to quit smoking.

The October 15, 1971, issue of *Nature* reported an alcoholic extract of the fresh oat plant being used to diminish the craving for nicotine in a number of smokers with pretty good success. One and a half parts of the crushed whole plant picked just before harvest is put into 5 parts by volume of 90% ethyl alcohol. Kept at room temperature, it's frequently shaken for up to 3 days and then filtered into another clean container.

This alcoholic extract (1 ml.) was then diluted to 5 ml. and given in oral doses of 5 ml. four times a day to a group of heavy cigarette smokers at Ruchill Hospital in Glasgow, Scotland. A second group of smokers were given a placebo intended to mimic the oat extract.

The first group taking this oat extract faithfully for nearly a month smoked an average of 19.5 cigarettes per day before the trial test began, but had dropped to an astonishing 5.7 cigarettes per day after the experiment ended. Whereas the nonoat group started out with 16.5 cigarettes per day and actually finished with a slight *increase* of 16.7 cigarettes smoked each day. Clearly, oats *do* help break the smoking habit when used as an extract in their fresh state! Old Amish Herbs of St. Petersburg, Florida (see the Appendix) has a nice antismoking oat plant extract which might help you to kick the habit. Between 7–10 drops 3–4 times daily *beneath* the tongue is recommended.

Rice Bran for Smooth Skin and Fractures

A chemist whom I met at the 28th Annual Meeting of the American Society of Pharmacognosy informed me that rice bran is used in his native country of India to keep skin smooth and to help bind fractures together so they can heal better. The bran is wrapped in cheesecloth and used in place of soap for daily washing of the face, neck, throat, arms, and hands to keep them nice and smooth. For fractures, cold water is added to rice bran and the combination well mixed by hand with a wooden ladle until a smooth thick paste forms. Some of this is then put into the palm of each hand and applied directly to the injured site. The hands are placed on either side of the fracture with the fingers spread apart. Gentle pressure is exerted in such a manner as to put the dislocation or fracture back together while the paste is still wet. This rice bran plaster is left on for several hours at a time, before a fresh one is needed. It helps to immobilize the injured bone. Equal parts of wheat flour and rice bran also make good plasters.

Rice Prevents Heart Disease

One of America's most successful diet programs, the Pritikin Diet, features generous quantities of cooked brown rice as one of the principal mainstays, along with steamed vegetables and baked potatoes. But can so much brown rice actually prevent coronary heart disease and help keep serum cholesterol levels down?

Well, the best evidence in support of this comes from the diet's originator, Nathan Pritikin himself. Prior to formulating his unique high-complex-carbohydrate, low-fat and low-cholesterol diet, Mr. Pritikin's serum cholesterol stood at an alarmingly dangerous 280 mg. per deciliter of blood in December 1955. About this time he was also experiencing some serious heart problems as well. *The New England Journal of Medicine* for July 4, 1985, published a summary of his serum cholesterol levels once he started on his now world-famous diet. In February 1958, he had 210 mg. of cholesterol per deciliter of blood; in November 1985, it was down to 94 mg.!

But even greater proof for the validity of his brown rice diet comes from a careful inspection of his heart at the time of his demise at the age of 69 in February 1985. An autopsy revealed Mr. Pritikin's heart to be virtually free of atherosclerosis with practically all of the arteries soft and clean.

Rice Water for Diarrhea

Donna Lee Ingram, R.N., who was a clinical nursing instructor at the University of Utah School of Nursing told me how she used an old wives' remedy to cure a bad case of diarrhea, even when the attending physician stood alongside of her snickering in disbelief.

Within hours after drinking from a polluted stream, a 63-year-old white male began to experience diarrhea which grew progressively worse by the afternoon of the following day. By the time he checked himself into the hospital a day later, his potassium count was dangerously low, which had produced heart fibrillations and some breathing difficulties as well. He was also pretty well dehydrated from his round-the-clock diarrhea.

"We first gave him an IV (intraveneous) potassium solution which brought his potassium count back up to normal again. He began to feel a little better after this, but the diarrhea problem persisted in spite of various antibiotics that we used on him.

"Now I remember my grandmother using flour and rice water to check diarrhea when we were kids. I mentioned this to the doctor on duty then, and he just laughed in my face, saying it was nothing but an unproven old wives' tale. Well, I went ahead anyway and got some rice from the kitchen and boiled up 1 cup of it in 3 cups of water, uncovered for 15 minutes on high heat. I then strained the water off and let it cool awhile before giving it to the patient. I gave the man a handful of flour first to swallow in small quantities, which he then washed down with sips of this rice water. Finally, I had him drink the rest of the rice water after all of the flour had been consumed.

"Within less than an hour, his heretofore uncontrolled diarrhea completely stopped, much to his own amazement, the doctor's complete surprise and my personal satisfaction!"

Rice Coffee and Tea Substitutes

In place of coffee or black and green teas, use these rice beverages instead. To make rice coffee, just put 10 tbsp. of raw rice into a slow oven on a cookie sheet and stir frequently with a ladle until it's well browned, but not burned. Then pass through a coffee mill and store in an airtight container. When using, put the usual amount you would for coffee grounds in a coffee-pot or percolator and prepare as you would coffee. But allow the rice to steep in a warm place for at least 30 minutes before serving. Rice tea is made the same way, except that the browned grains remain whole instead of being passed through the mill. Rice coffee and tea are good for relieving the terrible pounding headaches accompanying an alcoholic hangover.

Rye Protects Against Chemicals

In the Soviet Union, rye and whole wheat breads are recommended often in the diets of factory workers exposed to some toxic chemical occupational environments. Clinical nutritionists and doctors there believe that both of these grains, especially rye bread, have profound neutralizing and eliminatory effects on such harmful chemicals. Those laboring under similar conditions here might give serious consideration to adding more rye to their diets.

How Wheat Prevents Cancer

Whole wheat accomplishes many of the same things which other grains do; namely, to increase bowel movements, reduce blood sugar and cholesterol levels, and help prevent heart disease. It's also very useful for preventing cancer of the liver, small intestine and colon.

There are three possible ways that wheat may be able to do this. One is by actively binding cancer-causing compounds directly to grain fiber. Second is to reduce the length of time for a bowel movement, thereby limiting the potential damage cancer-causing agents could do. A third way is that wheat fiber seems to alter the microflora of the colon just enough so as to prevent chemical additives found in food from possibly producing tumors once it's consumed and thoroughly digested.

Relieves Rectal Itching

Wheat germ oil is the healing miracle for any kind of rectal or vaginal itching. It works far better than any known drug ointment does. Upon rising in the morning, bathe the rectum or vagina for several minutes with very hot water. Then apply wheat germ oil and leave for 3–5 minutes. After which you should wash the areas lightly with Aveeno or another brand of oatmeal soap. Rinse again in hot water and wash again.

Then mix together some concentrated vitamin C powder in a little water and apply to these areas, leaving on for just a minute before rinsing again. Thoroughly dry at once by using a heat lamp or hair dryer, concentrating on the sore areas. The drying is very important to the success of this treatment. Finally, apply some more oil and leave it on all day long. At night simply repeat the entire process again.

The Best Bread You'll Ever Eat

The following outstanding bread recipe is a composite of several different recipes from the isles of Corsica and Sardinia and the Roman Empire some 1,800 years ago. You'll find it's probably the best bread you've ever eaten.

All-Grains Braided Loaves

Needed: 2 cups whole wheat flour; 1/2 cup rye flour; 1/2 cup buckwheat flour; 1/2 cup millet; 1/2 cup rolled oats; 1/4 cup each cooked and dried split peas and navy beans; 1/2 cup cornmeal; and 3 1/2 packages active dry yeast. In a mixing bowl combine *only* 1/2 cup whole wheat flour, together with other dry ingredients.

Needed: 5 cups of canned goat's milk; 2 tbsps. each of molasses, maple syrup, and dark honey; 6 tbsp. butter; 2 tsp. salt. In a pan heat everything until lukewarm (115–120° F.), stirring constantly. Add to the dry ingredients above. Beat at low speed with an electric mixer for 1 1/2 minutes, scraping bowl frequently. Then switch to high speed and beat an additional 3 minutes.

Needed: 1 1/2 cups each chopped wheat berry and brown rice sprouts (see beginning of this section for instructions on making sprouts); 2 tbsp. each of toasted wheat germ, and bulghur and pot barley that's been previously cooked and allowed to cool. Using a ladle, stir in the sprouts, wheat germ, bulghur and barley, and as much remaining whole wheat flour as you can. Turn out onto lightly floured surface. Knead in enough of the remaining whole wheat flour to make a moderately stiff dough that's smooth and elastic (6–8 minutes total kneading time).

Next grease the inside of an electric slow cooker, turn the heat on low, and let the pot warm. Then unplug it and put your dough inside, covering the top. Turn once. Let rise until double in size (between 45 minutes and an hour). Really works like a charm and is a virtual foolproof way to make sure the dough rises quickly.

Punch the dough down and divide into three separate portions. Cover with a cloth and let rest 10 minutes. Roll each piece into a 10″ rope. Braid the strands together, beginning in the middle and working toward each end. Pinch the ends together and tuck the sealed portion under the braid. Then place into oiled 8″ × 4″ × 2″ loaf pans. (Grease them with olive oil or liquid lecithin from any health food store.) Cover and let them rise into braided beauties until nearly double. Make sure they are set in a warm enough place to nicely rise.

Bake them in a 375° F. oven for half an hour. Cover each loaf with aluminum foil the last 15 minutes, if necessary, to prevent overbrowning. Remove from their pans to wire rack. Brush the tops lightly with melted butter to which has been added a tad of mace and cardamom. Allow to cool before eating. Makes three incredible-tasting loaves!

Cornbread Fit for a King

This is the perfect companion for your favorite bean dish. And comes from La Rene Gaunt's *Recipes to Lower Your Fat Thermostat* with the kind permission of the publishers.

Golden Cornbread

Needed: 1 cup whole wheat flour; 1 cup cornmeal; 4 tsp. baking powder; 2 tbsp. dark honey; 1 cup milk; 2 egg whites; 2 tbsp. olive oil; 1/4 tsp. sea salt. Combine flour, cornmeal, baking powder and honey. Add milk, egg whites, oil, and sea salt. Mix well. Bake in a 9″ square pan which has been covered with liquid lecithin from any local health food store, at 425° F. for 25 minutes. Serves about 12.

Zesty Vegetable Rice Dish

Here's a dish that should really bring out the Nature lover in you with its unique "wilderness" appeal.

Wild Rice Delight

Needed: 1 cup uncooked wild rice; 1 tbsp. olive oil; 1/2 cup chopped Bermuda onion; 1/4 cup chopped green onion; 1/2 cup chopped green bell pepper (including seed center); 2 cups sliced zucchini; 3 small tomatoes, cut into eighths; 1 crushed garlic clove; juice from 1/2 lemon and 1/2 lime; 2 1/2 cups boiling chicken broth.

Sauté rice in oil until golden brown, over low heat, about 7 minutes. Spread in a greased, shallow, 2 1/2 quart casserole pan. Layer onions, green pepper, zucchini, and tomatoes over rice. Add garlic and citrus juices to broth. Pour over vegetables. Cover, and bake at 350° F. for 1 hour or until the liquid is absorbed and all the vegetables are tender.

Making Great Breakfast Cereals

To really liven up any cooked cereals, just add to them a little pure vanilla, some pure maple syrup, a dash of cardamom, and any fruit you like, then top with a little half-and-half for a truly memorable breakfast treat!

Wilderness Cereal

Needed: 1/2 cup each of cracked wheat and wild rice; 2 1/2 cups boiling *spring* or Perrier water. Combine everything and cook in a covered saucepan for 35 minutes, stirring frequently with a wooden ladle. In the last 10 minutes before the cereals are finished cooking, add $^1/_4$ tsp. cardamom, $^1/_8$ tsp. pumpkin pie spice, $^1/_8$ tsp. pure vanilla flavor, and 1 tbsp. blackstrap molasses for an unforgettable taste! Serve hot with ice-cold canned or fresh goat's milk.

A Most Nourishing Tea

An extremely nourishing tea for infants, children, the elderly, and anyone recuperating from a recent illness or surgery can be made from three different wholesome grains.

In a cast-iron skillet, *separately* roast $^1/_2$ cup *each* of barley, buckwheat and wheat grains on medium heat for 12 minutes, making sure you *stir* each of them *continuously* with a wooden ladle. One scant teaspoon of sesame seed oil may be used to lightly grease the bottom and sides of the frying pan so the grains won't stick. As each grain is done, remove it to a holding dish of some sort.

When all the grains have been thusly roasted, combine them together in a heavy stainless steel pot and add 1 gallon (16 cups) of spring or distilled water. Bring to a rapid boil, then reduce the heat to a lower setting and simmer, covered, for 25 minutes. Remove lid and stir in 1 tsp. pure vanilla and 1 tbsp. pure maple syrup. Cover again and steep until mildly warm. Strain and drink a cup several times throughout the day. A pinch of powdered cardamom may also be added when the vanilla and syrup are added, for extra flavor.

GRAPEFRUIT
(Citrus paradisi)

Brief Description

This fruit is so named because it grows in grapelike clusters. This large, yellow-skinned, tart-flavored member of the citrus family can weigh anywhere from 1 to 5 lb. White (pale yellow) and pink-fleshed varieties, like the Ruby Reds out of Texas, are the most common. However, there is also a special green-skin variety called Sweetie, which is less tart than the others. I've had juice made from a combination of the Ruby Red and the Sweetie; the taste is quite flavorful and not as tart as that of regular grapefruit juice.

The progenitor of the grapefruit is believed to be the pomelo (*C. maxima*), native to and long a popular fruit in India and other parts of Asia. The pomelo (also called shaddock) was first taken to London by an Englishman and then introduced to the West Indies later on. Through a genetic mutation of the pomelo seed, the grapefruit came into being and has proven to be one of the most popular citrus fruits in the Western Hemisphere.

Grapefruit doesn't improve with cooking as lime or lemon do. The tart flavor mixes quite well with fish, shellfish, liver, beef tongue, tripe, sweetbreads, or anything else with a pronounced flavor and rich texture. One of the best marinades I know of to bring out the full flavor in trout or salmon is to soak the fish in half a cup of grapefruit juice for about half an hour before broiling, baking, or frying it. Longer than this will spoil the flavor.

The tree that grows grapefruit is an attractive evergreen. Like other citruses, it is a prey to frost and hybridizes easily.

Award-Winning Actor Finds Cure for Stuttering

James Earl Jones is an award-winning actor, having won three Emmys, two Tonys, a Golden Globe, a Grammy, an Obie, and more. The only thing still missing from his trophy case is an Oscar, which he yet expects to win sometime in the very near future for work he still hopes to do on some unmade movie of notable distinction.

But this 64-year-old actor (as of 1995) is best known worldwide, not so much for his acting skills as for his distinctive, resonant voice. Any of us who've watched TV or gone to the movies have heard him during the station breaks on CNN, as the diabolical Darth Vader in the three *Star Wars* epics, and as Mufasa, the father lion, in Disney's runaway smash-hit animation, *The Lion King*, which by November 27, 1994, had earned $281,838,756. According to *Variety*, the show-biz daily, 3 of the 10 highest-grossing films of all time in the United States featured Jones: #3, *Star Wars* (1977); #6, *The Lion King*; and #7, *Return of the Jedi* (1983).

But, as this warm and gracious man with the explosive stage energy explained to a reporter one day in the summer of 1994 in his Sherman Oaks, California home, things weren't always so bright and happy for him in his teenage years.

By the time he was 14, Jones had grown into a gawky and shy kid at Dickson High School in Brethren, Michigan. "I had a stutter so bad that I never spoke out in class," he admitted.

This is probably why he never had any real friends to speak of. Fellow classmates avoided him as much as possible, thinking his self-imposed muteness had more to do with mental imbalance than anything else.

So Jones decided to find comforting refuge in poetry *and* grapefruit. He "loved to write the former and eat the latter," he claimed with a gentle smile. One time he submitted some of these poetic musings to his English teacher, Donald E. Crouch. Amazed by just how well crafted it was, Crouch wondered whether Jones might have copied it from a book of poetry. Jones denied he had done this. Crouch told him that the best way to prove that he himself had written it was to recite it by heart before the class.

Jones reluctantly got up out of his desk seat and slowly walked to the front of the room. He figured it was better to be laughed at for stuttering than to be wrongly disgraced. "I was scared out of my wits," he recalled. "But the moment I opened my mouth to speak, the words just flowed out ever so smoothly." His stutter had disappeared forever.

This African-American actor attributed his "cure" of stuttering to the many grapefruits he ate in his teenage years, as well as the memorization of the poem he had written.

Corrects Indigestion

A lady from Clearwater, Florida shared with me her method for relieving an upset stomach a few years ago when I was lecturing in that city. "I just grate the outer skin from an entire grapefruit right down to the white part," she told me. "Then I carefully spread these grated bits on a clean cloth to dry. When they get kind of crinkly-dry, I store them in a zip-lock plastic bag. And whenever I get an upset stomach or heartburn, I just put 3/4 tsp. of these peel gratings in my mouth and slowly suck them before chewing. Eventually my stomach is back to normal," she beamed with delight.

A professor from the National Naturopathic College in Portland, Oregon once told me that taking 2 tbsp. of olive oil every morning before breakfast, followed with 1/2 cup of grapefruit juice would help expel gallstones as well as stop any gall-bladder pains.

Curbing Hunger May Help with Obesity

It is claimed by some health food enthusiasts and so-called "nutritional experts" that a grapefruit diet will help a person to lose weight; but there's no scientific validation for this. At best, half a grapefruit every day may help curb hunger a little in some who are obese, but that's it.

A Grapefruit A Day Helps Prevent Heart Disease

Dr. James Cerda, professor of gastroenterology at the University of Florida in Gainsville thinks he has found a new use for pectin, the jellylike substance that forms the membranes of a single grapefruit's innumerable juice-containing cells. He discovered that when medium to high-risk patients of coronary heart disease took 15 grams of grapefruit pectin each for four months, their blood serum cholesterol levels decreased by almost 10%.

Furthermore, the ratio of so-called "good" cholesterol (high-density lipoproteins) to "bad" cholesterol (low-density lipoproteins) improved in 50% of the patients tested. And these benefits came without the patients having to change their dietary habits or life-styles. Another one-third of his patients taking grapefruit pectin showed an even more dramatic cholesterol drop nearing 20%.

In a recent telephone interview, Dr. Cerda claims that eating a grapefruit for breakfast several times a week can reduce individual risk for heart disease by as much as a third. When asked, though, whether drinking grapefruit juice would have the same effect, he answers without hesitation, "Absolutely *not*!" He says that it is the countless fibrous minisacs—the pectin—that hold the juice which brings health benefits for the heart.

Beware of Drinking Grapefruit Juice With Some Prescription Medications

It is the responsibility of an author such as myself, when writing a health encyclopedia like this, to inform my readers of the pluses *and minuses* of some healing foods covered in this text. There is a negative side to grapefruit and other citrus juices when taking antihypertensive drugs.

According to research reported in the February 9, 1991, issue of *Science News* (Vol. 139, p. 85), there is a potent "pharmacokinetic interaction between citrus juice and [these] drugs." If such medications are taken with grapefruit juice, for instance, a rash of side effects can occur; these include rapid heart rate, facial flushing, and dizziness. It is speculated by medical researchers that grapefruit juice may contain a substance that inhibits an enzyme that breaks down these antihypertensives, thereby leaving more of them circulating in the bloodstream.

Grapefruit juice should never be taken with these antihypertensive medications: hydrochlorothiazide, chlorthalidone, indapamide, captopril, enalapril, fosinopril, lisinopril, ramipril, quinapril, benazepril, cilazapril, perindopril, spirapril, diltiazem, felodipine, isradipine, nicardipine, nifedipine, nitrendipine, verapamil, acebutolol, antenolol, metoprolol, nadolol, penbutolol, pindolol, propranolol, timolol, and betaxolol. If you must take any of these, then drink water with them instead of citrus juices. However, you're better off finding natural alternatives to reduce existing hypertension. Kyolic garlic (4 tablets daily) and Kyo-Green (1 tbsp. in 8 oz. of water daily) from Wakunaga of America (see the Appendix) are two good alternatives. Magnesium (2,500 mg. daily) and potassium (2,000 mg. daily) are very useful minerals for this. All four supplements are readily available from most health food stores.

Grapefruit Seed Extract Cured Her Allergies Due to Bacterial Infection

The following true story was related to me sometime ago and used here with permission. Susan McKinney is a 38-year-old (as of 1994) special educator instructor who lives with her husband near Baltimore, Maryland. Awhile back they ate hamburgers at a restaurant somewhere in Pennsylvania. She had a cup of water with hers; her husband didn't.

Within hours she started feeling terrible. Severe diarrhea, abdominal pain, fatigue, weight loss, and other flulike symptoms set in. Over the course of the next two years she saw almost two dozen physicians who poked and probed her insides, not really knowing exactly what they were looking for. Varied opinions were given as to what she had; they ranged from multiple sclerosis and system lupus erythematosus to AIDS.

In time she developed food and environmental allergies. Finally, out of desperation, she turned to an alternative-minded nutritionist who managed to correctly diagnose her condition as a bacterial infection within her small intestine. He prescribed pills made of grapefruit seed extract. McKinney believes this remedy saved her life.

After taking the pills for seven days in April 1991, she felt 100% better than she had since first becoming sick in 1989. She is now fully recovered and stays away from greasy burger joints.

The Miracle Remedy for Sore Throat and Strep

Susan's story joins a growing number of similar testimonials nationwide for grapefruit seed extract. This liquid remedy comes from ground grapefruit seeds and was discovered about 15 years ago by a Yugoslavian physicist and medical doctor who specializes in finding natural drugs from plants. For the last decade, Bio/Chem Research in Lakeport, California, has been processing and supplying it around the world to a number of health food supplement companies.

Gargling with grapefruit seed extract and water will handle the most drug-resistant Strep germs imaginable. The bactericidal dilution of grapefruit seed extract against *Streptococcus faecalis* is only

1:80,000 in vitro. In vivo, doctors are recommending gargling with 2–3 drops in 5 oz. of water, for a ratio of about 300 to 500 part per million.

Unlike many currently marketed products for sore throat pain, after gargling with grapefruit seed extract, it is *completely safe* to swallow the solution.

Amazing Stories Told About Traveler's Diarrhea, Yeast Infection, and Dysentery Being Totally Cured

Grapefruit seed extract is one of the new *natural* "wonder" drugs of the 1980s and 1990s. Since this book first made its appearance some eight years ago, I've heard many true tales of remarkable healings with this citrus seed extract that are nothing short of incredible.

Jon M. of Miami went on a Caribbean cruise and suffered a bad case of traveler's diarrhea, which he blamed on the ship's food. By the time he returned home, he felt exhausted, dehydrated, and quite miserable. A colleague at work told him about grapefruit seed extract, so he started using it.

"I put 7 drops in half a cup of water and took that on an empty stomach twice a day," he stated. "And in just two days, all of my troubles were over!"

Liz O. of Kansas City, Kansas, went to her doctor for a check-up and was told she had Candida albicans. "I was listening to this guy on `The Jack Murphy Show' [a local talk show] one day," she said, "and he started talking about the wonders of grapefruit seed extract. I went down to the health food store and got me some and started putting 5 drops of it into every cup of coffee or glass of water or juice that I'd drink. Within a week, I felt more energetic than I had in a long time. I went back to my doctor and he told me my yeast infection had cleared up!"

A salesman who frequently goes to Mexico City on business told how grapefruit seed extract had cured him of the dysentery he contracted on one of his trips there.

Worried about Bad Water or Contaminated Produce?

Have you ever been in situations where the available drinking water brings many question marks to your mind when the urge to quench your thirst hits you? Certainly anyone who has ever traveled knows of what I'm speaking. And have you ever considered the safety on that fruit or lettuce salad you were about to dig into with relish?

Well, worry no more about such matters. For grapefruit seed extract comes to the rescue to solve these health dilemmas for you. Available water should first be filtered. At the very least, let suspended water particles settle. Retain the clear water and add 10 drops of grapefruit seed extract per gallon of water. Shake or stir vigorously and let sit for a few minutes. A slightly bitter taste may be noticed. This is just the inherent taste of the grapefruit seed extract.

The next time you're in a salad mood, first consider rinsing your produce with grapefruit seed extract. Its potency as a bactericide and fungicide will, quite simply, kill what is consuming/decaying/contaminating your lunch. For home rinsing, add 10 drops of grapefruit seed extract to each gallon of rinse water. You'll be glad you did.

GRAPES AND RAISINS
(Vitis species)

Brief Description

Grapes are readily identifiable as a fruit, with their trailing, climbing, tendril-clasping, wide leaved vines and pale green to reddish-purple fruit. Nearly half of this plant's innumerable grapes are native to North America. Many of our present species evolved from their cousins in the wild, such as the fox grape which kept the Lewis and Clark Expedition from near starvation. It's the ancestor of the now famous Concord grape. Apocryphal and ancient Jewish rabbinical sources all seem to suggest that the forbidden fruit consumed by Adam and Eve in the Garden of Eden was, in reality, a bunch of grapes and not the proverbial apple!

Wonderful Skin Moisturizers

The Dallas-based Neinast Salon is the beauty mecca for that city's top media personalities, business executives, bankers, attorneys, doctors, and real estate giants. The majority of patrons are men who come to receive a variety of services.

Owner Paul Neinast has shared a number of his beauty secrets for the first time in a book like this. Earlier some of these were mentioned under citrus fruits. Here we deal with his wonderful skin moisturizers from grapes.

"I find that the green Thompson seedless kind make the best facial toner for dry, sensitive skin," he began. "Just cut the grapes in half and slowly squeeze the juice on the lips and beneath the eyelids. Also rub some of the juice around the corners of the mouth and eyes as well. It's great for getting rid of crow's feet and the tiny cracks around the edges of the mouth. Or these grape halves can be cut into a small "x" and crushed right on the skin and left there for 20 minutes or so. Or they can be mixed up in a blender and lightly rubbed all over the face, forehead, throat, and neck and kept on for a little less than half an hour before removing.

"I also recommend champagne for tightening up loose skin and closing the pores. Champagne costing between $5 and $7 a bottle is good enough to use. Just splash some on the skin as you would after-shave lotion or cologne and let the air dry it out. We've also used this a lot on women's skin with good results. The champagne works especially well on middle-aged to older women who have slightly sagging, drooping skin problems beneath the eyes and around their throats," he concluded.

Wine Reduces Heart Attacks

Wines made from a special kind of grape (*Vitis vinifera*) have been proven clinically to reduce the chances of getting a heart attack and even help to reduce high blood pressure, when taken in moderation. Doctors at the Kaiser-Permanente Medical Center in Oakland, California, have surveyed the medical histories of over 200,000 patients and found that moderate alcohol users were 30% less likely to get heart attacks than nondrinking patients were.

Moderate alcohol intake increases good cholesterol (high-density lipoproteins), which, in turn, dramatically reduces the bad kind (low-density lipoproteins) that clogs arteries and eventually leads to heart attacks later on. Moderate intake of alcohol, especially white and red wines, should be about 2 fluid ounces or 1/4 cup per day for therapeutic benefits. Anything less or greater than this either won't work or can be harmful to health.

An effective concoction employed by some of the traditional *kanpo* doctors in Kyoto, Japan, to relieve the chest pains accompanying angina calls for a raw egg to be mixed with 2/3 cups each of sake or wine and canned apple juice. This concoction is then brought to a boil and taken internally after it has cooled awhile but is still quite warm. An average of 3 cups per day for 3–4 days is taken.

Raisin Tea for Strong Immunity

Two Canadian microbiologists working with the Canadian Health and Welfare Agency in Ottawa discovered that grape juice, red wines, and raisin tea showed strong antiviral activity against poliovirus, herpes simplex virus, and reovirus (an apparent cause of meningitis, mild fever, and diarrhea).

Not everyone drinks wine, and quite a few do not like too much grape juice on account of its tartness. But raisin tea is both a sweet and mild beverage that young and old alike can enjoy, while at the same time giving their immune systems a tremendous boost against future viral infections.

To make your own tea, just pour 3 cups of hot, boiling water over 1 cup of loosely packed raisins. Add 1 $1/4$ tsp. of blackstrap molasses, stir, then cover, and let set for an hour. Strain and refrigerate liquid. Drink 1 cup each day. The raisins may be saved and used a second time before finding some other way to use them such as in a rice custard pudding, bread, cole slaw, or carrot salad.

If you object to the chemical preparation of raisins, then just make your own. Wash a large bunch of grapes (any variety). Slice each grape lengthwise and remove the seeds, if there are any. Then arrange them in a single layer on flat cookies trays and slowly dry in a 165° F. oven with the oven door ajar until they are shriveled (somewhere between 13 and 21 hours.).

An Aluminum-Free Baking Powder

Currently about 2 million Americans suffer from Alzheimer's disease, a form of senile dementia that afflicts mainly the elderly. A growing body of medical evidence links aluminum with this horribly progressive destruction of the brain, which causes severe memory loss, extreme personality changes, and the inability to care for one's self. Aluminum is found in many things, ranging from deodorants, buffered aspirins and hemorrhoid preparations to baby food, baking powder, self-rising flour and cake mixes, processed cheese, and nondairy creamers.

Your chances of getting Alzheimer's later in life can be drastically reduced by not using aluminum pots and plans, nor cooking acidic vegetables like tomatoes or cabbage in fluoridated tap water in any aluminum cookware. And also by using an aluminum-free baking powder in recipes which call for this item. Some food scientists have estimated that regular commercial baking powder may contain 7–11% of pure aluminum in two forms, aluminum potassium sulfate and aluminum sodium sulfate, which can be injurious to human health over an extended period of time.

The simple recipe calls for mixing two parts of cream of tartar with one part of baking soda and cornstarch in a medium to large bowl. Blend thoroughly before storing in an airtight container to prevent moisture from getting in. Cream of tartar (potassium bitartrate) is a natural by-product of wine making, being a major part of the sediment left over. Arrowroot, the powdered starchy rhizome of *Maranta arundinacea*, may be substituted in place of cornstarch if you like. Arrowroot can be purchased in some larger health food or specialty food stores, while cream of tartar is found at most supermarkets as a rule.

Grapeseed Oil for Heart Disease, Impotency, Hypertension, Diabetes, and Obesity

A new use has been found for the seeds in a grape after the fruit has been submitted to different wine pressings. The oil obtained from these seeds now has a wide variety of culinary uses. But more about that later. Current medical research has proven that grapeseed oil has a beneficial effect on serum blood cholesterol levels.

Dr. David Nash, a prominent research cardiologist at the State University of New York Health Science Center in Syracuse, published a study revealing that just an ounce each day is all that is necessary for substantially increasing "good" cholesterol with the body. This "good" cholesterol is high-density lipoprotein, which prevents hardening of the arteries or atherosclerosis that can lead to heart disease. Dr. Nash discovered that this daily amount translates to a significant 39-56% decrease in risk of cardiovascular disease. As a bonus, triglycerides also decrease. (His study appeared in *Drugs,* Vol. 40, no. 1, 1990, pp. 13–18.)

A second report confirmed Nash's findings, adding that grapeseed oil reduced "bad" cholesterol (low-density lipoproteins) in just three weeks. (This was published in the *Journal of the American College of Cardiology*, March 14–18, 1993.)

"According to current knowledge, grapeseed oil—a high-linoleic product—is the only food known to raise HDL and lower LDL. Linoleic acid is one of two essential fatty acids people cannot manufacture themselves. Linoleic acid is an omega-6 fatty acid. Studies indicate that linoleic is sadly deficient in most diets. Grapeseed oil is 76% linoleic acid.

Interestingly, low levels of HDL are also related to impotence. In fact, the results of the Massachusetts Male Aging Study show that as HDL levels decrease, the probability of impotence increases.

(This was printed in the FDA *Consumer,* Vol. 28, no. 5, June 1994, p. 3.)

Grapeseed oil also

- Reduces platelet aggregation (or prevents cells from clumping together like bunches of grapes)
- Helps to prevent hypertension caused by sodium excess
- Helps to normalize lesions occurring from obesity and diabetes

How to Cook with Grapeseed Oil

Grapeseed oil is now available for all forms of cooking and is also ideal for salad dressings and mayonnaise. It has no fatty aftertaste and enhances the flavor of food, so it can be used for anything from a tuna salad to delicate party preparations. It has a nongreasy, slightly nutty flavor and will not cloud when chilled.

The recommended regular or deep-frying temperature is 360° F. At this temperature, there is no smoking, splattering, or burned taste because the smokepoint of grapeseed oil is unusually high (over 485° F.)—unlike the much lower smoke-point temperatures of other oils. Keeping the temperature at the normal range also prevents conversion to harmful trans fatty acids, as does not reusing the oil again. This is a major advantage over other oils.

Breaking a vitamin E capsule into stored oil is unnecessary with grapeseed oil because it is already naturally high in this nutrient: the presence of vitamin E gives grapeseed oil a shelf life of at least two years. For the same reason, maximum benefit can be obtained by using grapeseed oil raw.

Wouldn't it be wonderful if commercial food manufacturers would switch to grapeseed oil instead of the frequently used products laden with transfatty acids, designed for profit rather than health? Finally, potato chips, popcorn treats, and even baked goods would be healthy.

Grapeseed oil is available at many health food stores and nutrition centers under the brand name of Salute Santé Grapeseed Oil. If your retailer doesn't carry it, call the company that distributes it for the location nearest you that does: Lifestar International, Inc., 1-(800)-858-7477.

Recipes Made with Grapeseed Oil for Heart Health

Hot Rice Salad

Needed: 2 cups cooked brown rice; 1/4 tsp. pepper; 1 chopped onion; 1/4 cup grapeseed oil; 1 tbsp. lemon juice; 1 tbsp. oregano; 1 tbsp. parsley; a few olives.

Add pepper and onion to the hot rice. Then blend in the oil, lemon juice and oregano; pour over the rice. Garnish with parsley and onions. Serves 2–3.

Mayonnaise Substitute for Salad Greens

Needed: 2 tsp. chopped garlic; 1 egg yolk; 1 tsp. boiling water; granulated kelp to taste; 1 cup grapeseed oil; 1 tsp. lemon juice.

Add yolk to garlic and beat. Then, add water and granulated kelp. Next, add the oil and beat. Finally, add the lemon juice.

Fresh Tarragon Vinaigrette

Needed: 1/2 cup fresh tarragon leaves; 1/4 cup grapeseed oil; 3 tbsp. Dijon mustard; 2 tbsp. apple cider vinegar; 1/2 tsp. granulated kelp.

Whisk together the Dijon mustard, vinegar, and kelp. Slowly add the grapeseed oil while whisking, until thick. Add the chopped tarragon leaves. Serve over fresh greens.

Zucchini Escarpece

Needed: 3 medium zucchinis; 3 garlic cloves; kelp to taste; 3 tbsp. grapeseed oil; 3 tbsp. balsamic vinegar.

Cut zucchini into 1/4″ rings; slice garlic. In sauté pan, heat grapeseed oil and add zucchini slices. Sauté until golden brown on both sides. Place in serving tray; sprinkle with garlic slices. Pour balsamic vinegar over zucchini. Serves 4.

Curry Dressing

Needed: 1/2 cup grapeseed oil; 1/2 cup apple cider vinegar; 1/4 tsp. mustard powder; 1/4 tsp. curry powder.

Mix all ingredients. Refrigerate. Can be poured over any vegetable desired.

Olive Bread Spread

Needed: 1/2 lb. Calamata olives; 3 garlic cloves; 1/2 cup grapeseed oil; kelp to taste.

Combine all ingredients in a Vita-Mix whole food machine or equivalent unit. Purée for 2 minutes. Makes 3/4 cup olive paste. Serve on bread. Also good enough to use as a meal appetizer.

Sweet Potato Candy

Needed: 2 large sweet potatoes (about 1 lb.); 1 egg white, beaten (save yolk); 2 ripe bananas; 1 tsp. lemon juice; 1 tbsp. allspice; 1/2 cup ground walnuts; 2 tbsp. grapeseed oil.

Clean sweet potatoes (unpeeled) and boil under tender. Peel and mash with banana, egg yolk, lemon juice, and allspice. Refrigerate until thoroughly chilled and then form into balls. Refrigerate again for a half-hour. Roll in the egg whites and nuts and lightly fry in grapeseed oil or until golden brown. Serves 4–6.

GREEN (SNAP OR STRING) BEAN

(Phaseolus vulgaris)
(See also under BEANS)

Brief Description

Green beans belong to the same *Phaseolus* species kidney, navy and lima beans do, except that they are picked while still green in their pods. These green beans were first introduced by Native Americans to early colonists and became an instant hit with them. Green beans are one of the "safest" vegetables to serve guests with finicky tastes and picky eating habits.

Acne Problems Vanish

The bane of every American teenager is the dreaded problem of acne. Now there is an effective remedy for alleviating much of this. All a teen needs to do is wash his or her face morning, noon, and night with the pod tea of green beans.

For chronic acne, add 3 tbsp. of dried chamomile flowers to the pod tea after it's been removed from the stove. Cover and let it steep until the tea becomes cool, then strain and bottle. Wash the face every 3 hours if possible with the tea. A cup should also be drunk each day as well.

This same remedy also works for eczema, dermatitis, psoriasis, poison ivy rash, blackheads, herpes, cold sores, and similar skin afflictions.

Cosmetic Aid

A simple decoction made of the flowers of green beans is great for softening the skin, preventing it from badly peeling during recovery from a nasty sunburn, and sometimes hiding freckles by making them paler in tone. Bring $^1/_2$ quart of water to a boil, remove from heat and add 2 handfuls of freshly snipped bean flowers. Cover and let steep 50 minutes; then refrigerate, *leaving flowers in.*

Gently cleanse the face with this decoction or lay a wash cloth which has been soaked with the solution on the skin for about half an hour at a time. A tea made of stringbean pods and consumed daily is excellent for treating diabetes, claimed noted health authority Paavo Airola.

Recipe for Losing Weight

Jasmine Rice and Green Bean Almandine

This is a low-fat, low-carbohydrate diet that is very filling, but doesn't add extra pounds. It is recommended for those who are frequently hungry but wish to lose weight.

Needed: Vegetable cooking spray, 1/4 cup minced fresh onion; 1 cup (1/2") sliced fresh green beans; 1/4 cup water; 2 (10 1/2 oz.) cans low-sodium chicken broth; 1 1/4 cups uncooked jasmine rice; 1/4 tsp. salt; 3 tbsp. sliced almonds, toasted.

Coat a saucepan with a cooking spray; place over medium-high heat until hot. Add onion; sauté 3 minutes or until tender. Add beans; sauté 2 minutes. Add water and broth; bring to a boil. Stir in rice and salt. Cover, reduce heat, and simmer 40 minutes or until liquid is absorbed. Remove from heat; let stand, covered, 10 minutes. Stir in almonds. Serves 4.

Guava

(See under TROPICAL FRUITS)

H

Hawthorn

(See under BERRIES)

Hazelnut

(See under NUTS)

Hickory Nut

(See under NUTS)

Holleyhock

(See under ORNAMENTAL FLOWERS)

Honey and Other Bee Foods

Brief Description

It takes about 3000 bees three weeks to gather one pound of honey for your table. Most of the honey made by the bees remains in the hive for their own consumption, but one out of every 3 lbs. is carefully removed by the beekeeper. Using a heated knife, the beekeeper loosens honey, beeswax, and sections of honeycomb and collects them in a container. Law requires that honey be filtered or put through a nylon mesh. Honeycombs may be added later. FDA regulations prohibit additions of any sort. Honey is as pure as it is wholesome.

Honey keeps very well and never spoils. Honey is the ultimate preservative. So good in fact that wicked king Herod kept his murdered wife's body perfectly preserved in a tub of the stuff for seven years. And as late as 1943, the bodies of deceased Buddhist monks in Burma were still being preserved with honey until suitable funeral arrangements could be made months or even years later for them!

I've been to many countries and have sampled numerous kinds of honey. To me, the finest of them all has been Scottish heather honey—heavy and sweet and ecstatically fragrant. On the other hand, goldenrod honey yields a distinctly sassy sharpness to it. During my 1979 trip to the Soviet Union, I discovered the strong, macho qualities of very dark, heavy buckwheat honey, along with the almost dainty and angelic taste of beautiful white acacia honey. The lip-smacking honeys, though, come from fig blossoms and wild flowers found in and around the hills of Athens, Greece. They are undeniably the sweetest in the world.

Bee pollen is the male sexual grains of seed-bearing plants, which bees carry with them back to the hive. Pollen is comparable to spermatic cells in humans and animals. When bees enter the hive and brush it off, the substance falls to the bottom where it's later swept up by the beekeeper and sold to the health food industry for human consumption.

Propolis is a resinous substance, gathered by the bees, from the leaf buds or bark of trees (especially poplars). Bees use this stuff to seal any holes or cracks in the hive, to fix the comb of the roof and to protect the hive from outside contaminants.

When the honeycomb is constructed inside the hive, some of the cells are made to slightly larger dimensions. These are known as the royal cells. The eggs which are laid in these cells by the queen bee are designed to produce potential queens from the female grubs, which will be fed on royal jelly. This food is a sticky substance secreted by the glands near the mouth of the honeybee. It's a pearly-white, gluey kind of mass containing a high supply of protein and invert sugars like glucose and levulose.

Beeswax is the wax obtained from the honeycomb of the bee. After the honey is removed from the honeycombs, the combs are washed rapidly and thoroughly with water. They are then melted with hot water or steam, strained, and run into molds to cool and harden.

Great Wrinkle Remover

Honey adds softness and fresh beauty to the skin. Beekeepers' hands are often quite soft and wonderfully smooth during honey collection. Each day, splash warm water on your face and neck to open the pores, then apply a thin honey mask. The honey tends to soften and smooth away ugly, old wrinkles. Then just wash it off and finish with a dash of invigorating cold water to the face. Because of the composition of honey it causes the skin tissue to hold moisture. Dry skin cells plump up and wrinkles tend to smooth away.

By doing this often to the skin, you will notice a new glow of pink health returning to a heretofore "death warmed over" complexion. You may add dairy cream, whipped egg white, fresh lemon juice, apple cider vinegar, or any fruit juice to your honey mask before applying it.

Undisturbed Sleep

Have a hard time hitting the hay at night? Then try an old Slavic remedy for your insomnia. It calls for a combination of 2 tbsp. of honey with the juice of a lemon and an orange. Mixed in half a glass of warm water, this beverage often provides the shortest route to Dreamland. The darker the honey, the better it works!

Allergy Relief

Sometimes the very things that give us the most misery, are also the very things that help us over these same discomforts. Hayfever and related pollen allergies may be corrected or at least minimized to a great extent by taking honey at least a month before pollen season starts.

The best course of treatment to follow in preparation for this is to take 1 tablespoonful of honey after each meal. Then every other day get a small, waxy piece of the actual honeycomb and chew on it for a couple of hours in between meals. The ideal preventative, of course, is to eat honey and chew some of the comb from your own area.

Having been plagued with hayfever myself for years, I can attest to the effectiveness of this remedy when *both* courses of action are followed. Although they've never actually cured my hayfever as such, I can testify that they have reduced the misery and aggravation of watery eyes and runny nose by at least 80% during the allergy season.

Drinking Honey Stops Diarrhea

Colombian Indians residing in the northwestern part of South America often drink a lot of bees' honey to stop diarrhea and cure dysentery. Pediatricians at King Edward VIII Hospital in Durban, South Africa, gave 169 infants and children afflicted with infantile gastroenteritis a strong liquid solution of honey to drink each day. This helped to reduce the incidence and recovery time from bacterial diarrhea.

In 8 oz. of water, 4 large tbsp. of honey should be thoroughly stirred in before drinking the same. It does not work, however, on nonbacterial diarrhea. Also those with diabetes should be careful about ingesting so much honey at one time.

Keeping Cool in Hot Weather

When bees are flying through the air in exceedingly hot weather, they will usually hold a droplet or two of nectar honey in the folds of their tongues in order to transfer heat from their heads to their thoraxes. This passive shifting of heat between both body parts produces a remarkable cooling effect.

Based on these recent entomological finds, we decided to conduct some informal experiments ourselves with human subjects on a hot July day, here at our Research Center in Salt Lake City.

We found that chewing a small piece of honeycomb and keeping it in the mouth for as long as possible during periods of strenuous physical activity in hot weather produced a drop in body temperature. Another method is to take a small amount of stiff, granulated honey, wrap it in some gauze and put it in the mouth to occasionally chew on and hold next to the cheek as one would with chewing tobacco. This serves the same purpose as the honeycomb would. In either case, it will keep the body a lot cooler.

Great Food for Dieters

Dieters, in particular, can benefit from the dual properties of honey. The sugars in honey are almost completely predigested and can be easily absorbed into the bloodstream. Dextrose is assimiliated very quickly, giving that "instant" boost of energy the body needs. On the other hand, levulose is absorbed much more slowly and maintains the sugar level for some time. Honey's double-action sugars quickly satisfy a craving for sweets and tend to maintain that sense of satisfaction for quite awhile. For dieters, this is an important thing to know about.

Wonderful Throat Remedy

For a sore throat, an old-fashioned home remedy may be just as effective as a shot of penicillin from the doctor would be. So says Dr. Neil Solomon, a nationally syndicated columnist, in one of his June 1987 medical columns. "Every time a person has a sore throat, he doesn't have to run to a doctor," the column began. This famous Maryland physician then added, "The American Academy of Otolaryngology—Head and Neck Surgery, recommends warm tea and honey for a sore throat."

You can make your own throat syrup too, if you like. Just peel some garlic cloves and put them in a jar. Add honey a little at a time over a couple of days until the jar is full. Set in a sunlit window until the garlic has turned somewhat opaque and all the garlic flavor has been transferred to the honey. This is a great remedy for relieving hoarseness, laryngitis, or coughing. Take 1 tsp. every 3 hours as necessary. For a child, dilute each spoonful with a tiny bit of water. Bear in mind that garlic is decidedly hypoglycemic, so it should be used with care by those with known low blood sugar levels. An excellent honey-herb syrup is also available by mail from Great American Natural Products in St. Petersburg (see the Appendix).

Sores and Shingles Healed Miraculously

Throughout this book, I've been very careful to use adjectives like "magical" or "miraculous" very sparingly, and only where I feel it's appropriate and deserving enough. Well, this is one of those few instances where it can be said that herpes lesions, bedsores, and shingles are all miraculously healed within a very short time when honey is directly applied to them!

Either raw honey or garlic honey syrup can be applied two to three times daily as dressings with excellent results. Volume 62 of *The Chinese Medical Journal* for 1944 recommended mixing 80% honey with 20% Vaseline or Crisco shortening to heal chronic leg ulcers, small burns, and lupus erythematosus-like problems. The mixture is warmed and constantly stirred to make it easier to apply. The ointment is then applied in a thick layer to the afflicted areas and covered by sterilized gauze and bandages. When the infected sores show much discharge of pus, the dressing is changed daily or every other day. When it is fairly clean and dry, the change was made every 2–3 days instead.

Of 50 cases of skin sores treated this way, there were 38 or 76% complete cures right off the bat, with another 10 or 20% showing considerable improvement or at least partial cures. Only 2 or 4% failed to respond to this honey dressing. Such an incredibly high cure rate does indeed suggest miraculous healing powers for one of Nature's simplest and most abundant remedies.

Honey for Ulcers

Writing in Canada's *Medical Post* sometime ago, Dr. Basil J. S. Grogono claimed that the lowly honeybee may be able to do more for those who suffer from peptic ulcers than even health care givers could during past decades when they would often resort to drastic surgery. He noted more experts have come to recognize the role a tiny microorganism, *Helicobacter pylori*, plays in peptic ulcers.

While some have advised using drugs to combat this microbe, Grogono stated that these drugs all have nasty side effects and that the microbes may develop resistance to them. He cited, on the other hand, a recent study published in the *Journal of the Royal Society of Medicine* in which the antibacterial properties of honey were tested. One variety, which came from New Zealand bees fed on a plant called the manuka, was effective in fighting the ulcer-causing microbe.

Only honey that has *not* been pasteurized will work for peptic ulcers. One level teaspoonful per day is stirred into $^3/_4$ cup hot water and allowed to cool down to lukewarm. It is then slowly sipped on an empty stomach. Those, however, suffering from diabetes or hypoglycemia should *not* resort to this remedy.

Cures Major Wounds

Major injuries should receive prompt medical attention as soon as possible. However, sometimes circumstances warrant improvising when emergency medical care isn't readily available. A Wyoming housewife living on an isolated cattle ranch reported to me an incident that shows the remarkable curing power of honey.

> One day while keeping myself busy in the kitchen, I sustained a very nasty laceration with a French knife on my left hand. The gash I judged to have been at least 2 inches in length and very deep into the muscles. Blood was spurting everywhere.
>
> My husband was out on the range at the time and our other car didn't run. I screamed for my teenage daughter to come and help me. When she hit the kitchen at a running pace, she almost fainted by the amount of blood I'd lost.
>
> I told her to fetch me the jar of honey we kept in our storeroom. When she did this and had removed the lid, I scooped several handfuls of honey with my good hand out of the wide-mouthed jar and generously covered my injured hand with it. Then I had my daughter wrap the whole thing up in gauze and tape it down good.
>
> The next day when my girl unwrapped the dressing, we noticed that much of the badly cut flesh had already started knitting itself together again. We decided to apply a second honey dressing. My husband returned that night and upon being informed of the accident, insisted that I go with him into Buford to our local hospital. But I decided to wait until the next day to see how it was doing.

Imagine our surprise when the dressing was changed the next morning. Everything seemed to have come together of its own accord, almost as if the honey had acted like a glue of some sort to draw everything back like it used to be. We both decided to keep on using the honey dressing.

No stitches were ever required and my hand mended nicely on its own, without even a trace of a scar. I mentioned this to our family doctor later on and he just scratched his head almost out of disbelief.

The ancient Egyptian physicians used a combination of honey and grease to treat major injuries suffered in work or battle. Their usual mixture was $^{1}/_{3}$ honey, $^{2}/_{3}$ fat or butter. Some Harvard Medical School doctors tested this remedy out for themselves and found it to work quite well. The $^{1}/_{3}$ honey was just enough to make a nice paste of smooth consistency; too much honey made it quite sticky. And because no bacteria of any kind can thrive in honey for very long, it makes a dandy antibiotic as well. A little garlic juice added to this honey/fat combination would seem to be the perfect cure for most major wounds, when other medical treatment isn't readily available.

Pollen Is an Energy Powerhouse

If you're dragging your feet these days and can't seem to get enough pep and energy, then you should probably be taking bee pollen. About 15% of pollen is lecithin, a fat-melting substance. Another 20% is top-of-the-line protein. The balance of the ingredients are energy-producing carbohydrates and vitamins and minerals that work on certain strength-giving glands of the body.

In some European countries, beekeepers who inhale the pollen while routinely working with their hives during honey gathering periods, demonstrate more vibrant health and physical activity than at other times of the year. No wonder, since weight for weight, bee pollen contains more complete protein than accepted sources like steak, eggs, or cheese do.

A good quality bee pollen form Europe is sold in health food stores nationwide under the Nature's Way label. An average of 2 capsules per day is recommended for normal energy requirements. Caution should be exercised, however, by those with known pollen allergies before taking the stuff since adverse side effects are possible.

Propolis Combats Infection

During an outbreak of influenza in Sarajevo, Yugoslavia, several years back, Dr. Izet Osmanagic of the local university administered propolis to 88 students in a nursing college who were highly exposed to this epidemic. Another group of 182 students didn't take the substance. Of those who took propolis, only 7% became ill, while an amazing 63% of the nontakers came down with bad cases of the flu.

Dr. E. L. Ghisalberti of the Department of Organic Chemistry at the University of Western Australia has discovered that whenever propolis is administered along with penicillin or natural antibiotics such as garlic, propolis *increases* their effectiveness from as much as 10% to 100%. This helps to cut down drastically on the use of penicillin and potential drug side effects from its frequent use. Propolis is available from any local health food store.

Royal Jelly as a Sexual Rejuvenant

An amazing study on the wonderful sexual rejuvenating properties of royal jelly was reported in Vol. 56 of the *Journal of the Egyptian Medical Association*. Doctors in Cairo, Egypt administered royal jelly to a number of male patients known to be relatively sterile. Not only did the royal jelly increase their sperm count and make it more active, but also promoted growth of their genital organs as well. Ejaculations also seemed to be more frequent, too.

Royal jelly may be found at any local health food store. An average of 20 milligrams daily is recommended for infertility and poor sexual performance. An excellent combination of bee pollen, propolis and royal jelly is found in a unique food supplement called Aqua-Vite, available through Great American Products out of St. Petersburg, Florida (see the Appendix).

The Ultimate Lip Balm

Many years ago when I was still in elementary school, we had a favorite reprimand for those who poked their noses into other people's business where they didn't belong—"Mind your own beeswax!" Although the saying has faded with time, the importance of beeswax as a lip balm hasn't.

Both yellow beeswax straight from the hive and the commercially bleached white beeswax are used extensively in lipsticks. If you have beekeepers in your area, visit one of them sometime and purchase a little beeswax. It keeps well in a cool, dry place. Whenever you are out in the sun or wind too long, just carry some of this with you and rub a little bit over your lips every now and then. You'll find it keeps them nice and moist and soft.

HONEYDEW

(See under MELONS)

HORSERADISH
(Armoracia lapathifolia)

Brief Description

Horseradish is a perennial plant native to southeastern Europe and western Asia and occasionally is found wild but usually is cultivated in other parts of the world. The long, white, cylindrical or tapering root produces a 2- to 3-foot-high stem in the second year.

The dried, powdered root found in many herb formulas today is practically worthless. The real benefits lay in the freshly dug root. When grated, however, the strong volatile oils are released, so it's necessary to cover the grated root with apple cider vinegar and refrigerate it in a glass jar with a tight-fitting lid. It will keep for at least 3 months this way or the entire root can be packed in damp sand and kept in a cool corner of your basement or garage. Keep sand moist.

Great Massaging Oil

A very stimulating massage oil to relieve muscular aches and pains and help break up chest congestion can be made by steeping a small amount of freshly grated root in some cold-pressed oil of your choice (wheat germ, sesame, olive).

Cosmetic Benefits

Horseradish vinegar lightens the tone of the skin and gets rid of freckles and blotches. Also makes a great hair rinse and really enlivens a dead scalp. Cover grated root with apple cider vinegar and permit to set on a sunny window sill for 10 days. Vinegar is then strained and stored in an airtight glass bottle.

When using on the skin, dilute with at least 50% water. Or add to milk to bring more color to the face and to help relieve the itching of eczema. Soak 1 tbsp. freshly grated root in 1 cup of buttermilk for half an hour before straining. Dab on face and allow to remain for 15 minutes before rinsing with water. Refrigerate the rest for later use.

Warm up Tea

During the winter time or when an older person experiences cold sensations in the hands, legs, and feet due to poor circulation, a nice "warm up" tea can be taken to relieve some of this hypothermic feeling. Bring 1 quart of water to a boil. Add 1 tbsp. each grated ginger root and grated horseradish root. Cover and reduce heat, simmering for 10 minutes. Remove from heat, uncover and add 2 tbsp. each fresh or dried mustard greens and watercress. Cover and steep for an hour. Flavor with a pinch of powdered kelp and dash of lime juice. Drink 1 cup lukewarm every few hours.

Sauce Recipe

Here's a favorite horseradish sauce that's been in the Heinerman family for almost 200 years and was brought to America by my grandmother, Barbara Liebhardt Heinerman, from the old country (Temesvar, Hungary) at the turn of the century. It delivers full flavor and body, but without the unpleasant bite.

My Grandma's Basic Horseradish Sauce

Needed: 1/4 pint plain yogurt; 1/4 pint homemade mayonnaise (see recipe below); 2 1/2 tbsp. grated horseradish; 1/2 tbsp. each of lemon, lime, and grapefruit juices. Mix everything together in blender or by hand. Is a great accompaniment for fowl, fish, and beef.

Good Homemade Mayonnaise

Needed: 1 cup sour cream; 1/4 cup plain yogurt; 3 tbsp. of pressed onion juice; 1/2 tsp. pure maple syrup; 1/4 tsp. ground ginger root; 1 tsp. dillweed; 1 tsp. each very finely minced French tarragon and chervil (both fresh). Mix everything in blender until a smooth and even consistency has been reached. Makes 1 1/2 cups.

I

IRIS

(See under ORNAMENTAL FLOWERS)

JERUSALEM ARTICHOKE

(Helianthus tuberosus)

Brief Description

Of all the hitherto unknown foods which the newly discovered America bestowed bountifully upon Europe, the only vegetable of any consequence contributed by North America was the Jerusalem artichoke; all the others, including the "Virginia" potato, originated in South America. It was not an overwhelming generous contribution. The Jerusalem artichoke is of less interest to the gourmet than to the etymologist (a language specialist involved in the history of linguistic forms of words).

This vegetable originated in the central United States and Canada. A fast spreader, it had reached the Atlantic Coast before the advent of the white man and was diligently cultivated by many Native American tribes from Nova Scotia to Georgia. It was in such an Indian vegetable garden on Cape Cod, Massachusetts, that the famous French explorer Samuel de Champlain (1567?–1635) first saw it and so reported in 1605. In 1616 he rediscovered it in Canada and sent some of the edible tubers to France. Whether or not he had done so the first time isn't exactly known, but someone did, for Marc Lescarbot's *Histoire de la Nouvelle France* (Paris, 1609) mentioned that the Jerusalem artichoke was then being hawked by vegetable vendors in the streets of Paris under the name *topinambours* (which is what they're still called to this day there).

The taste of the Jerusalem artichoke can hardly be described as fine, delicate, or subtle, and even those who like it admit that satiety comes quickly. To many, its flavor doesn't even suggest artichokes, though it did to Champlain, who seems to have been the first to give it that peculiar and very inappropriate name. Nicholas de Bonnefons, gentleman-in-waiting to Louis XIV and a gourmet of sorts, wrote of this vegetable that "in salad, they have the taste of artichoke hearts, a little less firm, however; fried in fritter dough, they evoke salsify; boiled, they resemble, but from afar, potatoes"—by which he undoubtedly meant sweet potatoes.

A Food Well Tolerated by Diabetics

Those afflicted with diabetes mellitus have to be very careful of their carbohydrate intakes, since such starchy foods are quickly converted into glucose or blood sugar by the liver, pancreas, and spleen. But Jerusalem artichoke is one food that they can tolerate and eat as much of as they wish without worrying about their blood sugar levels soaring. This vegetable contains *no* starch, but instead has inulides, which can best be described in loose terms as a form of sugar or fuel energy that poses no problems for diabetics.

Survival Food to Prevent Starvation

The two famous American explorers sent by President Thomas Jefferson on an expedition to explore the vast Louisiana Purchase (1804–1806) were saved from starvation by Jerusalem artichokes. Meriwether Lewis (1774–1809) and William Clark (1770–1838) went up the Missouri River to its source, crossed the Great Divide, and descended down the Columbia River to the Pacific Ocean.

During one period of exploration, however, they and their entire company of men ran out of food and nearly starved to death. Some friendly Native Americans showed them how to dig up some edible roots; they quickly recognized them as being Jerusalem artichoke tubers, which they devoured *raw* with apparently great relish. The later consequences of such hasty action weren't reported in the men's journals, but knowing the vegetable as I do, it's safe to say that the expedition probably experienced a good deal of painful gas and terrible intestinal cramps.

Indian Remedy for Fatigue

Different tribes of the Great Plains relied a lot on Jerusalem artichoke in the early to midnineteenth century for food. The Omahas and Poncas called it *Pa-he*, while the Pawnee nation referred to this vegetable as *Kisu-sit*. During long buffalo hunts or fierce warfare with rival tribes or enemy white men, they would take some of this tuber, raw, boiled, or roasted, along with them in leather pouches of some kind. Whenever hunger would come over them or they would feel somewhat weak, they would just reach inside, pull out a piece of tubber, and nibble on it to give them extra energy.

Fixing Jerusalem Artichoke

As you've probably guessed by now, this tuber is neither from Jerusalem nor an artichoke. It is actually the sweet, crisp, and delicious thick root of a great yellow sunflower. It is a hardy perennial, and if you plant it, you'll have it on hand forever. Don't even bother to peel one, just scrub and slice it; these can be lightly steamed or sautéed for that matter. Or simply grate the tuber and add it, raw, to fruit and vegetable salads. If Jerusalem artichoke slices are marinated in equal parts of lemon and lime juice for two days and then dried out for a day, they make terrific snacks.

John McPhee, a former food writer for the *New Yorker* magazine, recounted an experience he had in the early 1970s with plant expert Euell Gibbons, during a time when they lived entirely on wild food alone.

> For dinner, we had boiled Jerusalem artichokes . . . They look like small sweet potatoes, since they are bumpy and elongated and are covered with red jackets. The flesh inside, however, is delicate and white. Boiled, it has the consistency of boiled young turnips or summer squash, and the taste suggests the taste of hearts of artichokes . . . [Euell] said he was sorry that Jerusalem arti-

chokes and a number of other wild foods . . . had been named for more familiar foods. "These things are not substitutes for tame foods," he went on. "They have flavors of their own, and it is not fair to them to call them by the name of something else. These are not artichokes. They're sunflower tubers." With a knife and fork, he laid one open and then scooped up a mound of the white flesh. "Boy!" he said. "That goes down very gratefully. Just eating greens, you can get awfully damn hungry." . . . For awhile we ate without speaking, because the [Jerusalem] artichokes were so good.

JICAMA
(Pachyrrhizus erosus)

Brief Description

Although relatively new to the North American palate in recent years, the jicama (pronounced hee-come-ah) has flourished in Mexico for many centuries. Those of Hispanic descent in California and the American Southwest also are well acquainted with this unique vegetable.

It was one of many produce treasures that the Spanish conquistadors discovered when they came to the New World. Called Mexican potato or yam bean, jicama is the bulbous root of a leguminous plant. It has a thin, patchy, light-brown skin, with a crisp, juicy, water chestnut-like interior.

Jicama is traditionally eaten in Mexico as a snack food, peeled, thinly sliced, and seasoned with salt, lime juice, and chili powder. Its sweet flesh and addictive "crunch" make it a natural for raw vegetable platters or a welcome addition to salads, soups, and stir fries. With 45 calories for a 3 $^1/_2$ oz. serving, jicama is a deliciously dietetic vegetable.

When selecting jicama in the market, choose one that is firm to the touch. Some scarring on the skin is common and doesn't indicate that the inside flesh is damaged. Before serving, peel off skin of the desired piece of jicama. The remaining, unpeeled portion can be stored, covered, in the refrigerator for several weeks.

Cure for Jaundice and Gout

Among the Yucatan Maya, a common cure for liver jaundice and gout is to take the sweetened juice of the root in $^1/_2$ cup amounts three times daily. One-half of a jicama can be peeled, cubed, and put into a Vita-Mix Whole Food Machine or similar food processor with 1 cup of ice cubes and blended for 1 $^1/_2$ minutes. The resulting thick "shake" can be sipped through a plastic straw.

Relief for Scabies

Scabies (mange in animals) is a skin disease caused by any one of several genera of skin-burrowing mites. Dog owners may be especially prone to contracting this disease from their beloved pets, if they're not careful to look for signs that indicate such. Animals infected with mange will continually scratch or bite themselves, shake their heads frequently, show continual ear drooping, or roll around on the ground or floor often.

In Mexico, a common remedy for man and beast alike has been to grind up the mature seeds of the jicama and then mix them with one part of tequila and two parts of castor oil before rubbing this mixture on the scabies/mange itself. This same remedy is also good for other itching skin conditions.

Mexican Treats

In Mexico, jicama is usually pickled. The very young seed pods are rubbed to remove their irritating hairs and then cooked and eaten like snapbeans are here.

The following award-winning recipe appeared some years ago in the Mexican newspaper *Excelsior* and was submitted by Señorita Maria Cortez Chávez of Mexico City.

MANCHAMANTEL/CHICKEN AND PORK STEWED WITH FRUIT

Needed: 3/4 lb. boneless stewing pork, cut into 1 1/2" cubes; 1 tbsp. salt; 1/4 cup safflower oil; 1 large chicken (about 3 1/2 lb.), cut into serving pieces; 25 almonds, unskinned; 1-1/2" piece cinnamon bark; 1 1/2 tbsp. sesame seeds; 5 chiles, cleaned of veins and seeds; 2 medium tomatoes (about 3/4 lb.), broiled; 1 1/2" thick slices fresh pineapple, peeled and cut into 1/2" cubes; 1 small banana plantain, peeled, and cut into thick rounds; 1 small jicama (about 1/2 lb.), peeled and cut into 1/4" slices.

Put the pork cubes into a saucepan, add water to cover and 1 tsp. of the salt, and bring to a simmer. Cover and cook for 25 minutes only, then drain, reserving the broth. Strain the broth, adding enough water to make 4 cups, and set aside.

In a heavy casserole, heat the oil and sauté the chicken pieces lightly, a few at a time. Remove and set aside. In the same oil, fry the almonds, cinnamon bark, and sesame seeds separately. Then drain each of them and transfer to a Vita-Mix Whole Food Machine or similar food blender.

In the same oil, dry the chiles lightly on both sides, then drain and transfer to the Vita-Mix container. Broil the tomatoes in advance. First, preheat your oven broiler to medium. Choose a shallow oven dish into which the tomatoes will fit snugly and can be turned easily; be sure to line the pan with aluminum foil. Place the tomatoes on the foil and broil about 2" from the flame. As they blister and brown on one side, turn them over, repeating this process from time to time during the cooking period, until they are evenly browned and soft inside. It takes about 20 mintues to accomplish this. Let them set until thoroughly cooled before adding to the Vita-Mix container. Include 1 1/2 cups of pork broth and blend for 1 1/2 minutes until smooth. Be sure the sauce isn't too watery.

Reheat the oil remaining in the casserole and fry this sauce for about 4 minutes, stirring and scraping the bottom constantly. Add 3 cups of the reserved broth and bring to a simmer. Add the chicken pieces, pork cubes, fruit, plantain, and jicama and sprinkle with some salt; then cover and cook over low heat for 1 hour 20 minutes, or until the meat, fruit, and vegetables are tender; be sure to stir from time to time to prevent burning on bottom of casserole. Serve hot, with corn flour tortillas. Yield: Serves 6.

Garden-Fresh Tuna Salad

Needed: 1 can (6 1/2 oz.) water-packed, low-sodium tuna, drained and broken into chunks; 1/4 cup plus 2 tbsp. fat-free bottled Italian dressing; 2 cups torn lettuce; 2 cups washed and torn fresh spinach leaves; 3/4 cup jicama, peeled, julienne-sliced; 1 cup sugar snap peas, strings removed and halved crosswise; 1 medium tomato, chopped; 1/2 cup julienne-sliced carrot.

Place tuna in a shallow nonmetal bowl; pour 1/4 cup Italian dressing over. Refrigerate 15 minutes to 2 hours before assembling the salad. To make the salad, in a large bowl toss together the lettuce, spinach, undrained tuna, jicama, sugar snap peas, tomato, and carrot with remaining 2 tbsp. dressing. Serve with corn tortilla chips. Yield: Serves 3.

Kale and Collards

(Brassica oleracea acephala)

Brief Description

It is believed by some food historians that kale could very well have been the first form of cabbage to be cultivated. It was certainly one of the best varieties of the versatile cabbage family known to various ancient cultures, though it isn't always easy matching their names for plants to the ones we use today.

Take Cato, for example. This Roman statesman (234–149 B.C.) was a champion of cabbage, but he lacked precision. He lumped all the kinds he mentioned under one word, *Brassica*, in his extant work, *De Agricultural*. Unless, of course, it happened to have been wild cabbage, in which case it became *Brassica erratica*. When Theophrastus (circa 225–287 B.C.), the Greek philosopher and scientist, wrote about curly-leaved cabbage in his treatises *History of Plants* and *Theoretical Botany*, it was undoubtedly kale of which he was speaking, and certainly the cultivated kind at that. We know this, because he made a clear distinction between it and wild cabbage. What Pliny the Elder (23–79 A.D.) referred to as *halmyridia* in his voluminous *Historia Naturalis* was actually sea kale, since he specified that it grew on the sea coast.

The Scots appear to be the world's champion kale eaters. The popularity of kale in Scotland is accounted for by the fact that it flourishes in the Scotch climate and therefore presumably reaches in it the summit of its somewhat limited possibilities. Kale grows well near the sea, never far away in Scotland, and it prefers cool climates. Scotch kale shows typical inbred Scotch stubbornness in being the least adaptable of all its many varieties to summer heat.

You'll notice that the Latin binomial given at the top of the entry heading, *Brassica oleracea acephala* implies that it is a non-heading cabbage. This scientific tag covers collards too, but collards and kale are not quite the same thing. Collards in general have much straighter, smoother leaves than kale, whose foliage is curled or crimped like some varieties of parsley. In fact, in French, kale is called *chou frisé*, which means "curly cabbage."

The general rule in Europe is that the curlier this vegetable is, the better for human consumption. Believe it or not, *more* kale is fed to livestock than to people throughout Europe, which is why it is sometimes called cow cabbage.

Kale Prevents Bone Fractures, Arthritis, and Osteoporosis

One of the more remarkable stories I've heard concerning kale and collards came from an elderly couple by the name of Merritts, who worked as volunteers for the LDS Church in the Family History Library located just west of Temple Square in downtown Salt Lake City. While their complete story is told in one of my other books, *Heinerman's Encyclopedia of Healing Juices* (Prentice Hall, 1994, pp. 149–151), a synopsis is worth giving here.

In her nice letter to me, dated August 11, 1993, Mrs. Merritt reported how she and her husband had slipped several times during the winter on icy sidewalks outside their apartment or else by accident in the bath tub. Yet when their respective doctors took X-rays of both, they could find *no* fractures or other bone injuries. "Our doctors were, quite frankly, amazed by this," she reported to me in a followup phone conversation later on. "In fact, one doctor insisted on doing the X-rays all over again, figuring his machine wasn't working right! The only things we suffered," she added, "were some bruises and having our dignities wounded," from such minor falls.

How did this couple, with an average age of 83, manage to retain such strong skeletal frames, when most other seniors their ages are notorious for suffering broken bones and hip injuries due to similar slips or falls? They did it by drinking a combination of kale, collard, and pineapple juice every day, made fresh in their juice machine at home. When fresh pineapple wasn't available, Mrs. Merritt would used canned or frozen pineapple juice; sometimes she would add a ripe apple to change the flavor a little.

An Ancient Roman Cure for Indigestion

Pliny the Elder mentioned that *halmyridia* or sea kale was good to take for indigestion; unfortunately he gave no specific directions for this. So I experimented one time on myself when I had overeaten and was suffering from an upset stomach. I cooked some frozen kale in double the amount of water called for for about 20 minutes on low heat. Then I let everything cool before placing it in my Vita-Mix Whole Food Machine (see the Appendix); I blended everything up for 45 seconds and drank the green liquid. Pliny was right: within minutes my intestinal agonies had subsided.

A Soup Worth Remembering

Kale and Spud Soup

Needed: 1 $^1/_4$ cups chopped medium onion; 6 small Pontiac (red) potatoes, scrubbed and diced; 9 medium kale leaves, chopped; 2 $^1/_4$ cups water; 1 tsp. each basil, marjoram, rosemary, sage, and thyme; $^1/_4$ tsp. finely chopped licorice herb root (*Glycyrrhiza glabra*), if available fresh or dried from any local health food store or herb shop; pinch of chili pepper for flavor; dash of granulated kelp for flavor; 2 cups soymilk.

Place the onion, spuds, kale, spices and water in a pressure cooker. Cook over high heat until the pressure regulator starts bouncing around; then reduce the heat to low and cook an additional 2 minutes. Remove the pressure cooker from the stove and let it cool on its own. Blend the cooking mixture in a Vita-Mix Whole Food Machine or similar food processor along with the remaining ingredients. This will have to be spaced out in several batches. Reheat and serve.

To make this soup without a pressure cooker, use a large, covered pot and follow the instructions previously given. Simmer the vegetables until the kale is tender, or for about 35 minutes. Add more water if needed; soups cooked the conventional way require a lot more liquid than those made in pressure cookers. Yield: Serves 4.

KIWIFRUIT
(Actinidia chinensis)

Brief History

In ornithology, a kiwi is a flightless bird with a Maeori name. But when in the twentieth century, the kiwi also became a fruit. It was back in 1904 when a traveler visited the Yangtze Valley in China and noticed the local inhabitants' fondness for a small, hard, wild Chinese berry, which they called minhoutao or monkey peach. Its electric green, tart flesh tasted roughly like a European gooseberry to this sojourner, who stuck it with the incorrect moniker of Chinese gooseberry.

He brought back some seeds for Alexander Allison, a nurseryman on New Zealand's North Island. In the next 30 years he and other gardeners developed superior kiwifruit vines through careful selection, pruning, and grafting. Most of these early fanciers were as much interested in the vine's showy white blossoms and attractive fan-shaped leaves as in its berries.

Early on it was renamed kiwifruit, but it took the late James MacLoughlin some courage and imagination to ultimately expand it into a worldwide multimillion-dollar business. After he lost his job as a shipping clerk during the Great Depression, Jim's wife's aunt invited them to stay on her lemon orchard at Te Puke on the North Island's east coast. Eventually, the bottom fell out of the lemon market and they faced financial ruin. But a neighbor convinced Jim that there was a future in kiwifruit, having himself sold the fruit from a single plant for 5£ (then worth about $20 United States). Jim quickly picked up on the economic significance of that, and so he risked putting in half an acre of them.

Lucky for him that the warm, wet climate and volcanic soil at Te Puke favored his vines. Neighbors soon launched their own commercial orchards, which further expanded during World War II when American GIs stationed in New Zealand developed a taste for kiwifruit.

Then Divine Providence intervened again. In 1952 an English fruit importer ordered a shipment of New Zealand lemons. To fill spare space in the ship, Jim decided to include ten cases of kiwifruit. A longshoreman's strike delayed the ship almost 1 1/2 months; by the time it finally docked in London the lemons had been reduced to stinky oblivion but the kiwifruit were in perfect shape. They sold well, and New Zealanders suddenly realized that they had just opened a world market.

Not all first reactions to kiwifruit, however, were so memorable. In 1976 a case of kiwifruit was delivered by a grateful client to the famous insurance house Lloyds of London and left in their foyer. It was during a time of Irish Republican Army bombings; someone became suspicious, looked in the box, and immediately judged the firm green kiwifruits to be grenades. The bomb-disposal squad of Scotland Yard was quickly notified, the entire building emptied, and the harmless produce blown to smithereens!

Kiwifruits are stunning to look at when peeled and sliced so that their delicate, edible black seeds glisten in an inner ring within a gorgeous green field. Those who've tasted kiwifruit enough declare it to be a succulent blend of strawberry, banana, melon, and pineapple flavors all rolled into one.

Food for Common Cold and High Blood Pressure

One of the kiwifruit's attributes is a high vitamin C content (approximately 75 mg. in a medium-sized fruit). The fruit also contains substantial amounts of niacin and vitamin E and is low in calories (46 in an average fruit) and fat (1 gram or less depending on fruit size).

But where the kiwifruit really shines is in its very high potassium and crude fiber contents. An average kiwifruit will contain 250 mg. of potassium. This much potassium makes the fruit a good way for people on hypertensive medication to get their potassium. It's also good for people who are looking for a little more variety in the high-potassium fruits they eat.

Nathan Moscher is a New York stockbroker. His line of work is very stressful. In the early part of 1993 he began getting frequent headaches. He went to see a doctor practicing in midtown Manhattan, who took his blood pressure. It showed that Nathan had a systolic blood pressure of 161 mm Hg and a diastolic reading of 115 mm Hg. He was immediately put on a sodium-restricted diet and prescribed the following medications: Hydrochlorothiazide 15 mg. 4 times daily; Ramipril 8 mg. 4 times daily; Nifedipine (slow-release system) 30 mg. 4 times daily; and Timolol 20 mg. daily, equally divided between morning and night.

Once he started this drug therapy, his headaches immediately ceased. But a different type of stress remained. "I began wondering to myself," he wrote me in a letter, "what on earth these pharmaceuticals were doing to my body. I worried about that night and day. Then I got some literature through the mail advertising your *Encyclopedia of Fruits, Vegetables, and Herbs.* I ordered a copy and started looking through it to find if there was a better solution to my problem. That's when I came across your entry for kiwifruits being good for hypertension. I thought to myself, `What have I got to lose?' And so I started buying kiwifruits by the dozens from a Korean market close by Wall Street where I work. I'd eat them in the morning for breakfast, snack on them for lunch, and nibble on them in the evening while riding the subway home."

The more Nathan did this, the more confident he felt that his problem of hypertension was being addressed on natural terms rather than with synthetic means. "I stopped all my medications cold turkey after just 3 weeks on the kiwifruits," he wrote. "When I went back to my doctor in a couple of months and had my blood pressure checked again, it was 130 over 80, clearly an improvement over what it was before. The doctor remarked, 'I'm glad to see those medications are working so well for you.' I pulled a green kiwifruit out of my pocket and handed it to him, saying, 'Doc, this is the only medication I've been on for weeks now, and I think you should be prescribing it more to other patients with the same problem.' He looked down at the fruit, then up at me, and blinked in disbelief. I winked and with a smile said, 'Have a good day,' and left and haven't been back since. That's how well kiwifruits have worked for me."

A Delicious Aid for Heartburn and Indigestion

Kiwifruits contain wonderful enzymes that soothe the stomach. Eating several kiwifruits following a heavy meal will help relieve a bloated feeling, heartburn, or acid indigestion. A simple recipe that doubles as an effective food digestant can be made from kiwis. It is adapted from one I picked up from a restaurant I visited while in Tokyo, Japan several years ago.

With agar-agar (a seaweed), this gorgeous—and low-calorie—dessert medicine is a snap to make. Look for kudzu, a nutritious root starch, in the macrobiotic section of any natural foods store. It helps give body to the mold and also makes a wholesome drink when mixed with warm water and squeezed kiwifruit juice.

Royal Kiwi Mold

Needed: 2 cups natural grape juice; 2 tbsp. agar-agar flakes; 1-2 tbsp. pure maple syrup, to taste; 1/4 tsp. sea salt; 2 tbsp. kudzu, diluted in 1 cup cold water; and 2 kiwifruit, peeled and sliced.

In a large saucepan, combine the grape juice, 1 cup of water, and the agar-agar. Bring to a boil, then reduce the heat and simmer for 5 minutes. Add the pure maple syrup and sea salt and bring to a boil again, then reduce the heat to low and simmer for 10 minutes. Whisk in the diluted kudzu and cook over low heat until the mixture begins to thicken, about 5 minutes. Then pour into a glass bowl and arrange the kiwi slices around the edge and in the center of the gelatin. Let cool to room temperature, and refrigerate until firm, at least 1 1/2 hours. Yield: 6 dosages or servings.

KOHLRABI
(Brassica oleracea)

Brief Description

Emperor Charlemagne ordered his gardeners to start growing kohlrabi in his domains 12 centuries ago, after he found out just how delicious it was. Food historians suspect that the ancient Romans before him also knew about and enjoyed it very much.

Kohlrabi is "a most underrated vegetable that more people should grow and eat," wrote Richard Gehman in *The Haphazard Gourmet*. The fact that he wrote "should grow" gives away one little-known detail connected with it: it is seldom found in grocery stores or supermarkets. So if you want to eat it, you most generally have to grow it yourself, which is what my family has done in past years.

There is a lot of misinformation connected with this vegetable. It has been variously described by some writers as a hybrid between a cabbage and a turnip, and others have outright declared it to be a turnip, unaware, perhaps, that it is a member in good standing of the cabbage family.

The edible bulb of kohlrabi is, in fact, not a root like the turnip, but instead a swelling of the stem just above ground level. It certainly doesn't have the bite of a turnip, but is blander and sweeter when consumed raw.

Potential Treatment for Tuberculosis and Asthma

Thirty-seven years ago a medical doctor from a small town in North Dakota compiled a list of folk remedies he had used for decades on many of his patients. Some of them were herbal based, while others involved food therapy. His papers were never published, but fell into the hands of a distant relative who preserved them in her files for a number of years. She came to one of my lectures at a National Health Federation convention held in Seattle, Washington, awhile back.

Afterward, we got to talking about natural remedies. She bought a copy of the old edition of *Heinerman's Encyclopedia of Fruits, Vegetables and Herbs*. In thumbing through it, she noticed on page 54 the mention of kohlrabi with other members of the cabbage family. This prompted a discussion of the vegetable and her telling me about this doctor's papers in her possession. Later, she sent me photocopies of some of the man's remedies.

Here is the one for kohlrabi that he used for patients suffering from tuberculosis. "Take one kohlrabi, wash it under running water, and chop it up, top greens as well as bottom tuber. Simmer in 1 1/2 quarts water without a lid until only half the liquid remains. Strain, cool, and bottle. Have patient drink one-half cup five times daily."

The lady who gave me this information said she had tried it on herself and with several friends for asthma and found it worked quite well. She said if it worked for that, it probably also works for TB, as the doctor claimed it did in his notes. It is reproduced here for the very first time, but I was asked not to mention her name nor the deceased doctor.

Healthy Stew

The following recipe involves a number of different vegetables, one of which is kohlrabi. It is a substantial meal by itself, but takes a little bit of preparation. The effort and wait, however, are well worth it.

TURKEY-VEGETABLE STEW

Needed: 1/2 cup all-purpose flour, divided; 1 1/2 lb. turkey tenderloin, cut into 1″ pieces; vegetable cooking spray; 1 tbsp. vegetable oil; 1/2 cup chopped onion; 2 cups cubed, peeled kohlrabi; 1 1/2 cups coarsely chopped cabbage; 1 cup sliced carrot; 6 small red potatoes, peeled and quartered; 1 tsp. dried thyme; 1/2 tsp. dried sage; 2 (13 3/4 oz.) cans salt-free or no-salt-added chicken broth; 2 cups small fresh broccoli florets; 2 (14 1/2 oz.) cans salt-free or no-salt-added whole tomatoes, drained and coarsely chopped; 1/2 cup water; 1 tsp. granulated kelp.

Place 1/4 cup flour in a large zip-top heavy-duty plastic bag. Add the turkey; seal bag and shake well to coat. Evenly coat a cast-iron Dutch oven with some cooking spray; add the vegetable oil, and place over a medium-high heat until hot. Then add the turkey and onion; cook 6 minutes or until the turkey loses its pink color. Add the kohlrabi and the next 6 ingredients; bring to a rolling boil. Cover, reduce the heat, and simmer 20 minutes or until the potato is tender. Add the broccoli and tomatoes; cook, uncovered, 5 minutes or until the broccoli is tender.

Place the remaining $1/4$ cup flour in a bowl. Gradually add $1/2$ cup water, stirring with a wire whisk until blended, and add to stew. Stir in the kelp (a seaweed). Cook over medium heat for 5 minutes or until thickened, stirring frequently. Yield: Serves 4–6.

Kudzu
(Pueraria thunbergiana)

Brief Description

Everyone has heard of the comic book hero Superman, who has been immortalized for more than half-a-century in print and in celluloid. As the old TV series starring the late George Reeve in the title role, boasted, Superman is "faster than a speeding bullet, more powerful than a locomotive, and able to leap tall buildings in a single bound!"

Well, while Superman may be a figment of some writer's imagination, kudzu the superplant *isn't*! This vine grows like a prairie wildfire without the benefit of fertilizers, pesticides, irrigation, cultivation, replanting, or even care. Writer James Dickey has called it "a vegetal form of cancer." Many now equate this once much-vaunted plant with Japanese honeysuckle, Russian olive trees, quack grass, and killer bees. (Kudzu is also spelled as kuzu.)

Kudzu spread unbelievably fast through much of the South due to a combination of a long growing season, a warm climate, and plentiful rainfall. Under ideal conditions, "the damned thing" (as one horticulturist called it) is capable of growing a foot a day and up to 100 feet in a season. Channing Cope, the Georgian who originally promoted it as "God's gift to Dixie," once calculated that a single acre of kudzu, left uncontrolled for a century, could expand to over 13,000 acres!

I've traveled enough throughout the South in the past 15 years to know just how fast it proliferates. Kudzu can cover an abandoned car in a few weeks and completely engulf an empty house in the course of a single summer, growing in one side and out the other. "The stuff will smother anything in its path," Ed Laws, head of Georgia highway maintenance, told me one year. "It'll even smother you if you stand around long enough to let it," he said with a sly wink of one eye. Not too long ago a group of tourists from New England were visiting Georgia and totally awe-struck by its lush green beauty. They asked an agricultural official if they could transplant some of the vine to their own state; the official *volunteered* to send them as many *dump truckloads* as they wanted *with the State of Georgia willing to pay ALL transportation costs.* But, in more somber tones, he warned them they would rue the day they ever did this and that their children and grandchildren would hold their names in infamy forever!

Japan, where kudzu first started, doesn't have this problem, however. Japanese farmers practice a kind of agricultural judo on the vine, turning its overflowing energy to their own advantage. Also, the country has a less ideal environment, with plenty of insect predators and disease to keep the vine in check.

This "green menace of the South" is actually a revered and *protected* plant in the Land of the Rising Sun, where it has long been used for soil erosion control, livestock fodder, as a honey source, as a lovely ornamental, as a primary food, and important natural medicine.

Kudzu has beautiful wisterialike flowers that emit a sweet fragrance resembling the aroma of crushed grapes. It has a woody stem, broad leaves, and a root that is usually about 6 to 7 feet in length and averaging between 320 and 440 pounds!

Kudzu Root Tea's Health-Giving Miracles

Since ancient times in both China and Japan, fresh kudzu roots, gathered from November until April, have been washed, peeled, diced into $^1/_8$" cubes and dried in the sun until they become firm and tan. In my trips to both countries I've noticed that these roots form an integral part of Oriental herbal medicine called *kanpoyaku*. They are generally dispensed through natural pharmacies—either by themselves or mixed with other herbs—in small brown paper envelopes or sealed jars. A typical dose is 4 to 11 grams. Within the last decade, though, imported kudzu root has begun showing up in some American herb shops and health food stores alongside more traditional herbs like goldenseal root and comfrey leaf.

In the hard-to-find book *Healing Ourselves* (New York: Avon Books, 1973), author Noboru Muramoto gives a great deal of valuable information about medicinal kudzu root teas and root creams. He recommends for maximum benefits that it be used in equal parts with ephedra (mahuang), cinnamon, unpitted jujube dates, dried ginger root, licorice root, and peony root.

To make the tea, combine the above ingredients in equal amounts with 4 cups warm water in a ceramic or enamel pot; simmer uncovered for 1 $^1/_4$ hours, or until about 2 cups of liquid remain. Take 3/4 cup 30 minutes before each meal, 3 times daily.

The many Chinese and Japanese alternative medical practitioners with whom I've spoken in past years concerning this remarkable root tea have all sworn in favor of its incredible curative powers. Collectively, their recommendations for this tea have been for a wide variety of ailments:

chronic headache
stiffness in neck, back, shoulders
measles
mumps
colitis
upset stomach
breast-feeding problems (promotes better milk flow)
sinus troubles
sore throat/tonsillitis
general respiratory problems/lung congestion
coughing
hangovers/migraines
constipation

Many Other Medicinal Benefits

White chunks of kudzu powder, when used directly in their dry form, closely resemble in both taste and appearance milk of magnesia and can be just as effective in relieving heart burn, acid indigestion, and other gastrointestinal discomforts such as stomach pains, abdominal cramps, diarrhea, gas, dysentery, and fever.

Actually, the application of some kudzu powder for any of these problems is incredibly simple. Just take a small chunk of kudzu powder about $^1/_2$″ in diameter, place it on your tongue, and permit it to slowly dissolve in your mouth. Within minutes any of these painful complaints will vanish almost like magic.

In the event you can't easily procure kudzu root or powder from local sources in your area, you may contact The Fmali Herb Co., 831 Almar Avenue, Santa Cruz, CA 95060, (408) 423-7913.

Making Your Own Kudzu Powder

If you are unable to locate a source from which to buy kudzu powder, you can always make your own. Drive down to Georgia between early December and the end of March to dig up about 5 pounds of kudzu roots. Ideally, each root should be from 1 1/2″ to 3″ in diameter.Bring your harvest back home and wash the roots thoroughly in a large plastic 5 gallon bucket, using a scrub brush and plenty of cold water. With a sharp, heavy knife, trim off 1 1/2″ of each root at the stem end and discard. Now cut the remaining root lengthwise into 1″ discs.

Half-fill a Vita-Mix Whole Food Machine or similar blender with water. Running the machine at medium speed, add 4-6 kudzu discs one at a time and puree until the mixture is quite thick. Pour this puree through a large wire strainer (such as you might use to drain cooked spaghetti) set over a 5 gallon plastic bucket. Squeeze the fiberous residue in the strainer to expel as much liquid as possible; then transfer to a second bucket of equal size. Repeat this procedure 6 to 8 times, or until all root sections have been used.

Add about 1 gallon of water to the bucket containing the squeezed fiber. Stir well, then lift out the fiber and squeeze by hand over the bucket to extract out as much liquid as possible; discard the fiber. Now pour the extracted liquid through the strainer into the first bucket. Squeeze, then discard, any fiber caught in the strainer.

Stretch a cotton dishtowel over the mouth of the now-empty second bucket and pour in the brown kudzu liquid from the first bucket. Gather the ends of the cloth, twist closed, and squeeze to express all liquid. Discard the fibrous residue afterward. Place the bucket containing the starchy liquid in a cold place and permit to set undisturbed for 24 to 30 hours while the starch settles. Now carefully pour off all the brown liquid that has arisen to the surface and discard it.

Using a knife, break up the claylike starch in the bottom of the bucket. Refill the bucket with cold water, stir until the starch is completely dissolved, then let stand for another 24 to 30 hours. Again pour off and discard all liquid that has accumulated.

Now break up the kudzu starch and place the pieces in a 1 gallon wide-mouthed jar. Rinse out the bucket and pour rinse water into the jar too. Stir the starch until it dissolves; then fill the jar with cold water and permit to stand for 50 hours in a cold place.

Pour off the water, and carefully scrape off and discard the upper gray layer of impurities until only the pure white claylike kudzu starch remains. Cut this starch into chunks. The kudzu is now ready to be used for culinary or medicinal purposes. Or place it on a tray (or several layers of white butcher paper) and set in a cool place to dry for at least 1 to 1 $^1/_4$ months. When sealed in a plastic container and stored in a cool dry place, the crumbly chunks of well-dried kudzu powder should last an indefinite length of time. (I am indebted to William Shurtleff and Akiko Aoyagi for this information, which I've adapted here for any readers who might be interested in making their own kudzu powder.)

Some Kudzu Recipes

VEGETABLE STEW WITH KUDZU

This recipe is extremely versatile and can be applied to a large number of vegetables during the sautéeing process. Bamboo shoots, daikon (Japanese) radish, turnip, burdock root, and cabbage can be substituted for zucchini in nearly equal amounts. Or several such vegetables can be combined together in reduced amounts.

Needed: 2 tbsp. sunflower or olive oil; 1 clove garlic, crushed; 1 small Bermuda onion, diced; 1 lb. zucchini squash, cut into pieces 2 $^1/_2$" in length and about $^1/_4$" square; 1 cup distilled water; 1 tbsp. natural soy sauce; $^1/_2$ tsp. grated ginger root; and 1 $^1/_2$ tbsp. kudzu powder, dissolved in $^1/_2$ cup water.

Heat a skillet and coat it with the oil. Add the garlic and sauté 45 seconds; add the onion next and sauté for 1 $^1/_2$ minutes. Then add the zucchini and sauté for 5 additional minutes, or until almost tender. Stir in the remaining ingredients and bring to a rolling boil. Simmer for 1 $^1/_2$ minutes or until thickened to stew consistency. Yield: Serves 4–6.

JEWEL SOUP WITH KUDZU

When I was in Japan some years ago, I discovered that kudzu powder is used just about everywhere to provide a delicate body to the very popular clear soups called *osumashi*. It is sometimes added to miso soup as well. All of the foreign ingredients cited here can be obtained from any Oriental food store in larger cities, especially those catering to Japanese clientele.

Needed: 1 1/2 cups dashi; 1 tsp. natural soy sauce; 1/2 tsp. granulated kelp; 1/2 tsp. sake (Japanese rice wine); 1 tsp. kudzu powder, dissolved in 1 1/4 tbsp. water; 6 oz. tofu, drained and cut into 1 1/2" cubes; 1/2 leek or 2 scallions, cut into thin rounds; 4 slivers of lime peel.

Bring the dashi just to a boil over moderate heat. Reduce heat to low, add the next four ingredients and cook, stirring constantly, until the soup begins to thicken. Add the tofu and simmer until the center of the tofu is well heated. Remove the tofu carefully with a slotted spoon and divide among two soup bowls. Garnish with sliced leeks and lime peel, then carefully pour in the simmer broth. Yield: 2 servings.

Kumquat
(Citrus japonica)

Brief Description

Kumquat is a small tree native to eastern Asia but cultivated throughout the subtropics, including southern California and Florida. It reaches 8 to almost 12 feet in height. The branches are mostly thornless and have dark-green, glossy leaves and white, orangelike flowers. This tree, along with peach blossom, are auspicious symbols for the Chinese lunar New Year when they are sold in the bustling street markets of noisy Hong Kong. Kumquat symbolizes gold and good fortune to the Oriental culture.

The fruit is no bigger than a large olive and somewhat similar in shape. The skin closely resembles an orange, although technically, it is not a member of the citrus family; but the outside surface is much thinner and edible. You can stick the whole fruit into your mouth and eat it, skin and all, spitting out the seeds along the way. It makes a refreshing mouthful after a rich meal, being both astringent and sweet. It has an intense, deep fragrance as you bite into the flesh and release the essential oils. There are hints of orange, orange blossom, bergamot, and lime.

In some of my trips to Hong Kong, mainland China, and Taiwan, I've discovered kumquats sold as candied fruit. And they make delicious jam and jelly, especially when spread on a lightly buttered and toasted English muffin.

When cooked with poultry or fish, kumquat enhances the finished dish because of the flavor coming from the scented oils in the skin.

A Liqueur to Lower Cholesterol

Kumquats can be preserved. Rinse, scrub, and dry them; prick each fruit all over with the tines of a fork or using a toothpick; then pack them into a Mason quart jar. Half-fill the jar with honey and top it with good Russian vodka. Place in your refrigerator and leave there for about 3 1/2 months.

By then you will have amazingly delicious fruit to add to fruit salads. And a most wonderful liqueur which can be taken every day (one tablespoon) to help lower your serum cholesterol.

Kumquats for Obesity

Since kumquats are such an intriguing cross-blend of sweet and sour, they have a natural appeal to the taste buds of those who are in the habit of frequently snacking on junk foods, which just add unwanted pounds. Kumquats by themselves won't help an obese person lose weight, but by snacking on a few every day, they will certainly satisfy the sweet-sour cravings for those junk foods that contribute to the problem.

Kumquat Liqueur Stills a Nagging Cough

The aforementioned liqueur obtained from kumquats that have been preserved in the refrigerator for over 3 months is very helpful in quieting a nagging cough. Simply warm up 1/2 cup of this liqueur and slowly sip it; the incessant barking will soon stop as the throat is gradually calmed with this soothing drink.

L

LEEKS

(See under GARLIC AND ONIONS)

LEGUMES

(See under BEANS)

LEMON

(Citrus limon)

Brief Description

Lemon is a small tree that grows about 15 feet tall, has thorny branches, and features purple-edged white blossoms; it requires a mild climate. Lemon is believed to have originated in India a long time ago. The Europe crop is centered on the islands and coasts of the Mediterranean. In the United States, lemons are grown chiefly in California, especially in the southern sea coast areas, and in Florida.

The trees are prolific, producing ripe fruit practically all the year. In the United States the fruit is cut from the tree while green, at a standard size, and the good lemons are placed in cool, dark rooms to ripen gradually; the skin grows yellow, thin, and pliable, and the quality of the fruit is better than when ripened on the tree itself. The imperfect fruit is manufactured into lemon oil, lemon juice, citric acid, pectin, and other useful products. Lemons have better preservative qualities than do other citrus fruits and are thus more easily transported.

The fruit is high in vitamin C and has long been known as a preventive of scurvy. Lemons have a refreshing, acid flavor; they are in great demand for use in summer drinks, such as lemonade and punch and are often preferred to vinegar as an ingredient in sauces and salad dressings.

Lemon juice is the main source of citric acid, which is used by calico printers to keep the fabric clear of rusty stains from the machinery; it is also a domestic remedy for rust, ink, and mildew stains. Lemon oil, or the essential oil extracted from the skin, usually while green, is manufactured mostly in Italy and France. It is used in the making of flavoring extract (essential oil combined with alcohol), perfumes and cosmetics, and furniture polish.

Knocks the Common Cold and Sore Throat

A terrific remedy for knocking the common cold and sore throat is to squeeze the juice from several lemons into a small saucepan and gently heat it on a low stove setting. When quite warm, stir in 1 tsp. dark honey or blackstrap molasses. Keep stirring until thoroughly mixed in with the juice.

Strain, if necessary, to remove any seeds and bits of pulp. When lukewarm, drink 2 tbsp. of this mixture, but do not swallow right away. Instead, sit down in a chair, tilt the head back as far as possible, and gargle with it for 2 minutes. Then it can be swallowed. *Do not drink anything else after this.*

I've never known this warm lemon juice and honey combination to fail. It may burn the throat a little in the beginning, but this discomfort will soon clear up as the person gets better.

Lemon Peel for Bleeding Gums

If your gums bleed quite a bit after brushing, then cut a small section of lemon peel from the fruit. Next turn it inside out so the white part is facing you. By wrapping enough of the peel around the tip of your second finger and holding the edges firmly with your thumb, you can rub your gums for a few minutes each day with the white part. Bleeding should stop in a few days.

Skin Beauty Tips from an Exclusive Salon

Paul Nienast is unique in the world of beauty care. His salon in Dallas is probably one of the most exclusive of its kind in America. This is not due to the fact that it rivals salons in Paris or New York City for professional hair and skin care, even for society's elite who flock to it from all over the country in search of eternal youth. His salon is exceptional because it probably uses more fresh fruits, vegetables, grains, oils, and dairy products on its clientele than any other single salon in America.

Paul consented to share for the very first time with the public in a book such as this some of his other remarkable "secret" techniques for incredibly beautiful skin that could fool many people into thinking you're 15–20 years younger. His wrinkle-remover formula is one of the most popular and sought after by the thousands who flock to his salon every month.

Take 2 slices of a lemon and put them in a wooden bowl (never metal or plastic). Add just enough half-and-half which has been previously heated to lightly above warm to cover. Cover the bowl and let set for about 3 hours, after which the solution is strained and gently massaged directly into the skin in a rotating fashion with the tips of the middle three fingers. Allow it to dry on the skin, then remove with a wet wash cloth and a little olive oil.

By doing this morning and evening, wrinkles should begin to disappear within a matter of just a few weeks or even less, depending, of course, on just how deep and old the skin lines are. If the skin is very oily, use 4 slices of lemon instead; if the skin is somewhat dry, use only 1 slice of lemon and double the amount of half-and-half.

"My wrinkle-removing formula is also terrific for rough, chapped 'dishpan' hands," Paul added. "From personal experience, I can tell you that there's probably not another application around that is so perfect for whitening and softening the hands with. I suggest that a couple of orange slices be added as well to the two lemon slices before the warm half-and-half is poured over them."

Paul finds a lemon juice compress to the skin is great for clearing up discoloration problems. "Just soak a wash cloth or small hand towel in a solution of the juice from 3–4 lemons and 1 lime diluted with a little hot water," he said. "Then apply directly to the skin for up to half an hour. It really gets rid of any blotches quite nicely and sort of evens the color out more."

"I make another blotch-remover by applying a banana purée with the juice of 1 or 2 lemons mixed in. It's also very effective evening out those areas that may be light and dark colors in the same place."

He briefly ranked the different citrus fruits according to their usefulness on the skin. "Only fresh fruit juice must be used in all instances, he warns. "Anything less than this simply won't do. Lemon is a good astringent for closing the pores and helping to tighten things up a bit. Because it's very acidic, I recommend diluting it with a little water. Orange juice is the best skin softener I know of in the fruit juices. Like lemon, it's good for normal to oily skin. Orange really seems to perk up dull, lifeless skin. While grapefruit isn't as dramatic an astringent as lemon, the juice is good for neutralizing fatty acids on the skin and doesn't require any diluting."

At Paul's salon in Dallas, a black fig purée with a little tangerine juice is used as a facial for half an hour to help tighten skin and close loose pores. After which it's removed with cool water. For blemishes, a paste is made by soaking a slice of white bread in 1/4 cup of warm milk along with 1 packet or cube of yeast. This is then applied to the skin for about 40 minutes and later removed with water.

A two-step treatment recommended for vibrant-looking skin is this. First, take some cotton balls and completely saturate them in a solution of equal parts of apple cider vinegar and hot water. Then, carefully cleanse the skin with them. After which some fresh orange juice is then applied and left to dry. A moist wash cloth can then be used to lightly wipe the face after this for fabulous looking skin.

Lemon and Migraines

Elana Russo of New York City wrote about a remedy which her grandmother used for getting rid of migraine headaches. A peeled lemon skin was turned white side down and placed on a handkerchief. Then the handkerchief was put against the forehead with the yellow side of the peel against the skin. When a burning sensation became evident, it was removed, and generally, the headache would be gone.

Another remedy recommended to me by my Indonesian friend Dr. Auzay Hamid is good for headaches. When women in Jakarta are bothered by migraines, they just go and wash the dishes or some clothes by hand to bring quick relief. A couple of lemons are cut and the juice squeezed into the hot, soapy water as well. The hot lemon water seems to transfer excess blood from the head down to their hands, besides reducing the swollen blood vessels in the top of the head. For severe migraines that don't seem to go away, standing barefoot in two pans of hot lemon water by the sink while washing dishes in hot lemon water is sure to work just fine. An upright, standing position is necessary, though, for guaranteed success.

Insect Remedy

If you just squeeze some lemon juice on a mosquito or centipede bite or a bee or wasp sting, within a matter of minutes the itching and pain will completely stop! This remedy was recommended by Thor Heyerdahl in his book *Fatu-Hiva*. And to keep ants out of your kitchen, just squeeze a little lemon juice along the baseboards and window and door sills and watch them head for the hills.

Heavenly Foot Care

Practically everyone suffers from sore, aching feet or foot problems of some kind at one time or another. Well, there's nothing quite like a nice lemon or lime juice–chamomile cream foot rub to help ease those pains away. The feeling to be derived from something so special as this is nothing short of pure ecstasy so far as hot, tired feet and sweaty toes are concerned.

Simply squeeze a little CamoCare Soothing Cream into the palm of one hand with a little lemon juice. Mix well by rubbing around in your palm with two fingers. Then rub the same on the bottoms of your soles and work well in between the toes. The sensation created is something akin to standing in ice-cold, minty water.

CamoCare is distributed by Abkit, Inc., New York, to most local health food stores. Lemons and limes are available from any produce stand or local supermarket.

LENTILS

(See under BEANS)

LETTUCE
(Lactuca sativa)

Brief Description

A Burpee & Co. seed catalog from 1894 proclaimed this about iceberg lettuce: "There is no handsomer or more solid Cabbage Lettuce in cultivation." That may well be the case, but it has been the butt of jokes and the bane of all true "gourmets." I have even spoken in disdainful terms of it in some of my past health lectures, equating "iceberg lettuce with Twinkies and Coca-Cola as some of the ultimate junk foods!" At other times I've sarcastically remarked that "even rabbits fail to thrive on iceberg lettuce, so nutritionally depleted in food value is it."

But this stalwart leafy vegetable is quite cosmopolitan and is equally at home on the aristocratic or the peasant table. The year 1994 marked the 100th birthday of this lettuce. It was discovered, quite by accident, by some of the Burpee people on their Fordhook Farm in Doylestown, Pennsylvania. What happened was that they were growing a large plot of "cabbage-head" or Batavia lettuce (a variety that apparently has ancient Celtic roots), when they noticed one particularly handsome specimen, large-headed and well-blanched at the center. More important, when it was tasted, it was found to be unusually crisp and "sweet" (in the sense of having a mild flavor).

Well, the Burpee folks knew a good thing when they saw it, and, together with other American vegetable growers, set about improving it still further. Today, there are countless varieties of iceberg. Incidentally, the California Iceberg Lettuce Commission recently corrected me in a letter, informing me that it should be called "crisp-head" lettuce.

Relieves Heartburn

While iceberg lettuce, for the most part, is very difficult to digest and not on my recommended list of foods that are good to eat, it, nevertheless, does possess at least one medicinal virtue. Any kind of lettuce is wonderful for treating acid indigestion and heartburn in place of antacids such as Tums.

In a blender whip up half of a cut lettuce leaf with 8–10 oz. of ice cold water until you have a nice semithick, green purée of near milkshake proportions. Drink slowly to relieve stomach upset. A little honey and a touch of vanilla may be added for flavoring, if desired.

Wonderful Poultice for Bruises

When I first wrote this book sometime in 1987, the aforementioned remedy was the only one I was familiar with using iceberg lettuce. Truth of the matter is, I was probably so prejudiced against it as a near worthless food that I didn't really look hard enough for more remedies involving it. Fast forward to the end of 1994 and a revision of his work. In the intervening period, my tolerance for iceberg—oops!, I mean "crisp-head"—lettuce has grown enough to allow me to more serious focus on some of its other medicinal applications.

A Mrs. Lutz from Corpus Christi, Texas informed me awhile back that a nice "cold pack" made from chilled lettuce leaves "really helps to take away the discoloration caused by bumps and bruises." She explained how to take half of a whole piece of round lettuce leaf, which has been chilled in the freezer compartment of your refrigerator for an hour and crumple it up in your hand. Then, she instructed, "place this cold, wet mass on any black-and-blue spot and tape down with some adhesive to hold. Works every time," she finished. "Tender soreness will ease up and the skin return to normal within a few days."

Culinary Functions

Iceberg lettuce is a close friend of every sandwich of any notable distinction—hamburger, peanut butter, cheese, club, tuna, devil egg, and more. It figures in the popular Chinese preparation of minced squab in lettuce leaves and in petits pois à la francaise, among other imports.

There are several different recipes which feature this vegetable as the main centerpiece. One is what I laughingly call my "Honeymoon Salad," which consists of "lettuce alone" ("let-us-alone"). On the more serious side is this imaginative little number:

"Crispy-Head" Lettuce With Velvet Dressing

Needed: 1 wedge of "crispy-head" lettuce (thoroughly rinsed); 1/2 cup sour cream; 1/2 cup low-fat or no-fat mayonnaise; 2 green onions, minced; 2 1/4 tbsp. of lime juice; 1/8 tsp. grated lime peel; 1/2 cup blue cheese, crumbled; a pinch of granulated kelp (seaweed from any health food store).

Plunge a cored head of iceberg forcefully into a large bowl of cold water. Drain, wrap in towel, and stick in the refrigerator overnight (so it can live up to the name of "icebox").

Next day combine the sour cream, mayonnaise, green onions, lime juice, and grated lime peel, and mix everything thoroughly. Stir in the cheese; then chill about 5 hours.

To serve, cut the lettuce into 2 or 4 large wedges, spoon on this velvety cream dressing, and lightly season with kelp. Serves 2–4.

LIME
(Citrus aurantifolia)

Brief Description

This small shrublike tree is one of the citrus fruit trees, similar to the lemon but more spreading and irregular in growth. It is native to Southeast Asia and has been introduced into Southern Europe, the West Indies, Mexico, and Florida. Chief production is in the tropical regions of both the Old and New World; most American limes originate from the West Indies or Mexico.

The lime is the most susceptible to frost injury of all citrus fruits; this confines its commercial culture in the United States to the southernmost tip of Florida. Here some of the varieties often do well in sandy or rocky soils—conditions generally unfavorable to most other citrus fruits.

The bright green fruit is smaller than the lemon, more globular, more acid, and with a thinner rind. It has the vitamin value and other properties of the citrus fruits. Limes are used to flavor drinks, food, and confections. Limeade and other lime-flavored drinks have a flavor and bouquet quite distinct from those made from lemons. In fact, limes are very popular in Yucatan, Guatemalan, and West Indies cuisines.

Limes are loaded with vitamin C and were formerly employed in the British Navy in the eighteenth and nineteenth centuries to prevent scurvy; hence the nickname of British sailors in the past as "Limeys."

Stomachache Remedy from a Calpyso King

Calypso is definitely linked to an African heritage. This rousing musical style of the West Indies originated on the west coast of Africa as *kaiso*, which is a Hausa word meaning "bravo." The *kaiso* had choral refrain, a dancing chorus, and a call-and-response structure. The style was closely linked to the Nigerian traditions of festivals when the entire community turned out and judged song contests performed by masqueraders. These African musical and festival traditions were the early heritage of calypso.

When slaves were brought to the New World to cut sugar cane, they were encouraged by the plantation owners to sing *kaiso*-style work songs to increase productivity; the lead singer of the winning work gang bragged in song about his group's prowess and scorned other slave gangs. The leaders adopted names such as Elephant and Thunderer, which were precursors to the colorful monikers of the later calypsonians. Eventually, the word *kaiso* became transformed into the noisy and colorful calypso of today, which is very popular throughout the West Indies.

Daniel Quevedo is a descendant of an early Calypso king who went by the title of Attila the Hun when performing in public, but privately was always known by his given Christian name of Raymond Quevedo. This Attila invented a special type of ginger beer which became very popular for curing the worst kind of belly ache.

I met Daniel in Kingston, Jamaica, some years ago and he gave me the "secret" recipe of Attila's:

ATTILA'S STOMACH REMEDY

Needed: 1 lb. fresh ginger, peeled and grated; the juice and "meat" of 2 limes; 1 cup of sugar; 1 tbsp. cream of tartar; 4 quarts boiling water; 2 tsp. active dry yeast. (I know, sugar isn't good for you, but in this recipe it is essential to making a good ginger beer remedy; honey won't work the same way sugar does.)

In a large *wooden* (no metal) bowl, combine the ginger, lime juice, squished "meat" or fruit flesh, and cream of tartar. Pour the boiling water over the top and permit to rest overnight. Add the yeast in the morning and stir well; let stand for 15 minutes. Strain the mixture, then add the sugar and stir until it is thoroughly dissolved. Store in quart jars. Makes roughly 4 quarts.

When a stomachache feels imminent, pour yourself a cup of this Calypso king brew and *slowly sip* it through a plastic straw. As Daniel said, "Take your time; there's no hurry; that gut ache ain't about to run away." Relief will be realized within minutes.

Excellent Skin Rub for Erysipelas, Sores, and Itching

Daniel demonstrated for me how to use half of a cut lime to relieve various skin irritations. Holding the lime half in one hand, he rubbed in a circular motion on an area of his dark skin which was itching badly due to several mosquito bites. He claimed it brought "instant relief." I tried it on my own mosquito bites and was astonished to see just how quickly this simple remedy worked! It also does wonders for erysipelas and sores.

Quit Smoking and Drinking and Lose Weight the Easy Way with Limes

After 15 years of cooking traditional south-of-the-border fare, Chef Armando Palacios of Armando's—a trendy eatery in Houston—began holding down the fat in the dishes he prepares. For instance, he cuts the fat but retains the flavor by steaming brown rice rather than frying white and by cooking black beans with celery and carrots instead of the usual fat and salt. He uses nonfat sour cream and less liberal doses of low-fat cheese.

Chef Armando discovered the miraculous key to his own health turn-around in limes, of all things! "Limes helped me to quit drinking, kick a heavy-smoking habit, lose a bunch of extra pounds, and lower my cholesterol," he bragged. "Before I didn't even know I had a body. It was just a torso that went everywhere I went and had no real goal in life. Now it has a goal and I'm in it for my health."

This chef, who stands 6 feet 2 inches and was 46 years old in 1994, started drinking his own limeade in place of alcoholic beverages and added fresh-squeezed limes to his V-8 and tomato juices whenever the craving for nicotine came up. And instead of adding salt to the Southwestern ethnic foods he loves so much, he started squirting generous quantities of lime juice on his black beans and tortillas. His blood pressure went down and he lost a lot of weight—he's now a svelte 170 pounds! He thinks everyone should be using limes more often.

Loquat
(Eriobotrya japonica)

Brief Description

This subtropical tree of the rose family is related to the apple and other well-known fruit trees of the temperate zone. Ornamental in appearance and rarely more than 33 feet in height, the evergreen loquat is frequently planted in parks and gardens, where it makes a lovely ornamental. The leaves, clustered toward the ends of the branches, are thick and stiff. The small, fragrant, white flowers are arranged in dense terminal panicles.

The fruits are borne in large, loose clusters; individually they are round, obovoid, or pear-shaped, and between 1 to 3 inches in length. They are quite fragile and their pale yellow to bronze, plum-like skin often looks spotted with brown bruises. Beneath the waxy-appearing skin is a delicate soft apricot-colored flesh surrounding 3 or 4 large seeds. The flavor is agreeably tart with a slight resinous hint, not unlike a mango.

Asthma and Bronchitis Relief from Japan

In Japan two systems of medicine peacefully coexist side by side. The one is orthodox medicine with all its technological advancements; the other is a Japanese adaptation of regular Chinese folk medicine known as kampo. Believe it or not, both systems are routinely practiced by licensed M.D.s. Therefore, it is common to see many physicians practice their surgery and drug-oriented medicine in hospitals each day for so many hours, and then leave to go to their clinics where this alternative kampo medicine is practiced.

In my trip to Japan several years ago, I had the good fortune of interviewing some of these doctors who practice dual medicines quite effectively. They shared with me some of their remarkable formulations. One of them consists of combing five crushed loquats, *including the pulverized seeds,* with a little dark brown rice syrup (about 2 tbsp.; honey may be substituted), and $^1/_2$ cup of sake (Japanese rice wine). The mixture is allowed to set for several days before being strained; the liquid is then given to patients suffering from asthma and chronic bronchitis, in $^1/_2$ cup doses on an empty stomach 3 times a day. It helps to evacuate phlegm from their respiratory systems so they can breathe a lot easier.

Soothes Cough and Eases Heart Pains

The same medicinal liquor, when warmed up a little and slowly sipped through a straw, will relieve the worst cough. And taken in 1 cup amounts, very slowly, on an empty stomach should help to ease heart pains, these kampo doctors told me.

LYCHEE
(Lycium chinensis)

Brief Description

Lychee is a small deciduous shrub growing throughout China. It is often found on the outskirts of villages, along roadsides and railroad tracks, in a wild state or in cultivated form in numerous garden plots. The stems are clustered with the slender branches bearing short thorns. The lower parts of the branches show several leaves clustered, ovate, or ovate-lanceolate in shape. During the summertime purple flowers appear from the leaf axils.

The shrub yields berries that have outer brittle, rough, pink shells that make them mistaken for nuts; hence the misnomer "lychee nuts." When the shells are pressed between thumb and finger, however, they can be peeled off cleanly to reveal a pearly white fruit inside. The shells themselves are elliptically shaped and bright red when fully matured. After they are removed, the fruit inside can be consumed raw; it has a sweet flavor and grapelike texture but is a little more chewy and delicated scented. Deep within the flesh is a long, shiny, oval, brown, inedible seed.

Tonic for Liver, Kidney, and Lung Problems

In Oriental medical theory and practice, lychee berries are designated for their tonic influences on three major organs of the body. They strengthen the functions of the liver and kidneys, which filter blood, thereby making this substance more pure as it flows throughout the body. Where consumption, asthma, or bronchitis may prevail, lychee nuts will stimulate improved breathing of the respiratory system.

Stabilizes Blood Sugar Levels in the Body

When I went to mainland China in the summer of 1980 with the American Medical Students' Association (part of the A.M.A.) to study the unique medical system of that nation, we learned some new things. First and foremost was the lesson that you can still achieve good health care for people without a lot of fancy medical technology or high-powered pharmaceuticals. Clinic after clinic and hospital after hospital that we visited, we saw this same principle at work with amazing results.

For instance, in some of the smaller clinics in large communes in rural China, we saw bare-footed or sandaled "doctors" with minimal medical training treating cases of diabetes and hypoglycemia with lychee berries. We were told that these "nuts" (or fruit if you will) helped to stabilize blood sugar levels very nicely. Upon my return to America, I began implementing this in my own recommendations to people afflicted with either problem. In a number of instances, where treatment was begun early enough by having the individuals consume up to 10 lychees a day, improvements in blood sugar levels were observed. The lychees suggested for this can only be obtained from Oriental food stores.

Helpful for High Blood Pressure, Yeast Infection, and Cancer

Since lychee berries are a proven tonic for the kidneys, it stands to reason that they will reduce elevated blood pressure levels. And for those afflicted with *Candida albicans* or yeast infection, lychees will have some merit. In a number of hospitals we visited while in China, lychees were used in conjunction with Schizandra and other herbs for cancer.

M

Macadamia
(See under NUTS)

Mango
(See under TROPICAL FRUITS)

Mangosteen
(See under TROPICAL FRUITS)

Marigold
(See under ORNAMENTAL FLOWERS)

MELONS

CANTALOUPE
(Cucumis melo cantalupensis)

CASABA
(Cucumis melo indorus)

HONEYDEW
(Cucumis melo species)

WATERMELON
(Citrullus vulgaris)

Brief Description

CANTALOUPE. A variety of muskmelon having a warty rind and reddish-orange flesh.

CASABA. A variety of muskmelon or winter melon with a yellow rind. First introduced to Smryna in Asia Minor.

HONEYDEW. A sweet, smooth-skinned, white variety of muskmellon.

WATERMELON. Large oblong or roundish fruit of a vine of the cucumber family. It has a hard green or white rind and a pink or red pulp with a copious sweet juice.

Some scientists who specialize in the evolution, migration and use of plants by indigenous cultures believe that these melons were brought to the Western Hemisphere around 2000 B.C. by emigrants from central Iraq, who crossed the ocean in unique vessels.

Melons as Laxatives

Besides making good diuretics, melons also help to correct constipation. Generally, a quarter or half a melon eaten by itself should induce a modest bowel movement several hours later.

Cure for Jaundice

An effective treatment for jaundice of the liver in some Caribbean countries calls for fresh green melons that are slightly unripe to be consumed along with some parsley tea on a regular basis. A handful of freshly chopped parsley steeped in 1 pint of hot water for 30 minutes, strained and drunk, makes a good tea.

Watermelon Remedies

The seeds of watermelon are universally acclaimed for their marvelous diuretic activity and soothing effect on bladder inflammation. Sometimes even the rind is dried and used the same way. In the Bahamas, freshly discarded seeds are pounded with a heavy object just enough to bruise them, after which they're boiled for about 45 minutes (6 tbsp. seeds to 1 quart hot water) on low heat, then strained and the liquid drunk at least three times daily in 1 cup amounts each time. The same tea also is good for expelling some intestinal worms as well.

In some South American countries, the thick rind is bound around the forehead and temples to relieve excruciating migraine headaches. Or else the rind is mashed into a pulp and applied as a poultice directly over the liver or gallbladder to relieve pain and suffering in either of those organs.

A piece of watermelon consumed immediately after eating a meal of beans usually helps to relieve and reduce the embarrassment of intestinal gas which frequently occurs later on.

Great Cooler for Intense Heat

Nothing seems to quench thirst quite like fresh watermelon, especially if the flesh has been remoeved from the rind and mixed up in a blender with some crushed ice. My oh my, but does that ever feel good going down!

A most curious but seemingly effective remedy for poison ivy itch and rash was demonstrated to me one time by some folks back in the hills of Tennessee. One of their youngsters got into some ivy and came home a red mess. His parents gathered a fresh watermelon from their garden, split it open, and methodically proceeded to rub down those afflicted parts of his body with the flesh and the rind itself. Next morning as my colleague and I got ready to leave, they brought the lad to us for inspection. And darned if the little tyke's rash wasn't in remission already by at least 70%!

But what has to be one of the most incredible yet highly successful treatments for major first-, second- and third-degree burns was witnessed by me in the People's Republic of China in the summer of 1980.

FInally ripened watermelon pulp and juice were placed in a clean glass jar, tightly sealed and allowed to stand at room temperature for 3–4 months. After which, the juice was filtered, having accquired a sour plum odor in the meantime.

I witnessed major burns first being washed with a cold normal salt solution or just with plain cold water. Then a piece of defatted or specially treated cotton was next dipped into the clear fermented watermelon juice and applied directly to the burned area. This dressing was changed several times each day.

My interpreter told me in response to a question I'd asked the doctors, that first- and second-degree burns generally took 8–9 days to heal and third-degree burns 17–21 days or less, as a rule.

MILLET

(See under GRAINS)

MUSHROOMS

Brief Description

Botanists seem to think that there are some 100,000 species of fungi, but less than half of them actually pass for the mushrooms with which most of us are somewhat acquainted. To distinguish mushrooms from all other fungi such as yeasts, lichens, penicillin, and so on, it may be helpful to think of them as "*fruiting* bodies."

Mushrooms haven't fared too well in literature. Keats, Shelley, and Tennyson all took swipes at them. And even in our own time, the famous English novelist D. H. Lawrence (1855–1930) compared "the beastly bourgeois" (or greedy capitalists) to "old mushrooms, all wormy inside, and hollow" besides. Finally, Emily Dickinson likened the average mushroom to Judas Iscariot, the apostle who betrayed Jesus Christ to the disbelieving Jews for 30 pieces of silver coin.

Some dietitians charge that mushrooms aren't nutritious. But all one needs to do is read what famed naturalist, Charles Darwin, the father of the theory of evolution, had to say about this in his classic work, *Voyage of the Beagle*:

> In Tierra del Fuego [at the tip of the South American continent] the fungus in its tough and native state is collected in large quantities by the women and children and is eaten uncooked . . . . With the exception of a few berries, chiefly of a dwarf arbutus, the natives eat no vegetable food besides this fungus.

It is also equally significant to discover that for the aborigines of Australia mushrooms were once considered "blackfellows' bread."

Mushrooms have no sugars (and are therefore a good food for diabetics) and not much carbohydrate, and that mostly in the form of indigestible cellulose (and are therefore a good for for weight-watchers). They also contain numerous vitamins and minerals (varying with the species of mushroom). One trace element they are particularly high in is zinc. (Another food with equal zinc content is raisins. Yet the former grows best without the benefit of sunlight, while the latter needs all it can get to dry out. This is one of the nutritional peculiarities about both foods that still baffles scientists.)

Above all, mushrooms contain a good deal of protein (not all of it being assimilable, of course). This places them in a near equal category with nuts and beans as being closer to meat than other foods.

The most widely consumed wild mushroom in the world is the boletus, of which *Boletus edulis*, is the most admired species. In taste it vaguely resembles raw chestnuts. They are very similar in appearance, though, to death caps, which account for 95% of all deaths from mushroom poisoning. This explains why Roman Emperor Claudius kept right on eating his dish of boletus mushrooms, without noticing that his wife Agrippina had slipped in a few toxic *Amanita phalloides* in order for her son, Nero, to ascend the throne.

The first cultivators of mushrooms were ants, which were so good at it that the entomologist Peter Farb wrote that "thanks to the numerical superiority to which they owe to their prosperity," the mushroom-raising species are the lords of antdom, who measure their gross national product in terms of mushrooms, not money. The first human mushroom cultivators were the Japanese, who have been raising the famous shiitake for two millennia now.

But all shiitake are not created equal. The best (selected from hundreds of possible strains) suggest garlic, pine and autumn leaves and are firm and meaty; others are mild, lightly earthy, and slightly soft. Those grown in an environment that approximates the original habitat are memorable. When temperature and humidity are accelerated to speed growth, the result can be "soggy Frisbees," as one Japanese farmer described it to me on a trip to Japan several years ago.

The name shiitake comes from the Japanese word *shii* for a member of the beech family. The fungus is carefully cultivated on beech and oak tree stumps in countries such as Japan, Taiwan, and China. The best shiitake, my Oriental sources told me, are those mushrooms which are firm and fleshy, particularly those that are dark, domed, and dappled (a white bloom is normal). Thin, pallid mushrooms often have a taste to match. Symmetry and size have little bearing on eating qualities. Shiitake should be dry, but not leathery, and have a distinct aroma. The heavy, tough stems should be well trimmed or small. Among the most flavorful (and rare) are cold weather strains with small cracks and fissures. Occasionally, "baby" shiitake or "buttons" show up. Small, deep brown, tender (even the stems), and intense, they are certainly worth the extra dollars paid for them.

Mushroom Cure for Breast Cancer Given Through a Near-Death Experience (NDE)

I will be the first one to admit that in two decades worth of collecting unique and unusual remedies from all over the world, *nothing* comes close to fitting the description of *strange* as does the following true story. I'll admit I was pretty skeptical when I first heard it myself, but I soon changed my mind after I spoke to the woman's attending physician who confirmed some of the critical elements in her narrative. Both parties were willing to share information with me concerning this extraordinary incident, provided that I did not use their last names or give the city they're from.

My informants were a 52-year-old woman named Elaine C. and her surgeon, Dr. Tom M. They reside in the state of Minnesota. I met Elaine by way of correspondence in the early part of 1994; she had read my book, *Double the Power of Your Immune System* (Prentice Hall, 1991) and wrote to my publisher in Englewood Cliffs, New Jersey. Her letter was forwarded on to me some weeks thereafter. She had some questions pertaining to the functions of a healthy immune system. I responded in writing, giving my unlisted phone number in the event she had more questions.

Elaine called me sometime in June of that year; we spoke for almost 1 $^{1}/_{2}$ hours. During that lengthy conversation, she related the following episode to me. I took notes and asked pertinent questions. She also gave me the name and phone number of her oncologist with whom I later verified some of the medical elements connected to her very intriguing tale. What appears next is a combination of the information obtained from both interviews.

Elaine C. went to the oncologist, Dr. Tom M., for a thorough breast examination. Her own mother and one sister had previously died of breast cancer some years before this, and she was worried that she might have inherited the disease herself. In his careful exam, Tom found evidence of dimpling, erythema, early skin edema, nipple changes, and extreme tenderness in her left breast. He recommended an excisional biopsy to remove sufficient tissue for histologic and immunohistochemical analyses.

Sometime in March, 1991 Elaine entered a hospital, signed all the necessary release forms, and checked herself in as Tom's patient. On the following morning he showed up at her bedside for some brief bantering with his patient. His clever and witty ways put her at ease, and she was wheeled into an OR (operating room) in a relaxed frame of mind.

It was necessary to put her under general anesthesia since a certain amount of tissue had to be cut out. Following the excision biopsy, Tom had the samples sent to Pathology for routine analysis. The entire procedure is fairly routine and lasts about two hours.

Sometime during this period, Elaine suffered a cardiac arrest. The EKG machine showed a flat line indicating no heart beat. Elaine vividly remembered what happened next, although her body still lay lifeless on the operating table.

She perceived herself slowly rising up out of her tabernacle through the crown of her head. She stood in midair some 10 feet above the table and was able to look down on what was then going on. Tom, her oncologist, was barking out orders for defibrillators to be brought *immediately* to the table to jump-start the ventricular muscle of her heart to restore normal beat again.

While she watched this scene of organized excitement with some curiosity, she became aware of two other individuals standing next to her. She turned and perceived that it was her deceased mother and sister, both dressed in simple white apparel.

Her mother spoke but never actually moved her lips; it was more along the lines of mental telepathy, Elaine later recalled. The spirit personage informed her that unless she started using some natural remedies and changing her diet radically, she would not only lose both of her breasts but also her life as they had done.

Then her sister spoke and said that they had to go through a lot of trouble and effort just to get permission to come to her; that it was not in the economy of Heaven to grant such visits as this to mortals. They apparently had to exercise a great deal of faith and prayer to convince those in charge that such a short visit would be in the best interests of all concerned.

Her sister gave her a brief list of dietary "do's and don'ts." Chief among which was the exclusion of all red meat *and chicken*, with only fish allowed. Also no more carbonated beverages of any kind as well as total abstinence from sugary and greasy foods.

Then her mother spoke and declared that the two things that would help her the most in reversing this disease would come from Japan and be both food as well as medicine for her.

Elaine asked for the identification of both items, but was told that if she exercised faith in God and meditated on the matter she would be led to find them very soon.

Her sister's final words were for her to visit "the place where health is sold." With that, her two relatives departed as quickly as they had appeared. In looking back down on her lifeless body still laying on the table below, she noticed a long, thin, filmy cord connecting her spirit with her mortal remains. The thought immediately flashed into her mind that once this spiritual "umbilical cord" was severed, her death would be permanent, but so long as it remained attached to her body, she would come to life again.

She saw one of the surgical nurses placing gels pads on her chest. Tom then lifted a pair of what appeared to be metal paddles and yelled for everyone to stand clear of the bed. A powerful electric shock ran through her body, causing it to momentarily arc upward before dropping back down on the table. Almost immediately everything went black for her as she felt her hovering spirit pulled back down into her system.

Tom related that following the use of the defibrillators, the EKG monitor showed a return of normal heartbeat. After that everyone breathed a little easier, including his unconscious patient. The report coming back from Pathology not long thereafter indicated the mass was malignant. But he decided in light of what had just happened that it was best to bring her out of anesthesia and return her to her room for rest.

Elaine said it took "powerful persuasion" on her part to convince her doctor to release her from the hospital in spite of the fact she had breast cancer. He finally made her sign a form relieving him of all responsibility concerning her condition, so that he wouldn't be threatened later on with a potential medical malpractice suit.

As she mulled over what had her heavenly visitors had said, it occurred to her that the things she was looking for would be found in a health food store. Locating one several miles from her residence, she went there on the third day of her requested discharge from the hospital. Not knowing exactly what she wanted, she inquired of the clerk what things the store carried that might be good for cancer.

She was informed that federal and state laws prohibited store employees from dispensing such information, which could be misconstrued as "prescribing and diagnosing without a medical license." But she was escorted to a shelf full of health books and told she might find something in them. In looking over the available stock, two works in particular caught her eye. One was a book by James F. Balch, M.D., entitled *Prescription for Nutritional Healing* (New York: Avery, 1990), and the other volume was on the *Reishi Mushroom* by Terry Willard, Ph.D. (Issaquah, WA: Sylvan Press, 1990).

She bought both books and went home to read them through carefully. In Balch's book, she found mention of Kyolic garlic, a unique, specially aged garlic extract *from Japan*. For cancer treatment, Dr. Balch recommended 2 capsules 3 times daily. Dr. Willard, on the other hand, suggested that the shiitake mushroom (the more common name for *reishi*) be taken in amounts of "9 to 15 grams a day in equally divided doses." Her brief search seemed to be over, as now she felt she had finally found what she had been looking for. Both the Kyolic garlic and shiitake mushroom came from Japan and easily qualified for food as well as medicine.

Elaine put herself on a consistent program of herbal and vitamin therapy and good, wholesome foods. Within 5 $^1/_2$ months the tumor mass was completely gone. She revisited Dr. Tom M. again for another exam of her left breast. He very reluctantly agreed to do so, half expecting as he later confided to me, that she would have been dead by then or close to it. He was understandably "very annoyed" (his exact words) when he failed to find any traces of the cancer in her body. He wrongly assumed that she had gone elsewhere for chemotherapy or radiation treatment and was now back just to harass him with claims of divine intervention and inspired remedies. But he pronounced her free of the disease and gave her a clean bill of health, besides his own considerable bill! When I revised this book in the winter of 1994, I decided to include her astonishing account, which appears in print for the very first time.

Mushroom Tea is a Great Remedy

In midsummer 1987 I met a Dr. Yan Wu with the Institute of Botany at the Academia Sinica in Nankang (near Taipei), Taiwan, while both of us attended the annual meeting of the American Society of Pharmacognosy at the University of Rhode Island in Kingston.

Dr. Wu related the number of uses for mushroom tea in his country. The mushrooms used are those which grow on oak (*Cortinellis shiitake*). If used in dried form, several mushrooms are soaked for an hour until soft. If used fresh, however, they are then immediately cooked. The mushrooms are coarsely chopped into $1/4''$ pieces, put in 2 $1/2$ cups of boiling water with a pinch of kelp and simmered, covered, for 30 minutes on low heat until just 1 $1/4$ cups of tea is left.

Mushroom tea, he claimed, was effective in dissolving and getting rid of fat congestion in the body, stimulating kidney function, reducing fevers and relaxing tension.

A California physician, Lawrence Badgley, M.D., recommends shiitake mushrooms in the diets of those AIDS patients who visit his clinics often. Like figs, mushrooms contain the anticancer compound, benzaldehyde.

Culinary Uses

Here is a delicious, easy-to-make soup using regular mushrooms.

Good Cream o' Mushroom Soup

The soup is prepared by washing and skinning $1/4$ lb. of mushrooms and simmering the skins in $1/2$ cup of water. The mushroom caps and stems should be chopped into small pieces and 2 cups of water added to the skins. The simmering should continue until the skins are tender.

Melt 2 tbsp. of butter, adding 2 tbsp. of whole wheat flour and 1 tsp. salt or kelp. Two cups of milk should gradually be added over low heat, stirring constantly with a wire whip until the soup eventually thickens. The chopped mushrooms are then added and sprinkled with chopped parsley before serving.

Mushroom Ragout

The following basic recipe serves as a prototype, to be modified as desired; try substituting mushroom stock for wine, thyme for rosemary, or adding garlic or shallots. Also a dash of liquid Kyolic aged garlic extract gives added zest.

Needed: $^1/_2$ lb. fresh shiitake mushrooms; $^1/_2$ cup defatted reduced-sodium chicken stock; $^1/_4$ tsp. dried rosemary; 1 thick slice bacon, diced (2 oz.); 1 stalk celery, finely chopped; 2 tbsp. dry red or white wine or dry sherry; granulated kelp (seaweed) to taste; and 1 tbsp. chopped fresh parsley.

Separate the shiitake stems from caps. Rinse the stems and slice very thin. Combine the stems in a small saucepan with stock and rosemary. Simmer, covered, until tender, about 10 minutes; do not drain. Meanwhile, clean caps with a soft brush. Break the caps into bite-size pieces or leave whole, as desired.

Cook the bacon in a 10" heavy skillet over medium-low heat until barely colored. Drain the bacon on paper towels. Pour off the fat; return the bacon to the skillet. Add the celery and increase heat to medium, stirring until softened or about 2 minutes. Add the shiitake caps; toss until barely softened, about 1 minute. Cook over medium-high heat until most of the liquid evaporates, about 5 minutes. Pour in wine or sherry, increase the heat to high, and toss until the liquid has almost evaporated, about 1 minute. Season with kelp and sprinkle with parsley. Serves 2.

Mustard Greens
(Brassica alba, B. juncea, B. japonica, B. nigra)

Brief Description

Gaius Plinius Secundus (also known as Pliny the Elder, 23–79 A.D.), a Roman scholar, remarked in his great work on natural history (*Historia Naturalis* in 37 volumes) that mustard grew everywhere in Italy without needing to be sowed. He stated that the Romans had only to gather it wild in order to use it. Pliny further mentioned that mustard greens have an ability to improve the taste of other plants with which they shared the cooking pot.

Mustard is often presented as the typical plant of the *Cruciferae* (so called because they bear flowers whose four petals are arranged in the form of a cross), a large family (over 200 genera and about 2,000 species) and a virtuous one (it adds nothing toxic to the system). A common characteristic of the family is that its seeds contain a good deal of oil—30 to 35% in the case of mustard—and the oil is often endowed with elements that produce a peppery effect in such diverse *Cruciferae* as cress, radishes, horseradish, turnips, and, of course, mustard itself. Among seasonings, mustard is second only to pepper in world trade.

Mustard is indispensable throughout the world. Or 'twas so in the mind of one anonymous French author who composed this jingle in Paris several centuries ago:

From three things may the Lord preserve us:

From valets much too proud to serve us;

From women smeared with heavy fard, good grief!

From lack of mustard when we eat corned beef.

Most *Cruciferae* are very high in the mineral sulphur, which explains much of their sharp tastes and distinctive odors.

Starvation Prevented with Mustard Greens

An elderly Russian gentleman inn his mideighties by the name of Oleg Serensky recently emigrated to America and settled with relatives in the ethnic neighborhood known as "Little Odessa" in

Brighton Beach, New York. When I was in New York City on business in 1994, I was introduced to Oleg through a nephew of his. With the nephew interpreting, I carried on an extensive conversation with this remarkable fellow.

He told me that he was one of 3 million people trapped in the city of Leningrad during the fateful winter months of 1941–1942. The Germans had completely surrounded the city and were determined to strangle the life out of its citizenry. Oleg remembered that the official ration of bread, heavily cut with cellulose and cottonseed oilcake, was 9 ounces a day for manual workers and troops and 4 $1/2$ ounces for the rest. The icy road across Lake Ladoga, the only lifeline for the city, was tenuous and rations were rare. Pharmacies had been emptied of cough drops, bad breath tablets, peach stone oil, castor oil, hair oil, and glycerin.

Oleg stated that by February of the following year, some 200,000 of his city's citizens had already died of cold and hunger. Oleg himself was in pretty bad shape, surviving on just about anything he could find. "Alcoholics died very fast," his nephew continued while the old uncle rambled on. "So did misers, who had been hoarding money and food for years; all too often their greed got the better of them, even on the brink of death. People pulled relatives to the cemeteries on sledges, but often became exhausted themselves and abandoned the corpses on the street. Horses were eaten early, and a dog or cat was worth a month's salary on the black market." Oleg recalled making soup from glue and boiled book bindings; of eating jellied calfskin mixed with cloves. Gangs of cannibals were common, though he swore he drew the line at eating human flesh. He remembered seeing dismembered fresh corpses in the recesses of unlit apartment houses. Water was dragged from holes cut in the Neva and filtered through gauze; it tasted awful, but it kept him from dying of terrible thirst. Fuel was exhausted, the trams stopped running, and there was no more light.

With the advent of spring, warmer weather melted the ice and snow. Nature began renewing herself in all of her fine greenery. Dispositions of those who somehow had managed to survive improved. It was at this time that Oleg had just about exhausted all of his options so far as finding more things to ingeniously convert into what might barely pass for food.

Then one day he noticed a large patch of wild mustard growing in a vacant field outside the northwest part of the city. Pulling up tender leaves by handfuls, he greedily ate them until he was stuffed. A few hours later he got a terrific bellyache, followed by a great deal of flatulence and diarrhea. Thereafter, he decided to boil them first in some water and then to only eat smaller amounts. For the next nine weeks, he subsisted only on these mustard greens, flavored only occasionally by an unlucky field mouse he happened to snare and throw into the pot of greens, skin and all!

Oleg said the mustard greens was a true Godsend and kept him from fully starving to death!

Grandma's Old-Fashioned Mustard Plaster

Mustard plaster is one of those reliable "old timey" remedies held over from grandma's era, due to its considerable value in treating a wide variety of disorders—asthma, bronchitis, pneumonia, fever and chills, sciatica, neuralgia, gout, bumps, bruises, sprains, tendonitis, common cold and flu, eruptive sores and boils.

A simple plaster still used by some farm folks in rural Indiana and by hillbillies in Kentucky, involves mashing the leaves and stems of fresh mustard plant into a pulp. The surface of the skin is then coated with Crisco or Vaseline before the pulp is applied and bound in place with some gauze and adhesive tape. By coating the skin with petroleum jelly, it prevents the mustard from causing serious blisters or raising welts on it. This plaster can be kept on for several hours or else left on overnight for best results.

Relieves Aches and Pains

If you're suffering from rheumatoid arthritis, lower backache, abdominal cramps, and so forth, then this remedy, however unusual, is for you. After cutting out the midribs from several large green cabbage leaves, just iron them with a steam iron until they're soft as velvet. Then rub a little olive oil on one side and put them on the areas of pain, covering them with a heavy towel. Leave for a while before changing again. *Guaranteed* relief, every single time!

Southern-Style Mustard Greens

Needed: 1 lb. each of collard and mustard greens; 3 strips bacon; $^1/_4$ cup chopped onion; $^3/_4$ cup hot water; 1 tsp. sea salt; 2 tbsp. apple cider vinegar; $^1/_4$ tsp. kelp. Wash greens well, removing tough stems and ribs, and shred leaves. Fry bacon in large pot over medium heat until crisp. Remove bacon and drain on paper towels. In the same pot fry onion until golden brown, then add greens, water, salt, and kelp. Cover and simmer for about 25 minutes until tender. Drain and reserve the liquid (called "pot likker" in the South), and transfer greens to a serving dish. Sprinkle with vinegar and more kelp. Crumble bacon over top and serve. Cornbread can be served on the side for sopping up the "pot likker." Makes a real tasty dish!

Nasturtium
(See under ORNAMENTAL FLOWERS)

Nectarine
(See under PEACH AND PEAR)

(Stinging) Nettle
(Urtica dioica)

Brief Description

A perennial plant found all over the world. In America it grows in waste places and gardens and along roadsides, fences and walls in the states northward from Colorado, Missouri and South Carolina. The square, bristly stem grows from 2–7 feet high and bears pointed leaves which are downy underneath, and small, greenish flowers that grow in clusters from July to September.

Stinging nettle may be considered a specific food plant as well as an effective medicinal agent. Euell Gibbons devoted an entire chapter to nettle in his book, *Stalking the Healthful Herbs* (New York: David McKay, 1966, pp. 132–138). He praised it accordingly: "This detested weed is one of the finest and most nutritious foods in the whole plant kingdom, a far better vegetable than many of those . . . [a] farmer's wife laboriously raises in her kitchen garden."

"Stinging nettles aren't everyone's cup of tea," archaeologist Michale Corbishley told me in an overseas phone call from London made during the Thanksgiving Day weekend of November 24–27, 1994. "But they make a pretty good cup of soup," he added with confidence.

Corbishley, age 50, is a former teacher and archaeologist. He is head of education at English Heritage, the state body responsible for preserving historic and architecturally important buildings, ruins, and sites. His job right now is studying the food that Europeans ate during the Bronze Age thousands of years ago. He gathers the raw ingredients, cooks the dishes, and eats them.

"Nettle soup is quite delicious and tastes a lot like spinach," he noted. "The Bronze Age people had no writing so all we know about them is what comes out of the ground." The Bronze Age diet was recently highlighted at a scientific conference by the Council of Europe at the British Museum in London, he casually mentioned.

"From the seeds or fruits preserved in boggy ground, sea shells, fish and animal bones and clues to how meat was handled by looking at butchery techniques showing in the bones, we can learn quite a bit about not only what they ate but also how they prepared it," he stated. Corbishley explained that, oddly enough, some of these ancient recipes still survive in modern times. One recipe still in use in the Shetland and Orkney islands off the northern Scottish mainland, is nettle purée, a sort of thick soup. "It is very nourishing and rather tasty," he noted.

English Heritage and the British Museum joined forces in 1993 to publish a book entitled, *A Taste of History*, about 10,000 years of food in the British Isles. Jane Renfrew, an archaeologist who wrote the sections on prehistoric food, said: "The main difference between prehistoric food and that of today is that our distant ancestors cooked rather simply; they did not go in for elaborate sauces and, having few ovens, did not bake pies."

"As a result they were a lot healthier and not plagued with the problems of obesity, diabetes, or hypertension as we are today," Corbishley was quick to comment. "Perhaps, if we ate more wild plant foods such as stinging nettle, we might all be better off," he mused in closing.

Fantastic Hemostat

Nothing seems to stop profuse bleeding more quickly and effectively than stinging nettle! Because the evidence to be presented is so incredible, the source from which it has been obtained needs to be cited as well in order to make everything more believable.

Francis P. Porcher was a surgeon and physician in the Southern Confederacy. His book, *Resources of the Southern Fields and Forests*, was an important medical text during the Civil War. In it, he related how he and another doctor deliberately cut open and laid bare a major artery of an adult sheep. Then just by soaking some gauze-like material in a strong cold tea made of stinging nettle and applying the same directly to the open wound, he was able to stop all bleeding within just a matter of minutes. More remarkable yet, was the fact that when the pressed juice of the plant was added to fresh blood poured out into the palm of the hand, it immediately began coagulating.

To make a strong solution for your own personal needs, bring 1 quart of water to a boil. Remove from the heat and add a generous handful of *freshly* chopped stinging nettle plant. Cover and let steep for an hour, before straining. Always best to use when cold. Dried plant materials may be used, but don't work quite as well. Tea may be used internally for bleeding ulcers or externally as a wash or poultice to stop any major hemorrhaging.

Remarkable Hair Tonic for Baldness

Stinging nettle lotion seems to help hair grow again where baldness may now be present. The following two lotions should be used at the same time every morning after washing and rinsing the hair as you normally would do. The alcoholic portion to be used should be diluted by half as much of the infusion prior to rubbing into the scalp good with the fingertips. And when doing so, be sure to bend your head down low, massaging the lotion in from the nape of the neck upwards towards the front. Afterward, allow the scalp to dry naturally without using a towel or hair dryer.

In 1 quart or 4 cups of gin, put 2 handfuls of washed and chopped freshly picked stinging nettle, $3/4$ handful of chopped fresh rosemary, 1 handful of chopped or cut fresh chamomile flowers, and

2/3 handful of chopped fresh sage. Cover the fruit jar with a good, tight lid and let stand exposed to *indirect* (shaded) sunlight for 2 1/2 weeks, making sure you shake the contents of the bottle good twice each day. Strain and refrigerate in a clean fruit jar with a lid.

Bring 1 1/2 quarts of Perrier or other bottled mineral water to a boil, adding half each of a small, coarsely chopped rutabaga and unpeeled potato, and 1 diced stalk of celery. Cover with lid, reduce heat and simmer for 25 minutes, then strain liquid into another pan, discarding the vegetables. Reheat to boiling point and add 1/2 handful of coarsely chopped, fresh stinging nettle, 1/2 handful of chopped, fresh garden sage, 1/4 tsp. grated horseradish root, and juice from half of a lemon. Cover and remove from heat, allowing to steep for 50 minutes. When cool, strain and refrigerate in a clean fruit jar with a lid.

When using the alcoholic extract, just remember to dilute 1 part of it with 1/2 part of the infusion. If used regularly for several months, new hair growth should become fairly evident.

Fun Recipes for a More Slender Image

Euell Gibbons once wrote that "stinging nettle is very efficacious in removing unwanted pounds!" Those obese individuals who've written to me in the past desperate for advice on how to reduce and whom I've put on a semidiet of stinging nettle have reported up to 32 1/4 lb. loss in just three months or less!

Nettles should be collected in the early spring when they are 4–8″ high. As the plant matures, it becomes pretty tough and quite unpalatable. A pair of good leather gloves is recommended to protect the hands when handling the stuff. And heavy paper bags are preferable to plastic ones for carrying the nettle in.

After you've gathered enough nettle it should be washed in cold water. A pair of kitchen tongs will be of considerable help when removing the washed greens from the water. Allow them to drain on paper towels for a few minutes before refrigerating. They'll keep for up to a week.

Nettles freeze very well. Place the rinsed nettles in a large kettle. Pour boiling water over them to cover. After 5 minutes drain the water off, pack them into freezer containers and freeze. In the frozen state they'll keep for up to 9 months.

Nettle Greens, Georgia Style

Needed: 2 quarts stinging nettles; $^3/_4$ cup stock from boiled chicken wings and chopped, cooked meat from those wings; 3 sliced green onions; 2 hard-boiled eggs; $^1/_4$ tsp. lemon juice.

Snip greens into bite-sized pieces. Put in pan with other ingredients except the eggs. Simmer on low heat for 20 minutes. Remove and serve, topped with sliced hard-boiled eggs. Season to taste with a little kelp, if needed.

Cream of Nettle Soup

Needed: 1 $^1/_2$ quarts nettle greens; $^1/_3$ cup Perrier or other mineral water; $^1/_4$ cup sesame seed oil; $^1/_4$ cup whole wheat flour; 3 cups canned goat's milk.

Cook the nettles with water in covered saucepan over medium heat for 10 minutes. Cool 15 minutes, then purée in blender or food processor. In a saucepan, warm up the oil and stir in the flour, mixing both well. Then slowly add the goat's milk and cook until the mixture thickens over a low heat. Add the purée nettles and heat thoroughly. Add some kelp to season. Serves 4.

Slenderizing Nettle Ale

This is a nice thirst-quenching ale that I've recommended highly to those seeking to lose weight. It's a healthier alternative to diet drinks and diet colas, not to mention helping shed unwanted pounds from around the thighs and hips in particular.

Needed: 4 quarts nettle greens; 2 gallons water; 2 thinly sliced lemons; 3 thinly sliced limes; 2 oz. grated fresh ginger root; $^1/_2$ tsp. powdered nutmeg; $^1/_4$ tsp. powdered mace; 2 cups light brown sugar; 1 cake of active yeast.

Boil the first 4 ingredients gently in a large open kettle or big pot for 50 minutes. Strain through 4 layers of cheesecloth before adding sugar. Cool to lukewarm. Dissolve the yeast in 2 cups of this ale liquid, then stir in the remaining ale. Bottle in quart fruit jars or clean gallon jars with narrow spouts. Let stand for one week in a cool place. Refrigerate 15 hours before drinking.

NOTE: I'm deeply indebted to Darcy Williamson, author of *How to Prepare Common Wild Foods*, for her nettle recipes, which have been somewhat adapted to fit the needs of this book.

Nuts

Acorn (Quercus species)
Almond (Prunus amygdalus)
Brazil (Bertholletia excelsa)
Butternut (Juglans cinera)
Cashew (Anacardium occidentale)
Chestnut (Castanea pumila)
Coconut (Cocos nucifera)
Filbert (Corylus avellana pontica, C. maxima)
Hazelnut (Synonym for filbert)
Hickory (Carya ovata)
Macadamia (Macadamia integrifolia, M. tenifolia)
Peanut (Arachis hypogaea)
Pecan (Carya illinoensis)
Pinenut (Pinus species)
Pistachio (Pistacia vera)
Walnut (Juglans nigra, J. regia)

Brief Description

ACORN. Small nutlike fruits found on many species of oak with cup shapes resembling saucers. Acorns from red oak are bitter due to the tannin they contain and must be processed before eating, while those of the white oak are sweet and edible raw.

ALMOND. Botanically classified as and related to fruits like peach and plum. The outer shell is leathery, but the seed inside the fruit, which can be either sweet or bitter, is the nut itself. Bitter almonds contain amygdalin or laetrile, and the oil contains mostly benzaldehyde, the same anticancer factor found in figs and mushrooms.

BRAZIL. This is one of the very few commercially available nuts which are never cultivated. It grows wild in the dense South

American Amazon rain forest, the trees often towering up to 150 feet or higher. The nut is contained in a pod similar in shape to a coconut which holds 12–30 of them. When ripe, these pods fall with such force they can bury themselves under the ground. Once removed from the pod the nuts are dried and put through a heavy brushing to remove their rough brown skin.

BUTTERNUT. The nut comes from the white walnut, a small tree with ash-gray bark, becoming separated into smooth ridges. The nut, found inside of a sticky, hairy husk, is thick and pointed and has very rough, obscure ridges on it.

CASHEW. This nut is the fruit of a tropical and subtropical evergreen, a species, interestingly enough, that's related to American poison ivy and poison sumac. The evergreen grows to about 40 feet tall and bears clusters of pear-shaped fruits called cashew apples. Below this fruit hangs the crescent-shaped cashew nut. The kernel has two shells, an outer one that is thin, flexible, and somewhat leathery and an inner that is hard like most nuts and must be cracked open. Between both of these shells is a brown oil that's so toxic it has an extreme blistering effect on the skin. For this reason, the oil must be burned off before it can even be touched.

CHESTNUT. This is a magnificent tree growing almost to 100 feet with a very broad spread. Longfellow's poem to this effect—"Under the spreading chestnut tree, the village smithy stands . . ."—aptly describes its hugeness. Barely 100 years ago, vast forests stood in many parts of the eastern United States; but a severe blight at the turn of the century and over the next 40 years practically destroyed all of them. What remains today are crossbreeds of Japanese and Chinese species. The nuts grow 2–3 together in a spiney burr about the size of a baseball. When ripe the burr opens and the nuts are removed. This is the only nut usually served as a vegetable.

COCONUT. This is the fruit of the coconut palm, a very important economic product found throughout the tropics. Unopened coconuts keep at room temperature for 2 months. The white meat inside the shell can be eaten raw or fried. Or it can be grated and squeezed into a very rich, fatty "milk" of sorts.

FILBERT AND HAZELNUT. These brown-shelled nuts are actually fruits of the same bush that differ only in their shape. To tell the difference between them, hazelnuts are shorter and rounder than filberts. They have the sweetest meats of all the nuts and are mainly used in desserts and candies.

HICKORY. A tall, slender, straight tree belonging to the walnut clan. Also known as shagbark hickory. Stern schoolmasters of early America "taught readin', writin', and 'rithmetic, all to the tune of the hick'ry stick," as the lines of one tune suggest. The outer husk of the nut is thick and woody, splitting to the bottom when the nut is ripe. The nut has prominent ridges, is quite sweet and edible.

MACADAMIA. Also called Queensland nut since it's native to Australia. The nut has a honey-brown shell that is extremely hard to crack open. The crisp, creamy-white nutmeat has a slightly sweet flavor to it.

PEANUT. Botanically classified as an underground pea of the legume (bean) family, the peanut is also called ground nut or goober in the South. Native to Brazil, the peanut comes in two varieties—the small, round Spanish kind used for candy, butter, and oil and the larger, oval-shaped Virginia type which is generally used whole. About half of the entire U.S. peanut crop is made into peanut butter.

PECAN. It's believed that the pecan tree was distributed from its original home in the South by the Iroquois and other Native American tribes centuries ago, throughout the northern part of the United States. When they made many canoe trips up and down the Illinois River and other tributaries, they deliberately planted the biggest and thinnest-shelled nuts at their main portages. The soft nutmeat is twin lobed and wrapped in thin, shiny, light-brown shells and was a mainstay in many an Indian diet.

PINENUT. Comes from various pines (mostly the piñon). A very sweet-flavored, high-protein kind of a nut that varies in size (1/3" to 2"), shape (cylindrical to round), and color (white to pale yellow).

PISTACHIO. The tree is part of the poison sumac family and contains green nuts with ivory-beige shells that split open upon ripening. Those with red shells have been colored with vegetable dye, and those with white shells have been coated with salt.

WALNUT. There are two types, black and English. The former have a strong flavor, and their dark-brown shells are somewhat difficult to open. The latter or more popular kind are white on the inside and golden tan to amber on the outside. Their light-brown shells are easy to open. Black walnut is chiefly used for medicinal purposes by the American herb industry, while the English kind are consumed raw or cooked.

Acorns and Oak for Burns and Liver Problems

The Iroquois had a nifty way of treating serious burns and rashes suffered by those unfortunate enough to have made contact with poison ivy and sumac. They would gather up a sufficient quantity of acorns, split them with a heavy stone or the blunt end of a tomahawk, then throw them into a large iron kettle full of boiling water.

After the mixture had boiled down to half of its original amount several hours later, it was strained and the strong tannic acid solution was saved for medicinal uses. Some of this healing water would then be applied to any severe burn or rash on the skin in the form of a poultice, as well as bathing the afflicted area often.

Today the same remedy can still be used with great success. Put two dozen or so cracked acorns in 1 $^{1}/_{2}$ gallons of hot water and boil down to half this amount, uncovered for a couple of hours. Then strain and store in sealed quart jars in a cool place until needed. Can this solution actually work in healing severe burns of any kind? Some medical doctors reading this might have their doubts, but clinical evidence suggests otherwise. The July 1926 issue of *Annals of Surgery* contained a lengthy and detailed report of some 17 pages by two Cleveland physicians who *successfully* treated extremely serious burns in children and adults with nothing but tannic acid. The before and after pictures speak for themselves. You too, in the confines of your own home, can do the very same thing for yourself and loved ones by using the above acorn solution or the following oak bark solution.

Simmer $^{2}/_{3}$ cup of coarsely cut, dried oak bark in 1 quart of boiling water, for 20 minutes on low heat. Then remove and steep an additional hour. Strain and drink 1 cup every other day for liver injury and to prevent hardening of the arteries. Or else use to wash and dress wounds and major burns when cold. Strips of clean gauze may be soaked in either solution to dress burns with and then changed every few hours.

Almonds for "Dishpan" Hands and Sore Lungs

Sweet almond oil is a super emollient for chapped, "dishpan" hands, diaper rash, herpes sores, shingles, psoriasis and lupus erythematosus. Just mix the juice of 1 lemon and lime with 4 tbsp. of oil and rub on thoroughly. Even helps to whiten or lighten dark skin somewhat.

"Almond Milk" is a very popular drink for throat and lung problems in Hong Kong and Canton, China. To prepare the drink, 10 parts of sweet almonds and 1 part of bitter almonds are soaked in water with a little rice. After both become tender enough, they are then ground into a paste by running them through a small food grinder or nut mill of some sort. The resulting milky mixture is strained to remove coarse particles and then diluted with a little more water and some honey before cooking on low heat until the consistency is somewhat syrupy.

A cup of the same is recommended 2–3 times daily to relieve hoarseness and the raspy or wheezing sounds common to heavy smokers and asthmatics. This is also good for scratchy, irritated throats and dry or hacking coughs.

Mental Energy from Brazil Nuts

All amino acids are valuable sources of energy, especially that needed for optimum brain function. Communication within the brain and between it and the rest of the central nervous system occurs through chemical "languages," called neurotransmitters. There are about 50 such languages; the amino acids, either as precursors or peptides, account for the majority of them.

Now Brazil nuts are one of the very few nuts or seeds that are so incredibly rich in both essential as well as nonessential amino acids. An Italian scientist who has studied the nut very carefully, calls it the "meat vegetable," because of its very high A, B-complex and C vitamin contents, not to mention the tremendous protein present as well.

University students in São Paulo stir in a teaspoonful or so of the meal into some juice and drink it for mental pep before taking their final exams. The meal can be purchased from some specialty food shops or else made by cracking open about 10 large nuts and running the meat through a nut mill.

Butternut for Intestinal Problems

Joseph Smith, Jr., a great American prophet, sometimes joked with his people (the Latter-Day Saints or Mormons) in a friendly way while sermonizing on topics touching human health. "If you have problems with your bowels," he was apt to say, "take some strong physic or go gnaw down a butternut tree somewhere to get things movin' again."

Several of his closest confidants were themselves trained Thomsonian herb doctors, who often used butternut meal or tea mixed with some salt to correct the most stubborn constipation or eliminate parasites such as tapeworm. Two teaspoons of ground butternut mixed with a pinch of sea salt and stirred into hot oatmeal or several cracked butternuts simmered in a pint of boiling water with a tad of salt added were two ways of taking this nut to give relief.

Cashews Make a Delicious Milk Substitute

Tired of milk or just can't drink it for some reason? Then try one of the most interesting and flavorful concoctions I devised some years ago for several adults and children who were allergic to real milk.

In a food blender, combine together at high speed for 4 minutes the following items: 1 cup of shelled, chopped raw cashew nuts, $^1/_2$ cup unsweetened pineapple juice, 1 tbsp. *each* of honey and pure maple syrup, $^1/_4$ cup chopped or crushed ice cubes, 2 $^3/_4$ cups spring or distilled water, pinch of powdered cardamon and $^1/_{16}$ tsp. of pure vanilla flavoring.

After blending well, refrigerate in a closed container. This makes one of the sweetest, most delicious milk substitutes I know of. The recipe yields about 4 cups and may be used on cooked or packaged cereals or consumed by itself. It's especially handy for those who are seeking to lose weight or are allergic to cow's milk in some way. In addition, this cashew "milk" furnishes a terrific protein drink for athletes and body builders.

Roasted Chestnuts for Bleeding Ulcers

A sure cure for bleeding ulcers is an old remedy I picked up some years back while lecturing among the Dunkard Brethren in North Carolina. This is an old-order religious group like unto the Amish, who advocate baptism by immersion; hence, their nickname "Dunkard."

One of their bearded ministers shared with me an effective treatment for stomach complaints. Roasted chestnuts are to be shelled and run through a nut mill until rendered into a fine meal. Then about 2 tbsp. of the meal is combined in a wooden salad bowl with 1 tsp. of dark honey and 2 tsp. of pure maple syrup, and everything is stirred well with a wooden ladle until a soft paste forms. This is then served on a metal soup spoon and taken internally to stop intestinal bleeding and help heal ulcers.

I've had occasion to work with this unusual electuary myself and find it works well for the problems intended. It may also be used with satisfying results by women who experience excess menstruation.

Jock Itch and Cataracts Disappear with Coconut

An admirer of my work with herbs from Valleja, California, wrote to me about her own curiously devised remedy for getting rid of cataracts. "Take the fresh juice from an opened coconut and with an eye dropper apply as much as the eye can hold. Then apply hot wet cloths that have been wrung out over the eyes and lay down for 10 minutes. One treatment is generally enough if done early on when they start to form," she claimed.

More remarkable still is an old Ayurvedic remedy for jock itch or vaginal infection, which I acquired from an old folk healer in Bombay. Dry an empty coconut shell in the sun for a week. Then break it into small pieces by smashing it good with a sledge hammer. Gather up the fragments and soak them in 1 quart of vodka or Scotch for 11 days, shaking the solution twice daily. Strain, then bottle.

Bathe the groin area as often as needed to get rid of jock itch. Douche regularly to eliminate vaginitis. Soak feet in a pan of the stuff to stop athlete's foot. Or soak fingernails in the same to clear up any fungal infection that develops beneath them. Works relatively quickly as a rule and seems to be effective in most instances.

Sweeten Your Breath with Filberts

The breath may be somewhat sweetened by chewing on a few filberts or hazelnuts. They work not so much because they're aromatic like peppermint would be as they do by absorbing much of the bad breath like a sponge does water. They're quite expensive, but worth investing in for this purpose and to help stop a throbbing toothache by crushing several into some powder and sprinkling on top of a little peanut butter, before applying directly to the site of pain for relief.

Hickory Oil Makes a Tasty Condiment

Long before we came to have such condiments as catsup, mustard, pickles, and mayonnaise, the Native Americans had devised their own tasty kinds from various nuts. The Cayuga (a tribe of the great Iroquois nation) used to crush the meats of the hickory, walnut, butternut, and chestnut and then slowly boil them in water, making sure to skim off the oil which floated to the surface and saving it. This oil was boiled again by itself and lightly seasoned with a pinch of salt.

This condiment was later used with bread, potatoes, pumpkin, squash, buffalo jerky, roast venison, and many other foods. The nut meats left after skimming off the oil were often seasoned and mixed with mashed potatoes or else added in with cornmeal to make an incredibly delicious bread. These same techniques ought to be used more often today as replacements for the stuff we put on our hamburgers, hot dogs, and salads that are unhealthy and harmful to the body.

Macadamias for the Liver and Alcoholism

A snappy cocktail that seems to satisfy the urge to drink and to help rejuvenate the liver is made from macadamias (preferably raw), ripe tomatoes (or the canned juice), a squeeze of lemon juice, a pinch of cayenne, and a thumbnail full of raw, grated ginger root. Combine 5 macadamias, 3 medium, very ripe tomatoes, $1/2$ tsp. lemon juice, the pepper, and ginger into a blender and mix well for 1 $1/2$ minutes, adding a little canned tomato juice, if necessary, to thin it down if it becomes too thick. Drink slowly by itself and E-N-J-O-Y! Beats the heck out of alcohol and tonifies a weak liver.

Peanut Butter for Hemophilia and Mouth Problems

A report appeared in *Nature* (Vol. 194, p. 980) for 1962 showing that peanuts have an unusual hemostatic factor, which can be of considerable benefit to hemophiliacs or those subject to chronic and prolonged hemorrhaging. Peanuts seem to have the ability to reduce this problem somewhat, and hemophiliacs are encouraged to eat more raw peanuts and to consume more peanut butter.

Peanut butter is also very good for helping to hold other medicaments in place within the mouth, when they're unable to be retained there of their own accord. For instance, when placing a crushed garlic clove by an aching tooth for relief, first put it on top of a small square piece of white bread covered with peanut butter before inserting. You'll find it stays in place a lot better.

Sores on the mouth or tongue are often difficult to treat with herbs, since they can't be retained for very long to do much good. But when a small slice of white bread covered with some peanut butter is sprinkled with either goldenseal root or white oak bark and then placed inside the mouth against the sores, better healing results can be shortly expected.

And a more convenient and pleasant way to take cayenne pepper so it doesn't burn the heck out of your gut is to mix a pinch of the spice with a little peanut butter and honey, before swallowing it whole. This is also a good way to persuade children to take bitter or burning herbs.

Pecans Relieve Migraines

As strange at it might seem, a poultice made out of raw meats of pecans and English walnuts will relieve a headache! In a blender mix (well) 2 tbsp. each of both nut meats and a little water until a thick purée forms. Then spread on two squares of gauze and tape to either side of the temples while reclining. Remain this way for several hours or until the throbbing ceases.

An Apache Pinenut Beauty Mask

Back in the early 1970s, I was introduced to a former runner-up in the Miss Arizona Beauty Pageant, who just happened to be Apache by birth. Later I accompanied her and some others to her home, where we met her folks. The girl's mother, a full-blooded Apache squaw herself, had one of the nicest complexions I had ever seen in a woman nearing the half-century mark of her life.

After chatting with them awhile, I gathered up enough courage to ask her outright what she did to make her skin look so young and beautiful. It was then that she explained to me about her nightly mask which she always put on before retiring.

Pinenuts, I came to find out, were her only secret ingredient. She would take a couple of handfuls and methodically chew up small portions of them very briefly in order to remove the shell, after which she spit them into a wooden bowl. Once this had been done she'd then run the nut meat through a small mill to make meal out of them.

Now pinenut meal is extremely greasy to say the least, but exceptionally wonderful to keep the skin from drying out. Before retiring each night she would mix some of this meal with a little goat's milk to make a soft paste, which was next rubbed on her face, forehead and throat. The rich amino acids from both the nut meal and the goat's milk had a chance to penetrate through the pores of her skin as she slept, and literally, rebeautify and enliven once more tired, old skin. By the time she awoke next morning and removed the mask with cold spring water, there was a supple youthfulness pretty enough to whistle at the draw and stares of many younger fellows.

A Nice Pistachio-Peanut Massage Oil

A good massage oil for dehydrated skin above the shoulders and for loose, flabby skin on the arms, buttocks, and thighs can be made by taking a handful of white, shelled pistachios which have been coarsely chopped and then simmering them in a quart of peanut oil on low heat for approximately an hour. After the oil has cooled down to lukewarm, it should be strained before massaging well into the skin on these areas of the body.

Useful Walnut Remedies

Both kinds of walnut, especially the black, including the nut shell and tree leaves and bark, have myriad uses. For one thing, the fat in English walnut kernels is a kind that the heart can handle without any problem.

An excellent herb tea for heart disease is made from the woody, interior walls of walnuts. Use the walls from 4–5 English nuts for each cup. Soak them overnight. Then boil them for 20 minutes the next morning. Take 3 cups a day. This tea alleviates pressure and pain in the chest. Tea may be sweetened with honey.

The bark and leaves of English walnut make a good mouthwash for preventing and reducing cavities. Bring 1 quart of water to a boil, then add a generous handful of well-pounded fresh English walnut bark. Cover, reducing heat, and simmer for 15 minutes. Remove from heat, uncover, and add 1 handful of finely cut leaves. Cover again and steep for 50 minutes. Strain and store in the refrigerator. Gargle and flush the teeth with a cup of this mixture 2 to 3 times daily or after every meal.

Certain skin blemishes, ringworm, and even warts just seem to disappear as if by magic whenever *green* or immature black walnut is used externally. Just make a couple of incisions into its outer shell and rub the juice on whatever you want cleared up. At first there may be a slight burning sensation, but don't worry. At other times, the juice could turn the skin where it's applied a little brown, but this will eventually wear off. Practically all kinds of warts, dark, ugly age spots, and ringworm have been successfully eliminated with this treatment.

The pericarp or ripened shell of the walnut is an effective remedy for fungal infection and to get rid of intestinal parasites as well. The best way in which to extract the juglone (the active substance), is to use boiled cabbage juice that has just been strained. Reboil about 1 quart of the juice, adding the green, broken, and bruised pericarps from 8 unripe walnuts. Cover, reduce heat to low, and simmer 15 minutes. Remove from stove and steep and additional 25 minutes. Strain and refrigerate. Take 3 cups daily in between meals to get rid of parasites, and soak hands or feet, gargle and flush the mouth, or douche the vagina to remove any kind of persistent fungus.

The walnut meat itself is fantastic for dissolving and eliminating kidney and gallstones. Clinical reports published between 1957 and 1961 in the *Chinese Journal of Surgery* related how hundreds of patients young and old were cured of stones. In all cases, 2 1/4 cups of raw walnut meat were deep-fried in pure olive oil until crisp. The meat was then ground to a coarse powder and mixed with 2 1/4 tbsp. of dark honey to form a paste. This paste was then given to each patient for two days, at the end of which most stones had partially dissolved, turned soft, and been eliminated.

For excessive accumulation of phlegm in the lungs and persistent coughing in the throat, a combination of walnut and ginger seems to do the trick every single time! This is a very ancient Chinese remedy, which calls for the meat from 3 walnuts to be ground in a nut mill and 3 slices of fresh ginger root to be finely grated. Both are then combined with warm liquid or hot broth and slowly swallowed before retiring. By the next morning all symptoms of the coughing and phlegm will have disappeared for good. This remedy is especially handy for asthma, bronchitis and those who smoke a lot.

The teeth may be occasionally polished by emptying the contents of a capsule of powdered black walnut on a wet toothbrush. Black walnut powder also helps to remove plaque and yellow stains. Use only occasionally, though, since frequent use might result in some loss of tooth enamel.

A tea made of walnut leaves is quite useful for drying up excessive milk flow once a child is past the nursing stage. Bring 2 cups of water to a boil. Remove from heat and add 3 tbsp. fresh-cut or dried, crumbled leaves and 1 tsp. chopped green outer shell which surrounds the actual walnut. Steep for an hour. Drink 2 cups each day.

Gray Hair Gone with Walnut

Gray can be easily removed from dark brown and black hair by dyeing them with walnut. A Texas grandmother explained to me once how she used to take 1 1/4 cups each of bruised walnut husks and coarsely chopped leaves and soak them both in a gallon of spring water (Perrier water may be substituted) for 24 hours. She would then pour the soak water into a large saucepan and boil until the liquid was reduced by half the original amount.

When it had sufficiently cooled, she'd then strain and add 1 1/8 tbsp. of either eau de Cologne or gin, bottle, and store in her pantry cupboard. A small amount would then be put on the hair daily with a sponge, wherever there was gray until a darker color had resulted. The only drawback to this, she said, was that everything the dye touched, including her scalp, turned somewhat dark as well. But she felt that a little inconvenience was worth putting up with to regain a more youthful looking appearance.

Ellagic Acid for Paralysis and Cancer

According to the *Journal of Pharmaceutical Sciences* (October 1968) and *Cancer Research* (May 1986), both black walnut and cashew nut shells contain a crystalline compound, that is of important therapeutic significance in treating muscular paralysis due to electrical shock, hypertension and skin cancer. Grapes, strawberries, black currants, raspberries, and other nut shells also contain this valuable plant phenol as well.

In the first study, two groups of mice were subjected to mild electrocution. Seven out of the 17 receiving black walnut shell extract remained alive, while 14 out of 17 rodents without the benefit of ellagic acid, perished in the other group. And in the second study, mice fed this substance from the above berries and nuts, experienced 45% less tumors when exposed to a chemically induced skin cancer and a prolonging of onset of the tumors by up to 10 weeks, than did another group without ellagic acid. It's strongly recommended that these nuts and berries be used in liquid forms (berry juices and nut shell teas) when treating any of the aforementioned health problems.

Nut Meats for Stuffings

Consider the meats of fresh almonds, cashews, filberts, macadamias, peanuts, pecans, and walnuts for use when making stuffings for roast poultry and fish, as well as for gravies and sauces. Nut meats really give them a dimension of culinary excitement, not to mention incredible, mouthwatering zest.

HOMEMADE NUT BUTTERS AND OILS

You can make your own nut butters at home, either separately or several put together for a very interesting blend with an intriguing taste. Simply remove the meats from the shells of a couple of handfuls of any kind of nut and put them in a blender or food processor after mashing them first with a heavy object. Add 1 cup of cold water to begin with and turn on to "liquefy," continuing to add a little more water as needed until a somewhat thick, creamy consistency has been achieved. Remove with a rubber spatula and refrigerate. Nut butters like this taste great on pumpernickel, rye, or cinnamon-raisin breads.

Hickory nut or black walnut butters of more runny consistency make very good sunburn lotions. Just follow the same instructions with the exception of adding fewer nut meats and more water. They can then be rubbed directly onto a bad sunburn with good results.

To make nut oils, just remove the shells from 2–3 double handfuls of any kind of nut and then pound them good with a hammer or heavy flat iron. After this put them in 2 quarts of hot water and boil until their oils melt and float to the top. Skim this off and allow to cool. What results is a very good and rare cooking oil that may be used any number of ways to create some exotic-tasting dishes you won't forget!

O

OAKS
(See under NUTS—ACORN)

OATS
(See under GRAINS)

OKRA
(Hisbiscus esculentus)

Brief Description

Okra is an annual herb with a tall, erect stem that grows 3–7 feet high and is covered with small hairs. The leaves are cordate, 3- to 5-lobed, and coarsely toothed, while the large flowers are yellow with crimson centers. Okra pods are anywhere from 5 to 12 inches long, horn-like in appearance, green or creamy green in color, and with ridges that are either smooth or hairy. The pods contain numerous seeds that are rounded, striate, and hairy. The entire okra plant is aromatic emitting an odor resembling cloves.

An unusual use for okra pods has been as blood plasma replacements. According to Volume Five of *The Wealth of India*, doctors used the mucilage in the pods as a blood volume expander in severely hemorrhaged mongrel dogs with good results. Further research some day may find this to be a useful agent in human blood transfusions.

Effective Burn Dressing

For major burns of any kind, liquid dressing made out of slippery elm bark, white oak bark, and okra can be effectively applied in serious cases where emergency medical facilities are not immediately available or economically affordable. The first thing to remember, though, when dealing with any critical situation like this is *TO REMAIN CALM.* Keep a cool head, let common sense prevail, and follow the simple methods outlined here.

The afflicted area should be covered with or soaked in ice water (or, preferably, clean snow) for some relief, until the dressing can be applied. In a stainless steel pot, bring 5 cups of spring or distilled water to a rolling boil. Reduce heat to a low setting and add 2 cups *each* of sliced okra and dried, cut slippery elm bark (*not* the powder), together with 1 $^1/_4$ cups of dried, cut white oak bark. In case the inside bark of slippery elm isn't available, the same amount of dried, cut comfrey root may be substituted instead. Cover and simmer mixture for nor more than 45 minutes. Mixture should be fairly thick and slimy by this time. Strain immediately while still hot, then again through several layers of muslin material.

For more rapid cooling, pour strained contents on a clean cookie sheet or layer pan covered with a clean piece of linen cloth. Set in the freezer for awhile to cool, but check it every so often to see that it doesn't become unduly stiff or frozen. Also setting the cookie sheet or layer pan on a table top and letting an electric fan blow cool air across the hot contents is yet another way of cooling it more quickly.

Before applying this cool mucilage combination to the injured skin, be sure that your hands are properly scrubbed and disinfected by soaking them in a small solution of Listerine Antiseptic Mouth Wash. Now there are basically two ways to apply this wonderfully healing herbal slime onto the burned skin. It may be lightly brushed on with a *sterile*, genuine camel hair brush with a fairly wide tip and then covered with gauze strips. Or else the gauze strips themselves may be first saturated with this herbal mucilage material and then laid onto the burned area by hand.

Five hours is about maximum before the dressing needs to be changed. Rapid healing will ensue and become evident in a day or so. The mucilage will not only reduce the pain and swelling of the severe inflammation, but the tannic acid present from the white oak bark should proliferate new cell growth and strongly discourage infectious bacteria from forming later on.

Poison Ivy and Psoriasis Relief

The foregoing treatment for major burns is also a handy remedy for relieving and promoting the healing of poison ivy rash and the misery of psoriasis. Follow the basic instructions, except to omit the white oak bark. The solution should be strained while still hot, but may be cooled more slowly at room temperature. The mucilage can then be rubbed on the skin by hand and allowed to remain for several hours before repeating the treatment as often as may be necessary.

Okra Recipe

A former "P.M. Magazine" chef and famous Encinitas, California restaurateur, LaMont Burns, highlights a unique chicken gumbo in his recent book, *Down Home Southern Cooking* (Garden City, NY: Doubleday, 1987), which has been reprinted here with the kind permission of his publisher.

SOUTHERN CHICKEN GUMBO

Needed: 1 (2 1/2 lb.) chicken; 3 tbsp. flour; 3 tbsp. butter; 1 tsp. brown sugar; 1/2 chopped, sweet red pepper; 4 cups cut okra; 1 large can tomatoes or 5–6 ripe, skinned tomatoes; 1 can corn or niblets or kernels cut from 3 ears of fresh corn; 2 sprigs chopped parsley; 2 snipped basil leaves.

Simmer chicken in water to cover for 1 hour. Remove chicken from stock and cool. Skin chicken and pick meat from bones. Cut chicken into bite-sized pieces. Brown flour in butter, add vegetables, sugar, and chicken stock. Simmer until tender and gumbo is thick. Add cooked and cut chicken and serve hot.

OLIVE
(Olea europaea)

Brief Description

The olive tree is an evergreen, commonly found in all the Mediterranean countries, but widely cultivated in tropical climates as well. The hard, yellow wood of the gnarled trunk is covered by gray-green bark. The branches extend upward to 25 feet or more. The leathery leaves are dark green on top and have silvery scales underneath. The tree yields fragrant white flowers and an oblong or nearly round type of fruit called a drupe that becomes shiny black when ripe. Other kinds of drupes would be plums, cherries, apricots and peaches. The oil which is produced from the fruit is quite valuable, having worldwide appeal for its excellent cooking and baking properties.

When it comes to classifying olive oil, consumers are often confused by the various terms used. So, here is virginity defined, plain and simple, and without a lot of hanky-panky either. "Extra-virgin" on the label refers to a low-acid oil that has not been refined by heat or solvents. Look for the words "cold pressed" also. Equally good is a label that lists the estate where the olives are grown, and the vintage year. In the absence of those claims, assume the oil is refined, that is, a low-grade oil that has been scrubbed of all flavor, odor, color, and acid and then blended with midgrade virgin oil to give it character. It may be a fine cooking oil and cheap besides, but it will taste awful in your salads.

Here are three separate brands of olive oil that I highly recommend; each is suited for different individual tastes. *Badia a Coltibuono* oil (about $12 for 8.5 ounces) has a peppery bite typical of Tuscan oils, which some love and others find too strong. Similarly fruity but less peppery is a Spanish estate-bottled oil, *Nuñez de Prado* (about $12.25 for 17 ounces), a beautifully rounded oil and a relative bargain. These are both cold-pressed extra virgins and oils for anointing favorite salads and drizzling on delectable breads. For cooking, I was mildly surprised as just how good Colavita's blend of 25% olive oil and 75% canola oil was; there is a noticeable olive flavor, lots of friendly monosaturates to keep your heart in great shape, and just $4.15 for a quart.

Prevents Heart Disease and Cancer

Considerable attention has been focused in the past on the virtues of such polyunsaturated oils as corn oil, which are known to help reduce serum cholesterol in the blood. The problem with polyunsaturates, though, is that they tend also to lower "good" cholesterol (high-density lipoproteins or HDL) *and* seem to promote tumor development as well.

But lately a lot of favorable attention has been given monosaturated fats like olive oil. Not only can olive oil lower "bad" cholesterol (low-density lipoproteins or LDL) just as well as corn oil can, but it doesn't adversely affect the good HDL or promote cancer either. These are two definite advantages which olive oil seems to hold over many other kinds of cooking oils.

Epidemiological studies conducted in four European cities—Uppsala, Sweden; London, England; Geneva, Switzerland and Naples, Italy—found only the southern Italian population to have both very low incidence of coronary heart disease deaths and low serum concentrations of the harmful LDL. And this is due primarily to "the almost exclusive use of olive oil as the only visible fat" in their diets.

Furthermore, food availability data in 30 different countries was analyzed by computer in relation to death rates for cancers of the breast, prostate, ovary, and colon. Mortality rates for all four cancers were related to total fat intake and animal fat intake, but not to vegetable fat consumption. Intake of milk, meat, animal protein, and calories from animal sources were positively associated with cancer death rates. Especially interesting, however, was the fact that inhabitants of countries in which olive oil is a major source of fat in the diet tended to have more reduced risks of all four cancers, particularly of breast cancer. This data appeared in the December 1986 issue of *Cancer*.

Since I wrote the preceding paragraphs almost a decade ago, additional research has more than confirmed the value of olive oil as a cancer preventive. Scientists have long been aware that Mediterranean women develop breast cancer at 50–60% the rate of American women. Now they say, based on more recent evidence, the reason could very well be the liberal use of olive oil.

Researchers at the Harvard School of Public Health analyzed questionnaires administered to 820 women newly diagnosed with breast cancer and an additional 1,548 cancer-free women whose age and area of residence paralleled those of the women with the disease. All the women were in Greece, where olive oil is widely used in cooking.

Dr. Dimitros Trichopoulos, author of the study, said part of the reason olive oil is better for the body is that it is less easily oxidized than polyunsaturated fats and contains plenty of antioxidant vitamins and other components, including vitamin E.

The analysis, published in the *Journal of the National Cancer Institute* for the week of January 15–21, 1995, said the data came from one of the largest studies that have examined the role of diet in the cause and origins of breast cancer.

According to Dr. Trichopoulos, the findings provide "an explanation for the paradox that Mediterranean women consume plenty of fats, and yet they have only 50–60% of the risk of breast cancer, compared with other women." Breast cancer is a major and rising problem around the world and especially in the United States, where 1 in 10 women can expect to develop the disease late in life.

The doctor's report strongly advocates the frequent use of olive oil in the diets of all American women. By doing this, it is concluded, their risks of breast cancer will be "substantially reduced."

Soothes and Heals Inflammation

Olive oil is terrific for ulcers and burns. To relieve heartburn, indigestion, and ulcers brought on by stress, spicy food, alcohol, coffee, and the like, just mix together 2 tbsp. of pure virgin olive oil with the white of one raw egg. Then take internally several times a day to experience rapid relief.

Severe burns on the surface of the skin can be effectively treated with just olive oil and egg whites, when nothing else is available. My own personal experience with this during a winter in the mid-1970s convinced me of its genuine healing effects. At that time we lived on a small farm in the central Utah community of Manti. I had put a can of granulated honey on the back of our coal-and-wood stove, which had a roaring fire going in it at the time. Worrying that the honey might get all over the stove and be hard to clean off, I immediately reacted before thinking and reached out with a couple of cooking gloves to remove the can. Just as I did so, a very hot geyser of honey spewed forth from the edge of the can, spraying the entire inside of my right forearm. I let out a considerable yelp of pain and dropped what I had on the floor.

Almost as if by sudden instinct, I ran outside and pushed my injured arm deep into a huge pile of snow to find temporary relief from the intense pain and burning. My father threw an old packing quilt over me so I could remain somewhat warm.

I stayed outside until I could no longer endure the cold. My father, meanwhile, prepared a simple remedy which his mother had brought with her from Temesvár, Hungary (now part of Romania), many years ago.

It consisted of mixing 2 cups of virgin olive oil with the whites from 6 farm-fresh eggs that our chickens had laid the previous day. He then took an old basting brush from one of our kitchen drawers, sterilized it under hot water, and dried it thoroughly before using it to brush on the oil-egg white mixture with. After this, he lightly wrapped my injured arm from elbow to wrist with loose-fitting gauze.

This dressing remained on for about 16 hours until it was changed again the next day. Although it took awhile for the pain to decrease, healing was already evident by the following afternoon. Surprisingly no scars remained, except for one little $3/4''$ × $3/4''$ mark about 4″ up from the wrist where the gauze had apparently become loose enough to expose the skin and probably cause some of the oil-egg white dressing to get wiped away on my blankets that night as I slept. In less than a week I was completely recovered and required no medical attention whatever.

Marvelous Degreaser

Olive oil makes one fantastic degreasing agent for getting other kinds of oil off the skin. An auto mechanic in Portsmouth, New Hampshire, told me once that he never washed his hands in soap and water after working on customers' cars. "Heck no," he said, "I just pour some olive oil over them and rub it good into the greasy grime. Then I take some paper towels and wipe it all off. And if they're not clean yet, I just repeat the same process again until all the grease has been removed. Then to remove the olive oil, I just run a little hot water over my hands and dry them with another paper towel."

Helps Tighten Loose Skin

For those who are bothered with loose, sagging skin around the face and throat areas or abdominal region due to recent loss of weight, an effective remedy used by Paul Neinast of Dallas in his famous beauty salon might just be worth trying. Neinast takes the yolks of 2 eggs and beats them with $1/2$ cup of olive oil. This then is brushed on the customer's face and throat and left there for 10 minutes. After which the stiffly beaten whites of both eggs are put over this, and the entire mask left for about half an hour. He claims that it really tightens up the skin!

How to Get Rid of Gallstones

For the complete removal of gallstones, the following remedy seems to have worked for several thousand people across the United States and in Canada. Of this estimated number, I've personally interviewed about 125 within the last decade during my extensive lectures from coast to coast and in both countries. In every single instance, the treatment with some slight variations here and there always seem to have met with success. From these different variations, I've put together a relatively simple and pretty basic program that is 90% guaranteed to succeed in getting rid of stones.

The first step involves a two-day, mild food fast and an easy internal cleansing that will help to prepare the body for the other steps later on. Only vegetables and fruits such as peaches, pears, soaked prunes (and the juice), figs, and psyllium seeds should be consumed. In a food blender, combine 1 cup of carrot juice (either fresh or canned), 1 cup each of diced peach and pear halves (either fresh or canned with their respective syrups), about 5 pitted, soaked prunes *and* $1/4$ cup bottled prune juice, 1 bunch of chopped parsley, and 2 tbsp. powdered psyllium seeds from any local health food store. Liquefy for 3–4 minutes. Makes a quart and can be refrigerated for several days. Drink 2 cups of this health juice cocktail every 4 hours during this two-day fast in which you should also be consuming plenty of soups and salads, but avoiding meat, bread, dairy products, coffee, soft drinks, condiments (catsup, mustard, pickles, mayonnaise), deep-fried foods, sweets, and the like.

The second step involves taking several enemas to cleanse the bowels. This can be done in the evenings of both days. Coffee enemas are recommended and the complete instructions for administering them given under COFFEE.

Only on the third, fourth, and even fifth day, if necessary, is the olive oil treatment to begin, continuing with your mild food diet at the same time. *On an empty stomach* morning, noon and night, drink a well-stirred or well-shaken mixture consisting of 8 tbsp. of pure virgin olive oil, 3 tbsp. unsweetened grapefruit juice, and 2 tsp. of apple cider vinegar, all of which is sweetened with 1 tbsp. pure maple syrup. Sometimes the response rate may be a little slower than usual, so the treatment may need to persist for a week or more. In such instances, a return to a more complete diet obviously is necessary for much needed strength and energy, but meat, animal fat, and refined carbohydrate intakes should be modified as much as possible to assure the greatest recovery success possible.

Intakes of lecithin (2 tbsp. of granules) every three days seems to be a good preventative from having them form again later on. This program works well when a little patience, common sense and modest dietary sacrifices are employed with the serious intention of getting rid of gallstones once and for all.

Sure Cure for Earache

Joel Bree, a friend of mine who lives in the orthodox Jewish section of Brooklyn, New York, came up with an effectively simple solution for treating earaches and inner ear infections. This came about as the result of a necessity to find an alternative to the antibiotic drugs which the local pediatrician was giving his kids during the many visits made to his office for ear infections.

His simple remedy calls for olive oil (Eden brand), vitamin E oil in the capsule (Schiff brand), and garlic oil in the capsule (San Helios brand). "I've experimented with other kinds of vitamin E and garlic oil, but never found them to work as effectively as these do," he remarked to me during a midsummer 1987 visit with him and his family.

First, get a small, clean glass container (an empty baby food jar is good for this). Then drop in it with an eye-dropper 13 drops of olive oil. Next, cut and squeeze the contents out of a single capsule of d-alpha tocopherol 400 I.U. vitamin E into the jar. Finally, cut and empty 1 garlic oil capsule so at least 7 drops are obtained. On account of the stiffness of the capsule itself, it may be necessary to squeeze rather hard to get the oil out.

Thoroughly mix all these oils together by shaking the jar from side to side. Then set the jar in a soup bowl and pour some very warm water around it. Leave in the bowl for about 1 minute until comfortably warm. It's advisable to first test the temperature of the oil by dropping a little of it on your wrist so as not to burn the child's ear.

Then, with the head tilted sideways, drop equal amounts of the oil in each ear. Remove the excess by gently dabbing inside of the ear with a wad of cotton. But don't rub hard. It's also a good idea while dropping the oil into the ear to gently rub the hollow space directly beneath the earlobe with the fingertips to permit more oil to enter the ear canal to reduce the pain and stop infection. This remedy is also good for ringing in the ear and water accumulation from swimming or taking a shower. However, Joel advises that this remedy shouldn't be used if there is a purulent discharge from the ear or if the eardrum has ruptured.

Relieve Pain in Teething Infants

When infant children begin their period of teething, oftentimes they are in varying degrees of pain. This seems to be most evident when they are fed with metal spoons, consequently a wooden one should be used instead. Another way of reducing their pain is to rub their sore, little gums with some pure virgin olive oil several times a day.

Dress Your Salads in Formal Attire

An infinite number of dressings for your favorite salads are in circulation these days. One which I'm particularly attracted to is a combination of several different recipes I've severely modified and borrowed from Frances Sheridan Goulart's excellent *The Whole Meal Salad Book* (Donald I. Fine, 1985), to whom I also express my gratitude for letting me use them.

Subtle Lime Liqueur Vinaigrette

Needed: 1 small crushed garlic clove; $^1/_2$ tsp. kelp; 2 $^1/_2$ tbsp. lime juice; $^1/_4$ cup unsweetened apple juice; 3 tbsp. of any fruit-flavored liqueur; $^1/_2$ tsp. paprika; $^1/_8$ tsp. thyme; $^1/_8$ tsp. rosemary; $^1/_2$ tsp. pure maple syrup; $^2/_3$ cup pure, virgin olive oil.

Mash garlic and kelp to a paste in a small bowl using the back of a heavy spoon. Add the lime juice, liqueur, paprika, thyme, rosemary, and maple syrup and blend thoroughly. Then gradually whisk in the olive oil and apple juice with a wire whip until the mixture is smooth and thick. Should make enough dressing for 4 whole meal salads.

Onion

(See under GARLIC AND ONIONS)

Orange

(Citrus sinensis, S. aurantium)

Brief Description

Although there are many varieties of oranges in the world, there are two basic kinds I've encountered in my numerous travels. The one we're the most familiar with is the sweet orange (*C. sinensis*), from which juice is made. The other is the sour orange (*S. aurantium*) available throughout tropical regions of the earth. Both are important in their own way.

The orange is native to China and Indochina and happens to be the most important fruit of international commerce. Its physical characteristics (especially the rich citric acid and vitamin content of the fruit) and history of cultivation are similar to those of other types of citrus fruits. Columbus brought the orange to the West Indies, and it is known that orange trees were well established in Florida before 1565 and were growing in California by 1800.

Flowers and fruits in all stages of development are on the tree throughout the year, although a large portion of the fruits ripen at one time. The orange is susceptible to many different insects and fungal diseases and is quite sensitive to frost. If the fruits are picked when still "green" (though fully mature), they must undergo a bleaching or degreening process to bring out the orange or yellow color in their rinds. Some oranges are artificially colored and waxed before marketing.

The navel is a winter orange and the Valencia a summer one; both are sweet. On the other hand, the Seville is a sour type, but much preferred in marmalade.

Lowers Cholesterol

At a local Chinese restaurant here on Main Street in Salt Lake City near our Research Center where I frequently lunch, I've noticed they always include fresh, unpeeled orange segments with their delicious buffet. I usually help myself to 10–15 of these segments to accompany large servings of their Mongolian beef with broccoli and sweet-and-sour pork. I find that these fatty foods digest a lot easier with oranges than without them. I highly recommend eating an orange after consuming meals high in fat and grease. This reduces your risk of incurring hardening of the arteries and a heart attack later on in life.

Relief for Hysteria, Palpitations, Convulsions, Epileptic Seizures, Acute Stomach Pains, and Bronchitis

Curacao is the largest and most populous island of the Netherlands Antilles, West Indies. When I went there in 1989 about 146,100 people resided there. The island has one of the world's largest oil refineries, receiving oil from the enormous reserves at nearby Lake Maracaibo, Venezuela. The next biggest industry is tourism.

Ramón Escotren is a local folk healer. I spent the better part of several days watching him working out of his dingy three-room dwelling on one of the side streets in the bustling capital of Willemstad. He used some herbs and spices, but a surprising amount of fruit to work his variety of cures with.

A middle-aged woman in hysterics was brought to him by her husband. She believed someone had cast an evil spell on her and was all worked up over this. Ramón gave her a glass of cold orange juice to sip slowly as he patted her hand reassuringly and spoke in low, soft tones with her for awhile. She eventually calmed down and left in peace with a very satisfied husband by her side.

A man in his thirties and of slight build was the next patient to call. He worked in the local ship-repair dry dock, which just happens to be one of the largest in the Americas. His problem was heart palpitations, occasional convulsions, and frequent stomach pains. After some conversation with the man, Ramón determined that he was suffering from lead poisoning due to the fumes of some of the paints and chemical solvents he had to work around every day. Ramón showed him how to make a tea from the grated orange peel: 2 tbsp. in 1 pint boiling water; simmer 5 minutes; steep 20 minutes; strain and drink 1 cup lukewarm every 5 hours. He also prescribed 1 glass of Seville or sour orange juice every afternoon. The man went away happy.

I asked my host at this point how successful this treatment was. He replied in his native Spanish, "Quite successful, for he had treated a number of similar cases from the docks. He couldn't explain exactly *how* the orange peel tea or orange juice helped, only that they worked! He noted, however, that the *sour* orange juice did better than the sweet kind.

Another man I judged to be in his early fifties came in that afternoon. His wheezing indicated acute bronchitis. For this Ramón had him sit over a pot of tea made from grated orange peel and cut orange tree leaves, and slowly inhale the fumes for 30 minutes. A towel was put over the man's head to retain the heat, and his face was kept a good 10 inches over the pot. The man left after that breathing much easier.

Sour orange juice will also help allay the frequency of epileptic seizures, I was told.

A Citrus Cooler for Poor Memory

Ever had one of those "dog day afternoons," when it seems that your brain has been put on "hold" and your body is in temporary limbo? Well, this little refresher is just the mental health break your brain needs to snap to attention again.

Fresh Citrus Cooler

Needed: 1 cup fresh orange juice; $1/2$ cup fresh lemon juice; $1/2$ cup fresh lime juice; $1/3$ cup pure maple syrup; 1 cup lime-flavored sparkling water, chilled; and lime wedges.

Combine the first 4 ingredients in a pitcher; stir until the maple syrup has dissolved. Add chilled sparking water; stir well. Serve promptly over ice. Garnish with lime, if desired. Yield: 3 cups.

Ornamental Flowers

Chrysanthemum (Chrysanthemum sp.)
Daffodil (Narcissus pseudo-narcissus)
Daisy (Bellis perennis)
Day Lily (Hemerocallis fulva, H. Flava)
Geranium (Garden) (Perlargonium sp.)
Hollyhock (Althae rosa)
Iris (Common) (Iris Germanica)
Marigold (Tagetes erecta, T. patula)
Nasturtium (Tropaeolum majus)
Pansy (Viola tricolor hortensis)
Peony (Paeonia officinalis)
Petunia (Petunia hybrida)
Rose (Rosa sp.)
Snap Dragon (Antirrhinium maius)
Tulip (Tulip sp.)
Violet (Viola cucullata, V. odorata)

Brief Description

CHRYSANTHEMUM. An ornamental plant belonging to the aster family. The flowers are of various warm colors such as white, red or yellow, with dark green leaves; both of which taste somewhat like a mild cauliflower. The Japanese and Chinese have used both for centuries in their remedies and recipes.

DAFFODIL. There are several kinds. Some have a crimson or reddish-purple circle in the middle of the flower, while others have a yellow circle resembling a coronet or cup in the middle. The common daffodil gets about a foot high, with leaves that are long, narrow, grassy looking, and of a deep green. The single, large, yellow flower at the top of the stalk presses down a bit due to its weight.

DAISY. A low-growing European herb of the aster family which has small white or pink rays and yellow disks in its flowerheads. It grows to about 18″ and has many broad leaves at its base with indented edges and finger-width size to them.

DAY LILY. This can be any plant of a genus of the lily family, which is characterized with long narrow basal leaves and showy yellow or tawny flowers in small clusters. Also any plant of a related genus (*Hosta*) bearing racemose white or violet flowers.

GERANIUM (GARDEN). A genus of South African plants, the species of which are widely cultivated in gardens everywhere on account of their very showy red or white flowers. There are also several kinds of scented geraniums as well, which some patients in the Soviet Union who suffer from hypertension and headaches sniff for 20 minutes every day in order to obtain relief.

HOLLYHOCK. A tall perennial Chinese herb belonging to the mallow family. This plant is cultivated in gardens as a biennial for its beautiful pastel flowers. Wee fairy folk were once thought to eat of its flowers a long time ago.

IRIS (COMMON). Cultivation has produced a great number of varieties, both among the bulbous or Spanish iris and the herbaceous or flag irises, which have fleshy, creeping rootstocks. The German or flag iris of modern American nurseries is a handsome plant with sword-like leaves of a bluish-green color that are narrow and flat. Flower stems are nearly 3 feet high with large, deep-blue or purplish-blue flowers that have an agreeable scent reminiscent of orange blossoms.

MARIGOLD. This can be any of several plants of the genus belonging to the aster family, especially African and French marigolds. A related species, marsh marigold, is known under the more familiar name of calendula for its marvelous skin-healing properties. All marigolds bear large yellow, orange, or red terminal flower heads.

NASTURTIUM. This is an annual native to South America, but cultivated in gardens all over the world. The trailing or climbing stems grow 5–10 feet long and bear small, almost round, radially veined leaves and are adorned with red, orange, or yellow flowers larger than the leaves themselves.

PANSY. An annual that's widely cultivated as a garden ornamental, but also occurs wild in fields and meadows and along the edges of forests in North America, northern Asia, and Europe. The angular, soft, hollow stem bears alternate, ovate to lanceolate, toothed leaves. The solitary, axillary flowers may be yellow, blue violet, or two-colored, the flowering time being from March to October.

PEONY. This is a perennial which grows wild in southern Europe and is cultivated elsewhere as a garden flower. The thick, knobby rootstock produces a green, juicy stem from 2–3 feet high. The leaves are ternate or biternate, with large, ovate-lanceolate leaflets. The large, solitary, red or purplish-red flowers resemble roses and bloom from May to August. This "queen of all herbs" was highly prized by the ancient Greeks for its miraculous properties. But not until it eventually lost favor with herbalists in the seventeenth and eighteenth centuries did it become a prize attraction of horticulturists.

PETUNIA. Any of a genus of tropical American herbs with funnel-shaped or tubular-spreading petals or corollas. The common garden petunia, as well as many forms and varieties, have all been derived from *P. axillaris* (with white flowers) and *P. violacea* (with violet flowers). Both are native to Argentina. Petunias belong to the same nightshade family (*Solanaceae*) as potatoes, tomatoes, tobacco, and chile peppers.

ROSE. There are over 100 species of rose in the genus *Rosa*, which consists of prickly shrubs found wild and widely cultivated in the temperate parts of the Northern Hemisphere. Their trailing, climbing or erect stems bear alternate, odd-pinnate leaves with familiar white to deep-red or, rarer still, black flowers that are single and five-petaled in wild species but mostly double in cultivated varieties. They yield fruitlike, fleshy hips rich in vitamin C.

SNAPDRAGON. Any garden plant of the genus having showy white, crimson, or yellow bilabiate flowers fancifully likened to the face of a roguish dragon. The nightshade family (*Solanaceae*) to which petunias belong is closely related to the snapdragon order (*Scrophulariales*) and is a connecting link between it and the phlox family (*Polemoniaceae*), which is confined primarily to the western United States.

TULIP. The name applies not only to any plant of this genus, but also to its flower or bulb as well. Tulips have been so long in cultivation that the common garden types cannot really be traced to any existing wild species to speak of. Holland is still the center of tulip cultivation though bulbs for the market are now also raised in the United States. Horticulturally, tulips are classed under two main divisions: early-flowering and May-flowering tulips.

VIOLET. Not only does this name apply to any plant, flower, or species of the genus, but also it pertains to actual colors resembling some violets. The common purple or hooded violet is common to the eastern United States and is the state flower of Illinois, Wisconsin, New Jersey, and Rhode Island. Violets vary in hue from reddish-blue to blue-red, are of medium saturation and low to medium in brilliance of color.

Mums for Hypertension, Angina, and Bloodshot Eyes

Clinical evidence found in Chinese and Japanese pharmaceutical and medical journals show that chrysanthemum flowers are excellent for treating high blood pressure and its associated symptoms of headache, dizziness, and insomnia. Snip enough mum flowers to equal 6 tbsp., then divide into 4 equal portions of 1 $^1/_2$ tbsp. each and set aside for use throughout the day. Beginning in the morning at 8 A.M. and every 4 hours thereafter, put 1 portion in a cup, pour boiling water over the flowers, and cover with a small saucer or piece of aluminum foil, allowing them to steep for 15 minutes before drinking. Repeat this same procedure 3 more times that day for up to one month. In one experiment where a total of 46 hypertension patients were thus treated, 35 of them showed fairly rapid improvement in their symptoms, with blood pressure returning to normal in less than a week. The remaining patients also showed varying degrees of symptom relief and dropping of blood pressure after 10–30 days of treatment.

This same procedure twice daily in identical amounts brought considerable relief from very severe constricting chest pains in 80% of a group of 61 patients suffering from angina pectoris. In both applications, the infusions should be taken on any empty stomach preferably for maximum effectiveness.

Tired and bloodshot eyes due to excessing reading, close-range precision work, lack of sleep, airborne allergies and related factors may be relieved with chrysanthemums. Steep approximately 2 tbsp. of whole flowers in boiling water for 5 minutes. The liquid can be drunk as a stomach tonic. While the flower heads are still very warm and relatively free of excess liquid, they should be placed over the closed eyes while in a reclining position and kept there for 15 minutes. They are then to be replaced by other hot ones, but not so hot as to burn the skin. Continue this procedure for an hour before retiring to bed.

Daffodils for Boils and Tendonitis

The purulent matter occurring in herpes sores, leg ulcers, boils, carbuncles, and nasty-looking wounds and abscesses can be effectively drawn out by applying a poultice of freshly grated daffodil bulbs and mashed chrysanthemum leaves, which both have been puréed together in a food blender. Turn the mixture out onto the wet side of a small clean linen cloth or white hand towel and then apply directly to the site, leaving there for about 45 minutes or so before changing again.

Put several chopped daffodil bulbs in a blender to purée. Then remove and mix with a little honey to form a rather stiff and sticky paste. Apply liberally to tendonitis, twisted ankle, dislocated shoulder, sprained elbow, or injured kneecap for incredible relief from excruciating pain and soreness.

Please Do Use the Daisies for Gout and Hernias

For the uninitiated, a past national best-seller by Jean Kerr, *Please Don't Eat the Daisies*, carried a somewhat clever prohibition against their use in its rather wry title. But here they're heartily recommended for gout, arthritis, and inflammations of the liver and bladder. To 1 quart of boiling water that's been removed from the heat, add a generous double-handful of cut daisy flowers and permit to steep uncovered for 1 hour and 10 minutes. Sweeten with a little honey, after straining. Drink 3–4 cups a day, as needed.

The same decoction may also be used as an enema to help relieve various kinds of intestinal inflammations, such as colitis and diverticulitis, among others. Equally valuable, however, are the daisy's usefulness in helping to heal any kind of internal burstings, like hernias, or even to some extent, appendicitis—*but only when emergency medical care isn't readily available.* In such cases, 2 double-handfuls of daisies would be added to 1 quart of boiling water and allowed to steep 1 1/2 hours uncovered, before drinking. About 5 cups per day, on the average, would be required in these instances.

Scalds and Burns Treated with Day Lilies

The root bulbs and leaves of day lilies are fantastic for sunburns, major burns, and accidental scaldings by either hot liquid or grease. While the severe burn or scald is temporarily resting in very cold ice water, the coarsely chopped bulbs and cut leaves and flowers of 2 day lilies should be puréed in a food blender with a little crushed ice added to form a nice, thick paste. This is then carefully applied to the injured skin tissue and leaf for several hours until another change is required. Adding 1/2 to 1 tsp. of olive oil while blending the bulbs and leaves will help to keep the poultice from drying out. Also, some wet pieces of muslin or linen cloth loosely laid over the poultice should help to retain moisture.

Geraniums Make a Good Astringent

Garden geraniums make a good astringent both internally as well as externally in helping to check diarrhea and hemorrhaging. Bring 1/2 quart water to a boil, adding 2 tbsp. grated or finely chopped root. Reduce heat and simmer uncovered for 7–10 minutes. Then remove from the stove, adding 1/2 handful finely snipped leaves. Cover and let steep 40 minutes. Strain, sweeten with honey to taste and drink 1 cup every 3 1/2 hours. Or make a poultice of the strong decoction for external purposes.

This ornamental flower also makes a handy styptic for nicks and cuts encountered while shaving in the morning. Just take 1 geranium left and fold it over and over again both ways until you have a small, fat green square of material. Then pound this good with a hammer or rock until it's well bruised. Apply to the cut or abrasion and hold in place with adhesive tape. Bleeding stops almost immediately. For smaller areas, use only 1/4 or 1/2 of one leaf instead.

Ulcers and Kidney Stones Healed by Hollyhock

Peptic and duodenal ulcers seem to respond very well to a tea made from the leaves of hollyhock. Simply bring 1 quart of water to a boil, then add 1 1/2 double-handfuls of freshly snipped hollyhock leaves. Remove from heat, cover with lid and let steep for 1 hour. Strain and drink 1 cup with every meal, sweetened with 1/2 tsp. of pure maple syrup.

A decoction made according to the foregoing instructions, with the exception of simmering on low heat for 5 minutes before removing and steeping for an hour, is very good for relieving the painful misery accompanying kidney stones. Two to three cups a day in between meals and on an empty stomach should also help to dissolve the stone as well. Another variation of this same remedy is to use only 1 double-handful of cut hollyhock leaves and 1/2 double-handful of cut, fresh catnip herb.

Edema and Bruises Disappear with Irises

Excessive accumulations of clear, watery fluid in any of the tissues or cavities of the body seem to disappear with a tincture of iris root. Grate enough fresh, clean iris root to equal 1 heaping tbsp. Add it to 1 pint of white wine, stirring well. Cap the bottle and let it set for 10 days in a cool, dry place, making sure you shake the contents twice daily. Strain, taking 2–3 tbsp. daily.

To remove black-and-blue marks, just make a poultice of iris root and rose petals to lay on the bruises until they clear up. Clean a freshly dug iris root by washing it well under tap water. Then cut half of it into small pieces which can be puréed in a food processor or blender. If the root is too tough for this, then pound it with a hammer until well mashed. Purée a handful of rose petals with a couple of tablespoonfuls of water and mix them in with the mashed root. Apply this entire poultice then to the injured skin, covering with a damp cloth or slightly wet towel. Leave in place for up to 4 hours. Repeat this process at least twice daily for a couple of days, at which time bruises should be gone.

Marigolds for Tetanus, V.D., and Constipation

In some parts of Central and South America, marigolds are extremely popular to *cure* tetanus or blood poisoning and to successfully treat venereal diseases such as syphilis and gonorrhea.

For tetanus just bathe the injured part frequently two or three times daily in separate hot and cold decoctions of marigolds. Bring 1 quart of water to a boil, then reduce the heat and add 1 $1/2$ handfuls of finely chopped, cleaned roots. Cover and simmer for 10 minutes before removing from heat to steep another 45 minutes. Strain and refrigerate this amount until it's ice cold. Next make another batch just like the first. Strain while still very hot and tolerable to the skin. Soak the afflicted part in the hot solution for 1 minute before switching to the ice cold one for another minute. Alternate this way for at least 1 hour. Repeat again 6 hours later. Use the same solution either as a wash or douche for treating venereal disease.

The ancient Aztecs of Pre-Columbian Mexico employed a tea made from marigolds for correcting chronic constipation. Aztec physicians would simmer a handful of chopped marigolds (flowers, stems and leaves) over a moderate fire in an earthenware vessel of some sort filled with a quart of boiling water for half an hour. When the brew was lukewarm, it was then strained through coarsely woven fabric of some kind and immediately given to the patient, several small bowlfuls at a time. This same brew was also used to relieve accumulations of serous fluid in the peritoneal cavity (ascites) and general fluid retention in other parts of the body, such as the legs, for instance. Aztec physicians also administered this brew while somewhat hot to induce sweating in their patients for the purposes of general body cleansing and ceremonial purification. According to *Science* for April 18, 1975, from which some of this information comes, better than 60% of the Aztec medicinal herbs evaluated by scientists proved to be as efficacious as the claims made for them from ancient native sources.

Nasturtium Is a Marvelous Expectorant

This winding ornamental really helps break up mucus congestion in the breathing passages and the lungs during colds and flus. A warm tea made of the flowers and leaves also acts as a disinfectant helping to kill unfriendly bacteria on contact, besides promoting the development of more white blood cells which fight infection inside the body.

The tea to be taken *isn't* made like your regular herb teas would be. Put a level double-handful of snipped nasturtium leaves and flowers into a food blender or processor. Then add enough hot *tap water* (*not* boiling) and liquefy to make a drink of smooth, somewhat runny consistency. Drink half of it and take the balance 3 hours later on an empty stomach, making sure both amounts are warm.

Pansies for the Heart and Skin

William Shakespeare referred to it in his plays as "heartsease," and several daily infusions while this plant's in bloom from March to October makes an excellent tonic for weak hearts. Certainly there is nothing wimpish about its lovely and gentle appearance when it comes to treating skin eruptions, particularly in children due to acne and contagious diseases like measles, mumps, and chickenpox.

Steep about 3 slightly heaping tbsp. of plant snipped into small pieces in 1 pint of very hot (but *not* boiling) spring or Perrier water, covered, for 35 minutes. Strain and drink 2 cups daily while pleasantly lukewarm (*not* cool). Sores on the surface of the skin can also be frequently bathed with this solution as well.

Jaundice, Allergies, and Peonies

An effectual remedy for jaundice, kidney and bladder problems is to make an alcoholic extract by steeping 2 tbsp. of fresh, minced rootstock in 1 pint of red wine for 9 days in a stoppered bottle, remembering to shake well twice daily. One tablespoon 4 times daily on an empty stomach helps the liver.

A decoction of the rootstock is also very useful in either preventing or else reducing the incidence of some allergies during those months in the summer and early fall when people are most prone to come down with them in some form or another. Bring 1 $^1/_2$ pints of water to a boil, then add 1 $^1/_4$ tbsp. finely chopped rootstock. Reduce heat and simmer 3 minutes before removing and letting steep, covered, for an additional 40 minutes. Drink warm in $^1/_2$ cup portions throughout the day as needed, or every 4–5 hours, for relief and protection.

Under no circumstance should the flowers or above-ground plant be used internally, since they contain toxic substances that could make a person quite ill and uncomfortable.

A Good Night's Rest with Petunia Tea

A popular remedy for restlessness and insomnia which was introduced to me by a Costa Rican folk healer, calls for 1/2 cup of freshly snipped petunia petals and a few leaves to be gently brewed in 1 1/4 cups of hot (not boiling) water, uncovered for the space of 25 minutes. After which, it may be strained, sweetened with a touch of pure maple syrup and a drop of pure vanilla extract, then slowly drunk while pleasantly lukewarm.

Eating Disorders Corrected with Rose Tea

Lately several eating disorders, often associated with personality quirks, have been in the news a lot. One of them is anorexia nervosa, an extreme aversion to food. Its counterpart, on the other hand, is a sudden urge for a glorious pigout at the nearest buffet line. This morbid appetite seizure is called bulimia, and often alternates with periods of anorexia.

The best treatment I've ever found for both conditions is a regular infusion of rose petals, taken regularly about 6 times a day in 1 1/2 cup amounts. The best kind of roses to use for this are the red ones known as hybrid perpetuals. Bring 2 quarts of pure spring water to a rolling boil. Then remove the pot immediately from the heat and add 2 generous double-handfuls of fresh red rose petals, along with 2 tbsp. of dried chamomile flowers obtained from any local health food store. Cover with a good, tight lid and permit to steep for only 25 minutes at the most! Remove the lid and allow to set an additional 15 minutes until lukewarm. Strain, sweeten with a touch of pure maple syrup, and drink according to the instructions previously given. Reserve the rest for later use in a cool, damp place, but do not refrigerate. Keeps up to 17 hours before a new batch must be made. Each time it's taken, the tea should be lukewarm to work the best.

Treat Sore Eyes to Snapdragon

I'm really surprised that no American or Canadian herbalists have discovered the fantastic therapeutic powers of this unique and clever ornamental for all manner of eye disorders.

This book, in fact, may be the only recent herbal encyclopedia that deliberately omits eyebright from its wonderful repertoire of plant medicinals. I have nothing against the herb personally. It's just that there's something much better for the eyes than this.

My Hungarian grandmother treated numerous eye disorders ranging from cataracts in the early stages and eye inflammations to conjunctivitis and you name it, with an eye wash made from snapdragon flowers, leaves and root (in other words, the entire plant)!

Bring 1 quart of distilled water (very important to use only this kind) to a rolling boil. Add 1 heaping handful of carefully cleaned and coarsely chopped fresh root. Cover with a tight-fitting lid, reduce heat and simmer 15 minutes exactly. Remove promptly from the heat, uncover and add the cut and chopped contents of 1 small to medium snapdragon. Cover again and steep for an hour. Strain twice, bottle, and refrigerate. Makes the best eye wash you've ever seen. Bathe eyes frequently with an eye cup as needed.

Tiptoe Through the Tulips for Pain

Certain kinds of pains seem to respond very well indeed to tulip bulbs. A Chinese pharmacist in Hanoi told me this while I was attending an international symposium on Oriental folk medicine and acupuncture in Taipei, Taiwan.

To relieve a toothache or painful insect bite or sting, simply crush a little bit of fresh tulip bulb and apply the mashed poultice to either for quick and effective relief. Fresh, crushed tulip leaves were also a very good poultice for running abscesses to help draw out the pus.

Violet Syrup Soothes Hacking Cough

For scratchy, irritated throats, raspy voice, or hacking smoker's cough, nothing quite compares for comfort with a delicious syrup of violet flowers and leaves. It almost feels like liquid gold going down and leaves the throat and lungs with silky feeling.

Add 5 cups of closely clipped violet flowers and 1 cup finely clipped leaves to either a stone crock, earthenware, or glazed china container (don't use metal or plastic). Over them pour 2 $^3/_4$ pints of boiling distilled water (or pure rain water) and cover with a tight-fitting lid. Steep for 24 hours before pouring off the liquid and straining it gently through several layers of clean muslin. It might be a good idea to shake the contents of the container several times during the 24-hour steeping period to slightly agitate things.

Transfer this liquid to a regular stainless-steel saucepan and warm up slowly on a low heat, making sure it never boils! At various intervals of the warming up stage, add portions of lukewarm honey in an amount equal to 4 cups total. Keep stirring with a wooden ladle until a syrupy consistency has been achieved. Bottle in clean fruit jars and store in a cool pantry. Use 1–2 tbsp. at a time as needed for throat and lung problems. Works like a charm every time!

Daffy and Daisy Duck Salad

Saturday morning TV cartoon fans would know that Daffy Duck is Bugs Bunny's crazy feathered friend, while Disney fans should recognize Daisy as Donald Duck's ageless heart-throb. Well a little of both, plus root and leaves from each respective ornamental, wind up in this rather comical treat.

Needed: 1 medium-sized head of dark Romaine lettuce; washed and chopped into salad-sized pieces; $^1/_2$ head of shredded purple cabbage; 2 tsp. finely shredded fresh daffodil bulb; 1 tbsp. finely minced onion; 1 crushed garlic clove; $^1/_2$ cup each chopped daffodil and daisy leaves; 1 $^1/_4$ cups finely snipped daffodil and daisy flowers; 2 cups coarsely chopped and boned cold duck; $^1/_4$ cup chopped, shelled walnut meats; $^1/_4$ cup finely chopped red apple (unpeeled); 1 pint plain yogurt; 7 pitted dates; $^1/_2$ cup raisins; and touches of sherry and pure vanilla mixed together.

First, rub a good-sized *wooden* salad bowl on the inside with the crushed garlic clove. Add the lettuce, cabbage, ornamental leaves and flowers, walnuts, apple, raisins, and duck. Mix everything very thoroughly. In a food blender combine yogurt, dates, daffodil bulb, onion, sherry, and vanilla. This is your dressing for what may appear to be a very daffy salad idea but with a daisy of a taste to it!

SCENTED GERANIUM HONEY

Bruise fresh, scented geranium blossoms. Place them in layers on the bottom of a small saucepan. Pour room temperature honey into the pan and cook over low heat. Stir the mixture just until the honey is warm—about 2 minutes. High heat will damage the honey. Pour the mixture into sterilized jars and seal tightly. Store the jars at room temperature for about 1 week to allow flavors to blend. Rewarm the honey over low heat and strain the blossoms out. Recap or use immediately.

PANSY

(See under ORNAMENTAL PLANTS)

PAPAYA

(See under TROPICAL FRUITS)

PARSLEY

(Petroselinum crispum)

Brief Description

Parsley is a nonhairy biennial or short-lived perennial with a much-branched stem. A thin, white, spindle-shaped root produces the erect, grooved, glabrous, angular stem that can reach a height of slightly over 2 feet. The plant is often cultivated as an annual for its foliage, especially in California, Germany, France, Belgium, and Hungary. There are numerous varieties. Parts used are the ripe fruits (seeds), the above-ground herb, and the leaves.

White or greenish-yellow flowers appear in compound umbels from June to August. Curiously enough, parsley is poisonous to most birds but is very good for animals, curing maladies such as foot rot in sheeps and goats. The wild parsleys found throughout the British Isles are closely allied to the celeries and were used by the Anglo-Saxons in ancient times to mend skulls broken in combat.

Removing "Dragon Breath"

Ever smell a dog's breath or someone with acute halitosis? They're bad enough to gag you. But now there's a simple cure for both extremes. The next time you feed your dog, mix several sprigs of parsley in with a little raw chopped or ground beef, then combine that with the animal's regular dry chow. You'll be surprised how well this works! And as for human breath problems, simply dip a couple of sprigs in vinegar and thoroughly chew them slowly before swallowing. The purifying effect should remove offensive odors for at least 3–4 hours.

An Ignored Cancer Preventative

Unbeknownst to most people, a few sprigs of parsley pack a wallop as far as cancer goes. For one thing, they contain about as much vitamin A as 1/4 tsp. of cod-liver oil. For another, they yield about two-thirds of the vitamin C of an *entire* orange! Furthermore, parsley ranks higher than most vegetables in an important amino acid, histidine, which strongly inhibits tumor development within the body, according to *Mutation Research* (Vol. 77, pp. 245–250). And vitamins A and C are now recognized as significant nutrients in the fight against cancer. Therefore, it seems we'd all be better off eating more parsley than just leaving it on our plates.

Yucatan Remedy for Kidney Problems

Throughout the Yucatan Peninsula of southeastern Mexico, a tea is made out of fresh parsley herb to treat kidney inflammation, inability to urinate, painful urination, kidney stones, and edema. Bring 1 quart water to a boil. Remove from heat and add 1 cup of coarsely chopped parsley. Cover and let steep 40 minutes; then strain before drinking. Take 1 cup of warm tea 4 times daily with meals.

Overcomes Sexual Frigidity

This same parsley tea also manifests mildly aphrodisiac properties for couples experiencing sexual frigidity of any kind in their relationships. The same directions would be followed, except that 2 cups of chopped parsley are used with 1 quart of water, steeping time is an hour and 2 cups of very warm tea are consumed by each partner at least 20 minutes before sexual activities begin.

Interestingly enough, parsley has been fed to sheep in Spain to bring them into heat in any season of the year. Some couples with whom I've spoken after they've tried the above remedy have reported to me somewhat increased stimulations in their sexual desires, although several wished that this herb worked more powerfully in this respect.

Clears Up Bruises

An old Romanian gypsy remedy called for several sprigs of parsley leaves to be crushed, then applied directly to any bruise on the skin and left there for awhile. Repeated applications would usually clear up any black-and-blue marks within a day or so.

Exciting Recipe

The following dish is part of a traditional Iranian New Year's feast that's both delicious as well as very, very tempting.

PARSLEY AND GREEN ONION RICE WITH FISH

Needed: 3 cups uncooked rice; warm water; kelp; 2 bunches chopped green onions; 2 bunches chopped parsley; 3 lb. fish fillets; sea salt as needed; pinch of turmeric; 2 tbsp. butter.

Rinse rice several times until water is clear; soak in warm water with salt added. Bring a large pot of water to a boil (about 8 cups water). Drain water from soaked rice and add rice to boiling water. Boil about 10–15 minutes until rice is not crunchy but still quite firm. Stir occasionally to prevent grains from sticking together.

Drain rice in a strainer; add chopped onions and parsley. Pour some cold water over rice, parsley, and onions. Cover the bottom of the pot with butter and some water. Sprinkle rice and these two vegetables into the pot a spoonful at a time, keeping them in the center of the pot so as not to touch the sides of the pot. Cover pot lid with paper or dish towel and place lid tightly on pot. Cook approximately 10 minutes over medium heat; then reduce heat to low. Allow rice to steam 30–40 minutes.

Cut fish into serving pieces; sprinkle with sea salt, kelp, and a bit of turmeric. In a skillet, brown fish, cooking until done on both sides in butter. Serve with rice.

Parsnip
(Pastinaca sativa)

Brief Description

Parsnips look like an anemic version of their cousin, the carrot. The parsnip's starchy root, however, is one of the most nourishing in the whole carrot family. This starch is converted to sugar whenever the root is exposed to the frost. Parsnip isn't a common vegetable anymore, even though most of us have heard of it. Americans usually serve parsnips glazed with brown sugar and fruit juice only on special holidays like Thanksgiving or Christmas. Refrigerated in a plastic bag, parsnips keep for nearly a month.

Fatigue Fighter and Cleanser

Imagine a food so highly concentrated with energy-giving properties that it is a remarkable internal cleansing agent as well. Such a one is parsnip, which is loaded with more food energy than most of our common vegetables except potatoes, yet is a relatively strong diuretic for helping to remove toxins from the body.

A diet of parsnips, either steamed or baked for lunch *and* dinner for at least a week, becomes an extremely valuable cleansing agent and even gets rid of some stones in the kidneys and bladder. And parsnips in the diet once a day or at least every other day is very useful for strengthening those who have hypoglycemia or are just recently recovering from serious illness or surgery or both.

Save the juice left from cooked parsnips and drink a glass morning and evening for up to 6 weeks to get rid of gallstones. This is an old remedy from colonial America, which was introduced by the renowned eighteenth-century religious reformer the Reverend John Wesley.

Parsnip Perfections

For an unusual flavor, peel and slice or dice parsnips and boil them in a small amount of apple cider. Remove the parsnips when tender. Then boil the cider until it becomes syrupy and serve as a glaze. Add chopped parsley for color.

The next recipe comes from La Rene Gaunt's cook book, *Recipes to Lower Your Fat Thermostat* and is reprinted here with the permission of her publisher.

Perfect Parsnip Patties

Needed: 6 cooked parsnips; 1/4 tsp. powdered cardamom; 1/4 tsp. powdered mace; 2 tbsp. whole wheat flour; 1/2 cup plain yogurt.

Wash, peel and quarter parsnips. Remove core and discard. Cook them in boiling water for 15 minutes until tender. Drain and mash. Use cooking water to adjust consistency. Season with spices. Whisk flour into yogurt. Stir into seasoned parsnips. Shape into 8 patties and brown slowly on a nonstick griddle or frying pan. Turn just once. They should have a crisp crust. Serves 5. Liquid lecithin from any local health food store can be used in place of oil for griddle or frying pan.

This last little recipe comes from Better Homes & Gardens' *Fresh Fruit and Vegetable Recipes*, which the author thanks them for letting him use here (see Appendix).

Potato-Parsnip Whip

Needed: 1 1/2 lb. medium potatoes, peeled and quartered; 1 lb. or 5 medium parsnips, peeled and cut up; 2 tbsp. butter; some heated canned or fresh goat's milk.

Place potatoes and parsnips in a large saucepan, then add water to cover. Bring to boiling. Reduce heat and cook, covered, for 25 minutes or until tender. Drain well. Add butter, pinch of salt and kelp. Beat well with an electric mixer or mash with a potato masher. Gradually beat in enough warm milk (about 1/4 cup) to make light and fluffy. Makes 8 servings.

Passion Fruit

(See under TROPICAL FRUITS)

Pasta
(See also under GRAINS)

Brief Description

Pastas are basically high-starch, low-protein foods with small amounts of vitamin enrichment to them. Pasta can be divided into two main groups: noodles and macaroni. Noodles are characterized by the addition of eggs to flour. This increases the protein, but also increases the fat content as well. Yogurt may be substituted for eggs when making your own homemade noodles as the recipe that follows indicates.

The macaroni group includes spaghetti, lasagna, macaroni, shells, and other shapes. These are usually enriched with vitamins or wheat germ. My suggestion is to use these products in small amounts with high-nutrition foods. Fill out soups with a handful of whole wheat macaroni or noodles. Top whole wheat pasta with creamed chicken or tuna or a vegetable-tomato sauce.

The average American consumed 12 lb. of pasta in 1985. Spaghetti leads the market share in pasta sales with 55%, followed by macaroni at 27%, egg noodles at 13%, lasagna and specialty shapes at 3%, and all other kinds of pasta at a mere 2%.

Really Lowers Cholesterol

Most dieters have dismissed pasta as too fattening. Nothing could be farther from the truth, however. In fact, clinical studies show that pasta is one of the best foods to help keep serum cholesterol levels low!

A team of international doctors from Minnesota and Italy worked together to study the diets and blood cholesterols of healthy Italian men in the age range of 40–55 years residing in Naples and their American counterparts residing in the Minneapolis–St. Paul area.

The great bulk of the Italian diet was bread, pasta, and local vegetables, with meat, fish, milk, cheese, and eggs being definite luxuries due to their costs. Some fruits and very small amounts of cheese were quite often consumed as well. The diet of the Minnesota businessmen was just the opposite, in that dairy products and meat were frequent meal favorites, along with sugary foods, potatoes, and greasy entrées. Some vegetables were consumed, and fruits were eaten occasionally. Refined bread and other white flour products were used more often than pastas were.

Cereal grains provided the Italians about 67% of all their calories with another mere 20% coming from fats. This probably explains why the Italians had 30 mg. of cholesterol *less* in 100 milliliters of blood, than did their American counterparts who obtained close to 50% of all their calories just from fats! And as a rule, Italians have considerably fewer heart attacks than Americans, although many of them can be somewhat rotund and overweight. If anything, then, pasta can be a true life-saver so far as cholesterol goes and should be included in our diets more often.

Pasta Pieces

Homemade noodles and spaghetti taste great and are better for you than some of the store-bought pastas might be.

Whole Wheat Noodles and Spaghetti

Needed: 1 1/2 cups whole wheat flour; 1/4 tsp. sea salt; 1 cup plain yogurt. Mix flour and salt. Add enough yogurt to make a stiff dough. Knead the dough for about 3 minutes. Heavily flour the countertop. Press dough out with hands. Sprinkle more flour on top of dough and roll out with a floured rolling pin. *It is absolutely essential that the dough be very thin!* Let rest 5 minutes. Cut into 1/4" slices for noodles and 1/8" for spaghetti with a very sharp knife. Spread on wax paper and let dry until hard, about 3 hours. Cook by dropping noodles in boiling water or bouillon. Cook 10–15 minutes. Serves 6. Note: These noodles make incredible chicken or turkey noodle soup!

Pasta Primavera

Needed: 1 tbsp. olive oil; 1 clove minced garlic; 1/2 chopped medium onion; 2 large chopped tomatoes; 1/2 tsp. oregano; 1/4 tsp. marjoram; 1/2 tsp. kelp; 1/2 cup white wine; 1 tbsp. olive oil; 2–3 cups favorite sliced vegetables (broccoli, carrots, zucchini, summer squash, etc.); 1 lb. hot, cooked pasta of your choice; some grated Parmesan cheese, if desired.

Heat 1 tbsp. olive oil in sauté pan. Add garlic and onion; cook until translucent. Stir in tomatoes, oregano, majoram, and pepper; sauté 2 minutes. Add wine; let simmer while preparing vegetables. In separate sauté pan, heat 1 tbsp. olive oil. Add vegetables and cook until crisp/tender. Add tomato-onion mixture and serve over hot, cooked pasta. Sprinkle with some grated Parmesan cheese if desired.

PEACH, PEAR, AND QUINCE

Brief Description

Peach (*Prunus persica*). The many varieties of peaches are divided into two basic categories: "freestones," with soft, juicy flesh that separates readily from the stone, and "clingstones," with firmer flesh that adheres tightly to the stone. Clingstone varieties like the Red Haven are generally used for canning; freestones like the Rio Oso Gem are eaten fresh or frozen. Peaches originated in China several thousand years ago and were venerated as fruits of immorality.

Pear (*Pyrus communis*). This is a delicate, aristocratic, temperate-zone fruit that exists in thousands of varieties—with new ones being constantly produced. Few fruits vary so greatly in color, texture, flavor, size, and shape. Pears are also an exception to the usual rule that tree-ripened fruits are best—they are picked when full grown but still green, and attain their finest texture and flavor (soft on the inside but still firm on the outside) off the tree. America's most widely grown pear, the Bartlett, is bell shaped, with yellow skin and a red blush when ripe. It's excellent for poaching, canning or eating raw in season, which is from July to mid-October.

Quince (*Cydonia cydonia*). These yellow, pear-shaped fruits originated somewhere in Asia Minor and have been cultivated for some four millenniums. In medieval times most Europeans ate them fresh as well as cooking and preserving them. Quinces were once thought to be a type of pear, and in fact pears are often grown on quince rootstock, but the two fruits simply cannot be hybridized. Until the late eighteenth-century, marmalade was usually made from quinces: the word "marmalade," in fact, derives from *marmelo*, which is Portuguese for quince.

Towards Beautiful Complexion

Because of their extremely high moisture content and delicate mineral balance, peach, pear and quince make ideal beautifiers for a more wonderful complexion. Paul Neinast, who runs a beauty salon in Dallas, Texas, combines peach with papaya, banana, and avocado in a blender until well puréed. This facial mask is then applied and left on 30 minutes, after which it is rinsed away with tepid water. Then he will saturate several cotton balls with any polyunsaturated oil (sunflower oil is good to use) and gently rub the skin in a circular motion. This keeps dryness out and moisture in and gives the skin more elasticity. The face may also be rubbed with a little juice from some freshly pressed green grapes before the oil is applied. This treatment appears to give the skin a much softer texture than it may have had before.

Also a morning cocktail consisting of these three fruits is a great way to help flush out the system of all the old debris which may have accumulated during the night. In a food blender, combine 1 fresh, pitted peach half, 1 fresh pear half, and 1 whole quince. Do not peel any of them, but wash thoroughly before liquefying. Add just enough broken ice cubes and a small amount of cold spring or Perrier water to make a nice, refreshing beverage that's smooth but not too thick.

If someone is troubled with boils, carbuncles, or similar festering sores that seem to refuse to heal, just have the individual mix together in a food blender about 4 fresh peach tree leaves, a couple of slices of raw, *un*peeled potato that are $^{1}/_{16}$" thick and about 3" wide and 1 $^{1}/_{2}$ cups of extremely hot, boiling water. When a nice, warm purée has formed, pour onto a clean, thick cloth and hold on the boil for awhile. In the event nothing is drawn out, it may need to be lanced first with a sewing needle, which has been sterilized over a flame for 30 seconds, before the other warm poultice can then be applied with good success.

An old, but very reliable remedy for removing the inflammation and discoloration accompanying bumps, bruises, and abrasions calls for 3–5 peach tree leaves to be first mashed by hand before simmering in about 2 cups of sweet condensed milk for some 25 minutes on low heat. After this the solution is allowed to steep by covering with a lid, it is then strained when cool. This handy lotion is applied to the injured area either by rubbing directly on the skin or else soaking a wad of cotton or gauze material in the solution and then holding it in place with some adhesive tape. By the way, this same lotion is also terrific for relieving intense sunburns as well.

You may want to gather a number of peach tree leaves during the summer months when the fruit is ripe and preserve them by freezing for later use in the winter time when they're no longer available. Now peach tree leaves have considerable enzyme activity which needs to be lessened before freezing, but *not* destroyed by boiling them in hot water. The best method is to suspend them *over* boiling water and steam for about 10 minutes, before cooling *over* but *not* in, ice-cold water.

Fill a large, deep pan with a good lid half full with water and bring to a rolling boil. Put freshly picked and washed leaves into a large wire strainer (the kind often used for straining cooked spaghetti noodles) and hang over the edges of the pan, making sure the bottom doesn't touch any hot water. Cover and keep pot on high heat, while the leaves steam for the allotted time. The same methods for cooling are next employed in another pot, which should be covered and placed in the freezer for just a couple of minutes to hasten the cooling process.

After this, the leaves should be carefully laid out on several thicknesses of paper toweling and covered with more paper toweling to completely drain them. Then they may be neatly stored in airtight plastic containers in the freezer until needed.

Fresh peach tree leaves that are crushed and then mixed with a little sweetened, condensed milk are an effective lotion to help clear up poison ivy rash, shingles, and psoriasis. An alternative to this would be for a handful of leaves to be added to 2 cups of boiling water and then covered to steep for awhile before using.

Another highly effective skin lotion for burns and inflammation is to combine the fruits of all three (peach, pear, and quince) in a food blender with enough shaved ice to make a thick, cooling purée, which is then spread on some clean muslin and laid on the surface of the skin. But no matter how you use them, they're great for a fantastic complexion!

Relieves Indigestion and Constipation

Peach tree leaves are sensational for digestive problems and constipation. A woman from Paris, Tennessee, wrote that she made a tea out of the leaves for her 15-month-old baby who was then suffering from intestinal gas. "He took right to the stuff," she said, "and it did the job I wanted it to." Furthermore, she, herself, drank several cups each day and found it to have marvelous laxative properties. The same tea also helps to relieve bladder inflammation in men.

Peach Syrup for Fevers and Congestion

An old standby often employed by some Native Americans and African Americans is a syrup prepared from peach kernel and peach bark to treat intermittent fevers, chronic bronchitis, and asthma and the common cold and flu. Add about 2/3 cup each of pounded peach bark and dried, split peach kernels to 2 cups each of apple cider vinegar and pure distilled water. Cover and let stand in a warm place for 5 days, shaking several times each day. Then simmer gently on low heat until the contents are reduced to slightly over a pint of liquid. Half a cup of brandy or whiskey is then added to preserve it and the solution stored in a well-sealed fruit jar until needed. One tablespoonful of syrup at a time is given every 3–4 hours to reduce fevers and eliminate accumulated phlegm.

This syrup works quite well for intestinal parasites, too, in 2 tbsp. amounts 2–3 times daily. The hydrocyanic acid present is thought to be a good substitute for the drug, quinine, which is obtained from Peruvian bark. This syrup may also be used as drops to relieve a painful earache. CAUTION: *No more* than just a couple of tablespoons of this syrup should be taken at any time, although the same amount can be spread over an entire day without any particular discomfort, if need be.

Fruit Tonic for Weak Constitutions

A very nice mucilage preparation can be made from pears and quinces as a soothing and strengthening tonic for delicate digestive systems, an aid for anemic conditions, and a speedy recovery from serious illness or major surgery. Remove the inner cores from several ripe pears and quinces before cutting them up and inserting into a food blender with about 3 cups of carrot juice. Mix for several minutes on medium speed. Makes a delicious tonic drink when taken in the morning and evening in 8 oz. portions. Chill before using.

Homemade Yeast

Back in nineteenth-century America, when baking yeast wasn't always available for making bread, women often made their own. You can do the same, if you like, with peach leaves. Just take 3 handfuls of peach leaves and 3 medium-sized potatoes and boil them in 2 quarts of water until the spuds are done. Remove the leaves and discard them. Next peel the spuds and rub them good with 1 pint of flour, adding enough cool water to make a paste. Then pour on the hot peach leaf tea and scald the floured spuds for about 5 minutes. If you add to this a little yeast, it will be ready for use in just 3 hours. Otherwise, if none is added, it will require standing a day and a night in a warm corner somewhere, covered with a cloth, before getting new yeast ready for use.

Making Good Fruit Jam and Jelly

Jam is nothing more than whole fruit, crushed or chopped, and combined with sugar, but will not hold its shape as resolutely as jelly does. Jelly, on the other hand, is a mixture of fruit juice and sugar that is clear and firm enough to hold its own shape very well.

The secret to making good preserves lies in properly blending four key ingredients: fruit, pectin, acid, and sugar. The fruit gives the distinctive flavor and color. Pectin and acid are found in all fruits, but to varying degrees; combined with sugar, they cause the product to jell. Sugar also serves as a preserving agent and, of course, adds its own sweet flavor.

For the health-conscious crowd who may object to white sugar, brown sugar may be substituted, but this will not give as clear a product as white sugar, especially in the case of jellies. Also $^1/_8$ to $^1/_4$ of the amount of sugar called for may be replaced with honey without seriously affecting the preserve.

Most fruits require adding high-pectin fruits such as apple, crab apples, or grapes or their juices, or adding instead commercial pectin such as Certo (a liquid) and Sure-Jell (a powder). The easiest way to make wild preserves is to use commercial pectin: cooking times are shorter and more predictable, and results are nearly always good.

Here is the basic procedure for making jelly. Crush the fruit and simmer it until the juices start to flow. Spoon the crushed fruit into a jelly bag. The best jelly bag I've ever seen was an old salt sack, whose heavy weave traps the pulp and seeds while letting the juices slowly drip through. About five layers of cheesecloth will also suffice very nicely as well.

After putting the fruit into the bag, be sure to collect all the juice that runs or drips out. Then measure out the required amount of pectin called for on the recipe sheet found inside the box it comes in. Don't get greedy and go for large batches; rather, be content to stay with making smaller quantities which are easier to handle.

Put the juice in a pot, add the pectin, heat, and stir in the honey/sugar. *Never* use iron, copper, aluminum, or galvanized steel cookware when making jelly. *Always* stay with stainless steel, enamel, or glass cookware. Boil, *stirring constantly*. Then administer the jelly test to see if it's ready to take off the stove. Dip your spoon in the boiling jelly stock, hold it above the heat to cool slightly and tip it sideways. If two drops form and then merge and slide off the spoon in a sheet, the jelly is done. If the liquid runs off like water, it needs to boil longer. As the late plant forager Euell Gibbons, once wrote, "No amount of instruction can take the place of experience in performing this task well . . . (but) you will be surprised how quickly you master this art."

When your jelly passes this simple test and appears to be ready for the next step, remove it from the stove. Then skim off and discard the accumulated foam and ladle the jelly into sterilized half-pint Mason canning jars with two-piece lids (a metal band and a metal cap with a rubberlike gasket surrounding the rim). These are better than the more conventional recycled jars (baby food, mayonnaise, mustard, pickle, and so forth).

After filling each jar, seal them good. Then submit all jars to a boiling-water bath for about 7 minutes, which helps to further sterilize the contents and tightens the lids more securely to the tops, thereby presenting potential contamination from occurring. Jellies and jams made thusly can be stored for up to 3 years in a cold, dry place. The metal ring portions of the lids should be removed after a year to prevent them from rusting. In the event, however, that mold should appear, under *no* circumstance eat them. Such molds can be highly toxic and have been known to produce cancer in lab animals. Discard such spoiled items at once!

PEACH-CRAB APPLE JELLY

Use the ripe, red-colored fruit from the Japanese flowering crab apple, a common ornamental shrub, and exceedingly ripe Hailstone peaches, if possible. August is the best month for obtaining both fruits in a mature state of development. Cut them in halves and remove pits and seeds. Cover with spring or distilled water. Bring to a boil, simmering for 20 minutes. Pour off juice; strain through coffee filter for clearer jelly. To 7 cups of juice, add 1 package commercial pectin. Bring to a complete rolling boil, stirring all the time. Boil for 1 minute, then take off the stove. Skim foam and pour jelly into containers. Seal with two-piece lids.

PEACH-PEAR-APPLE JAM

Wash, peel and core about 4 2 $^1/_2$" diameter peaches and 5 Bartlett pears. Make sure both fruits are fully ripened. Crush the fruits good and measure 2 cups of this prepared mix into a large saucepan. Wash, peel, and core 1 large apple; finely chop and add 1 cup to peaches and pears. Stir in $^1/_8$ tsp. *each* of powdered cardamom and cinnamon. Mix 6 cups brown sugar and $^1/_2$ cup honey with $^1/_3$ cup bottled lime juice into the fruits and bring to a boil over high heat, stirring constantly. Then stir in 12 tbsp. (6 fl. oz.) of commercial liquid fruit pectin. Bring to a full rolling boil and boil hard for 1 minute, stirring constantly. Remove from heat, skim foam, put into jars, and seal. (The author is indebted to Charlie Fergus for much of this information.)

PEA
(Pisum sativum)

Brief Description

Fresh garden peas are becoming extinct in U.S. markets because few customers want to shell them. Instead they seem to prefer canned or frozen. Unfortunately, those that are available are often large, starchy, and nearly tasteless. Besides garden or shell peas, there are two other edible-pod varieties: the small, flat snow—or sugar—peas often used in Chinese cooking and the plumper sugar snap peas that can be eaten raw or cooked and shelled when mature. Peas were Thomas Jefferson's favorite vegetable and he grew them at Monticello.

Dissolves Blood Clots

Clinical studies conducted by doctors in Calcutta, India showed that peas have the ability to dissolve clumps of red blood cells that are destined to become clots. This clot prevention property is due to the presence of special plant proteins called lectins. It is, therefore, suggested that peas be incorporated into the diet more often, especially in the diets of those more susceptible to clots due to poor circulation, thick blood, and coronary heart disease.

Bathing Skin Eruptions

In some parts of Europe, children afflicted with measles, mumps, or chickenpox are sponged with the water in which peas have been boiled. This apparently seems to keep them from itching so much and from forming permanent pit marks in the skin. A poultice made from dried peas, boiled until they are soft, is a wonderful remedy for boils and abscesses.

Safe, Gentle Laxative Soup

For an effective, yet easy to use laxative, turn to peas. Either fresh or dried peas, boiled as a vegetable or served as pea soup, get the bowels working.

GARDEN PEA SOUP

Needed: 1 tbsp. butter; 1 cup diced potatoes; 2 cups Perrier water; 1 cup distilled water; 1 cup freshly shelled green peas or frozen and thawed peas; 1 tbsp. chopped chives; kelp to taste.

Melt the butter in a 2-quart soup pot and swirl the potatoes around in it. Add both kinds of water and simmer until the potatoes are soft, about 20 minutes. Purée in a blender and return soup to the pot. Bring to a boil and put in the fresh green peas. Cook them until just tender—test by eating one—and stir in the chives before serving. Season with kelp. This recipe was adapted from *Eat Better, Live Better*, courtesy of Reader's Digest.

PEONY

(See under ORNAMENTAL FLOWERS)

PEPPERS
(Capsicum species)

Brief Description

The several voyages of the famous explorer Christopher Columbus helped to introduce the capsicums to the rest of the world. At the time of his arrival in the West Indies in 1492, these peppers were being cultivated and used extensively by native peoples of the New World from northern Mexico southward through South America.

In his very first letter to King Ferdinand and Queen Isabella of Spain, Columbus described the many wonderful things he had seen. Although he did not call peppers by name, there can be little doubt that he was referring to them when he penned, “In these islands there are mountains where the cold this winter was very severe, but the people endure it from habit, and with the aid of the meat they eat with very hot spices”; see the 1910 *Harvard Classics* Volume 43, pp. 22–28.

On his second voyage, Columbus was accompanied by the physician to the fleet, Dr. Diego Alvarez Chanca, who gave us our first written account of the peppers of the West Indies. The letter was written to the municipal council of Seville, Spain, and arrived there on April 8, 1494. Dr. Chanca’s observations had all been made in a three-month period. Concerning the voyage itself, he wrote, “By the grace of God and the good knowledge of the Admiral, we came as straight as though we were following a known and established route.” About the people eating peppers, he had this to say: “Their principal food consists of a sort of bread made of the root of an herb, halfway between a tree and grass, and the *agé* [the sweet potato], which I have already described as being like the turnip, and a very good food it certainly is. They use, to season it, a vegetable called *agi*, which they also employ to give a sharp taste to the fish and such birds as they can catch, of the infinite variety there are in this island, dishes of which they prepare in different ways.” This passage appears in *The Letters of Christopher Columbus*, ed. by R. H. Major, 2nd ed., pp. 19–71 (London: Hakluyt Society, 1870). As a result of his letter, Chanca gets the credit for the first written record of peppers.

Chanca's spelling of the name given to capsicums by the natives of the Caribbean Islands (*agi*) has been corrected to *aji* (pronounced ah-hée) by modern linguists. Today that is still the common name for capsicums in the West Indies and South America. So where on earth did the term *chile* come from, which is used to describe just about all these hot peppers today? Apparently it originated from an old word in the Nahuatl tongue, the language of the ancient Aztecs, the dominant group in Mexico at the time of the Spanish Conquest.

The Aztecs referred to capsicums as *chilli*. The stem "chil" pertains to the *chilli* plant. It also happens to mean "red." The "li" is a suffix without significance denoting closure, as was the custom in Nahuatl. Dr. Francisco Hernández (1615) was the first person to use the term *chilli* in print. In Mexico today, the term *chile*, which was derived from the ancient Nahuatl word, refers to both hot and sweet types and is used in combination with a descriptive adjective, such as *chile verde*; or a word that indicates the place or origin, such as *chile poblano*. The same variety can have different names in different geographic regions, in its various stages of maturity, or in the dehydrated states. Consquently, the names of peppers in Mexico can be very confusing, indeed.

What Makes Chiles Hot?

The fiery characteristic in chile peppers is due to a very potent chemical called capsaicin; this substance can survive both cooking and freezing processes. In general, the smaller the chile, the bigger the thermonuclear detonation it's capable of delivering. This is because smaller chiles have a larger amount of seeds and vein (or internal rib) relative to larger chiles, and these are the parts that contain up to 80% of the capsaicin in chile.

Capsaicin triggers the brain to kick out a flood of endorphins, those natural painkillers that promote a sense of well-being and stimulation. But they can become addictive, which probably explains why chile lovers become hooked on these tiny "vegetable volcanos" (as I prefer to call them.) Capsaicin has some medicinal benefits, which will be covered later in this section.

The best antidotes I know of to combat the fire of chiles are dairy products, such as milk, yogurt, or even ice cream. And starchy foods like white bread, mashed potatoes, or cooked white rice will also have a neutralizing effect on the natural alkaloids in capsaicin. Alcoholic beverages are just about the worst thing to drink when consuming chiles, because they tend to *increase* the absorption rate of capsaicin within the body, which then only makes the food they're in seem a lot hotter! An irony of fact, however, is that chiles are more popular in the hot and humid tropics than anywhere else. This is because their stimulating qualities raise in heart rate and induce the process of sweating, which becomes a natural form of air conditioning!

Rating the Chiles

I've compiled a short list of some of the most popular chiles found in the Western Hemisphere. They're arranged alphabetically by their common names, but keep in mind that they're all classified scientifically as *capsicums*. They are also given a ranking according to their heat potential: a count of 2–3 will get your stomach churning; a 5–6 grabs your tongue and gives it a mighty good shake at the roots; and a perfect 10 means the damn thing will blow your lips plumb off your face!

AJI. Originated in Peru, its heat capacity is 7–8. Aji is usually seen in its fresh green or red state or yellow dried form, with a tapered point, measuring 3–5″ in length and 3/4″ in diameter. Rather thin fleshed in appearance, it delivers a clear, searing heat and tropical fruit flavor. Aji is used in ceviches, salsas, and sauces, and is often pickeld.

ANAHEIMS. Long green or red chiles from California and the Southwest, anaheims are pale to medium bright green in its immature state and fire-engine red when fully matured. Tapered and measuring about half-a-foot in length and roughly 2 1/2″ in diameter, they are medium to thick fleshed. The green vegetable flavor improves with roasting. Heat capacity on both is between 2 and 3. The immature form is used in stew and sauces, and as rajas or else stuffed (rellenos); the same for the ripened form, plus being good pickled and grilled.

BELL PEPPERS. These peppers come in green, blond, orange, red, yellow, and violet hues. They are all pleasantly sweet to the taste and lack capsaicin. They can measure anywhere from 3 to 6″ in length and up to 4″ in diameter. All are thick fleshed and vary in sweetness. The range is mildly sweet for the green, blond, and violet bells to the very sweet orange, red, and yellow bells with their crisp, fruity tones. All are good in salads, salsas, stews, and pastas; and can be grilled or roasted in shishkabobs.

HABANERO. This little number with its tropical fruit tones comes from the Yucatan and the Caribbean. Heat capacity is a devasting 10, which means it's 30 to 50 times hotter than the jalapeño. Be careful to use gloves when handling it if roasting it in the oven, open all windows and doors, and *get out of the house*! Then before reentering the premises to shut the oven off, take a deep breath, hurry inside, and then back out again. if you think I'm joking, go ahead and throw caution to the wind, and find out the hard way! Habañeros are employed mostly in salsas, chutneys, and seafood marinades and can be pickled.

JALAPEÑO. Named after the town of Jalapa in the Mexican state of Vera Cruz, the jalapeño is green in its immature state, but crimson red when fully developed. This is the first chile ever to be taken into space by American astronauts in 1982. Heat capacity for both colors is 5.5. Jalapeños can be used in salsas, stews, sauces, tamales, dips, cheese, soups, cooked fish or meats, pizzas, or breads.

NEW MEXICO. Also known as the long green chile, because that exactly fits its description. Heat capacity is 3–5 and it measures between 6 and 9″ in length and up to 2″ in diameter. Is medium fleshed and flavored like no other chile in North America: sweet and earthy. I've been down to the chile festival held every fall in Santa Fe, where I've seen these chiles roasted in enormous quantities. They freeze well, and frozen are even better than canned. The New Mexico or long green chile is ideal for making, what else, but green chile sauces as well as other sauces, stews, and salsas. A sandwich made with crunchy peanut butter and several of these long green chiles, roasted and peeled, laid in between both slices of bread is a most enjoyable eating experience. The red version is commonly roasted and used in red chile sauces, barbecue sauces, chutneys, salsas, relleños, and tamales, not to mention pipián sauces and great rajas.

PIMENTO. This chile is scarlet, almost heart shaped, tapering to a point, and measuring roughly 4″ in length and 3″ in diameter. Very fleshy, wonderfully sweet and aromatic to the point of exhiliration, this chile barely heats up to 1 and is tastier than the red bell pepper. Most commonly used in its powdered form as paprika, the best of which comes from Hungary, salads come alive with fresh pimentos, but the canned kind are mostly used as garnishes.

POBLANO. One of the most popular chiles used throughout Mexico, the poblano is dark green, with a purple-black tinge, tapering down from the shoulders to a point. It measures up 5″ in length and 3″ in diameter, with a modest heat capacity of just 3. Thick fleshed as most of the others chiles are. The green poblano is always cooked or roasted and never consumed raw. Roasting gives it a fuller, smoky, more earthy flavor. Poblanos are highly favored for making chiles rellenos or any other stuffed chile dish because of their size and the thickness of their flesh. The riper form of poblano is a deep red-brown in appearance and much sweeter. It is best roasted and used for rellenos or rajas, as well as in soups, stews, tamales, and a variety of sauces.

SCOTCH BONNET. Common to Jamaica and other Caribbean islands, and coastal Belize, this chile pepper has a heat capacity between 9 and 10; it is an essential ingredient in the Jamaican specialty called jerk sauce and in Caribbean curries. When I asked Daniel Quevedo, the descendant of a famous Calypso king (see LIME) why they called it jerk sauce, he grinned a wide smile, revealing an even set of pearly whites. "it's 'cause it'll jerk your tongue right out of your head," he laughed heartily. "Or maybe 'cause the one eating it might be a jerk." Pale yellow-green, orange, or red in color, and smaller than the habañero though similar in shape, the scotch bonnet measures about 1 $^1/_2$″ in length and the same in diameter. It is very hot, fruity and smoky in flavor, and is also used as a condiment sauce.

SERRANO. The serrano comes from Mexico and the American Southwest; it heats up to 7. When ripe it can be either dark bright green or shiny scarlet. Both colored chiles are cylindrical with a tapered, rounded end, and measuring up to 2″ in length and $^3/_4$″ in diameter. The serrano is thick fleshed with a clean biting heat and rather pleasant high acidity; it is the hottest chile commonly found in the United States. Both are wonderful is salsas, pickled, or roasted and used in sauces. Green and red serranos are used interchangeably, although the red is apt to be a tad sweeter. Reds are often used for decorative purposes, besides for eating.

TABASCO. This little number, is thin fleshed and bright orange-red in color and hailing from Louisiana and Latin America, has a sharp, biting heat of about 9, with some stemminess and hints of celery and green onion to it. It was brought back to Louisiana in 1868 by a soldier returning from the wars in the eastern Mexico state of Tabasco, located on the Gulf of Campeche. It is probably the only chile in history to ever be copyrighted, being almost exclusively used in the famous McIlhenny Tabasco® pepper sauce. This sauce has many uses on the bar, at table, and in the kitchen and is one of the most important commercial sauces known the world over.

There are also many different kinds of dried chiles. Ounce for ounce they pack more whallop than their fresh counterparts. This is because the drying process intensifies and magnifies their flavors and gives them a much higher yield of natural sugars. Chef Mark Miller of Santa Fe's celebrated Coyote Café considers them absolutely indispensable: "We use them in large quantities, mainly in the preparation of sauces."

Really Relieves Arthritis

Amazing relief for rheumatoid arthritis sufferers can come from red pepper. It seems that the pain which arthritic victims suffer so much from takes place something like this: a unique protein called nerve growth factor (NGF) helps to produce a hormone known as substance P (SP), which transmits all pain signals throughout the body to the brain quick as lightning, producing the expected verbal response of "Ouch, that hurts!" or a facial grimace of pain. Now when cayenne is taken into the body on a regular basis, its main constituent, the fiery capsaicin does several things: (1) It blocks the supply of NGF; (2) it causes a massive release of SP from the hypothalamus, which at first increases arthritic pain but later diminishes quite a bit; (3) by producing such a depletion of SP from the hypothalamus, pain signals no longer are able to get to the brain. The first noticeable result is *no feeling of pain* in those with arthritis. An interesting side note to all of this is that some U.S. Navy doctors have been studying capsicum's usefulness in helping to alleviate the "phantom limb pain" which war veteran amputees often experience, according to *Science Digest* for September 1983.

The recommended dosage for effective pain relief from crippling arthritis is approximately 2 capsules, three to four times a day with milk or apple juice. This must be done on a regular, consistent basis in order for lasting benefits to be derived. Don't worry about the early increases in pain; it will diminish soon enough, leaving your body relatively pain free before long.

Brings Down Blood Sugar Levels

A report in the *West Indian Medical Journal* (Vol. 31, pp. 194–197) mentioned how a pack of mongrel dogs picked up off the streets in Kingston, Jamaica, were given powdered cayenne pepper. The result was a dramatic plunging of their blood sugar levels for up to several hours at a time.

Which is to say, if you're diabetic that an average of 3 capsules of Nature's Way or any health food store brand of capsicum will help bring down high blood sugar levels very nicely. If you're just the opposite and hypoglycemic, you'd better avoid cayenne altogether, both in food and in most herbal formulas, too.

Lowers Cholesterol

A group of rodents were fed high-fat diets, but given some cayenne pepper as well. There was an increased excretion of cholesterol in their feces and no rise in their liver cholesterol to speak of. So when you're consuming any kind of greasy food, be sure to drink an 8 oz. glass of tomato juice that has 1.8 tsp. cayenne and a squeeze of lemon juice in it with these meals.

Halts Bleeding Quickly

For any sudden gash, nick, or serious cut, just apply enough cayenne pepper or powdered kelp or both to the injury until the bleeding stops. I cut my hand between the thumb and first finger several years ago while dining in a restaurant. I drew a small crowd around my table when I requested some capsicum to dress the injury with. A few "oohs and ahhs" accompanied the success of my treatment when all bleeding ceased within a matter of minutes.

Prevents Blood Clots

The New England Journal of Medicine reported that residents of Thailand have virtually no blood clot problems because of their frequent consumption of red pepper. If you use capsicum on a regular basis you won't ever have to worry about getting blood clots! About 2 capsules a day is good for general health maintenance and eating more Mexican, Indian and other spicy foods laced with red pepper will virtually guarantee keeping your blood pretty thin and moving fairly good as a rule.

An Ulcer Healer

How can something so hot help something so painfully raw and sensitive as a stomach ulcer heal up quite nicely in the course of time? The internal consumption of capsicum stimulates the gut's mucosal cells which release more slimy mucous that neatly coats the walls of the intestines, including sore, bleeding ulcers. If you've ever watched a dog lick its wounds or held a burned finger in your mouth, you'll know about the kind of relief I'm talking about which comes to stomach ulcers covered by lots of mucous. That then is about how cayenne pepper helps to heal ulcers. Suggested intake is 1 capsule 2 to 3 times daily with meals.

Knocks Out Cold and Flu Miseries

Some Jewish grandmothers have relied upon a pinch of cayenne pepper and a finely chopped garlic clove in a bowl of hot chicken soup as the best way to fight the aches, pains, and fever accompanying colds and flus. Called "Jewish penicillin" by many, chicken soup is often recommended by medical doctors in place of antibiotics.

Cayenne also seems to work quite well with vitamin C. In fact, vitamin C doesn't perform as well unless cayenne accompanies. An Old Amish Herbs remedy called Super C has cayenne, ginger, and vitamin A in with vitamin C to make it more potent (see Appendix).

Keeping Your Toes Warm in Winter

Cayenne pepper sprinkled in your socks keeps your feet warm in winter. I known an old duck hunter from Malad, Idaho, who always put some cayenne in the bottoms of his woolen socks and also into the fingertips of his mittens or gloves, when staying out in the cold behind a duck blind for long periods of time.

Reduces Risk of Heart Disease and Tuberculosis

One thing that strikes as odd every time I go to Mexico, Central or South America, or the Caribbean Islands is the *low* incidence of heart disease among these area residents. Granted that they may have other problems of an infectious nature, but not the coronary diseases seen so often in this country.

After evaluating their life-styles and taking diets into consideration, I was finally able to isolate the chile peppers as being the main reason for this. Also, a virtual absence of tuberculosis in some of these places can be attributed again to the frequent consumption of different chiles.

Popular Talk Show Host Flirts with Chile Peppers to Avoid Obesity

According to various A. C. Nielsen ratings published periodically throughout 1994, the "Late Show" with David Letterman on the CBS television network was the highest-rated late-night talk show for that year (as well as the year before when it moved in midsummer from NBC to CBS). Letterman is known for his quirky, avant-garde style of doing things, and has brought his own late-night talk show to the "cutting edge" of television programming.

His show is taped every afternoon before a live audience at the Ed Sullivan Theater in New York City. But during the day when he and his staff are planning that day's show from his office on the 12th floor of the Ed Sullivan Building, he will frequently resort to a shot of hot sauce. He proudly showed off a shelf display of about two dozen brands of hot sauce to me and a friend of mine who is a reporter with *The New York Times*.

"I like food that can do something to you other than just make you fat," he exclaimed. "I prefer to interact with my food. My philosophy is simple with something like chile peppers: you bite them, and they'll hurt you back. Rob [Burnett], the head writer, and Robert [Morton], my producer, and other members of my staff . . . we see who can get the hottest pepper concoction going. Then we take turns eating it, just for the experience."

"We got hold of some Scotch bonnet peppers," he continued enthusiastically. "Without a doubt they've got to be the hottest little numbers in town." There was a brief pause; then he added, "Just about as hot as my show is right now." We both looked his way and noticed him flashing one of his famous skunk-eye winks in our direction, hinting of some mischief afloat.

"My first reaction when I tried one of them little babies," he said, "was that it didn't seem all that bad. But in a minute or two I started to sweat all over and I realized that I might actually . . . go . . . down." Demonstrating what he meant, he bent over slightly, placing hands on knees, and did this slow, weird, hyperventilating thing.

"But once the pain passes, it's not so bad. In fact, it's pretty great fun," he added with his trademark protracted cackle. "But I'll tell you this, there's absolutely no weight gain when you start eating concoctions of some of these babies," he said with a sweeping gesture of one hand toward the shelf containing all the small bottles of different hot sauces.

Others with whom we spoke on his staff also seemed to think that there is something in chile peppers that keeps weight under control, even though a person might eat a lot of things which otherwise might contribute to obesity. While Letterman's point of view may not be scientific, it merits some serious consideration by those who love to eat but don't want to put on the extra pounds. Try some chile peppers; they'll help you stay trim.

"Sure Cure" for Drowsiness

A Hispanic woman named Margarita Challes first introduced me to her *guaranteed* "sure-cure" remedy for drowsiness sometime back in 1977 when I made my first trip to Mexico City. I was doing some research at the National Museum of Anthropology in Chapultepec

Park. The weather for that midsummer was typically hot and humid. While I had the benefit of air conditioners for part of the day, the rest of my time was spent out in the open. It wasn't too long before I felt tired and groggy.

Now Señora Challes worked in the building that I was staying in at the time. She noticed my drooping energy levels and told me in Spanish that she had something for this. I asked her what it was and she produced some "pepper cookies," which I was reluctant to try but encouraged to do anyway. I took one, thanked her, and cautiously nibbled away on it as I pored over my notes at a desk in my upstairs room.

Containing some dried chiles, of course, they proved to be, how should I say, rather animated with flavor. But surprisingly enough, half of one of Margarita's cookies shook me out of my lethargy in no time at all, and I felt pretty peppy after that. I grew to like her cookies, but in very moderate doses, taking them as I would a potent medicine.

She graciously wrote down the recipe for me, which I took back home to Utah with me and modified over the course of time. These cookies will do away with the worst form of drowsiness and are ideal to *nibble* on while driving long distances, operating heavy equipment, or studying for a final exam.

MARGARITA'S "WAKE-UP" COOKIES

Needed: 4 dried Chiles de Arbol (a Mexican chile rated 7 on the heat scale from the state of Guerrero, but jalapeño or serrano chiles can be substituted, if necessary); 1/2 cup milk; 1 egg; 1/4 cup sunflower seed oil; 1/2 cup plain nonfat yogurt; 1/2 cup brown sugar; 2 tsp. vanilla; 1 1/2 cup white flour; 1/2 tsp. salt; 1/2 tsp. baking soda; 1 1/2 cups uncooked oats; 1 cup chopped cashews; 1/2 cup packed raisins.

Be sure to wear rubber gloves when discarding the stems from any of these chiles. Then break them into smaller pieces and place in a medium-sized pan. Add the milk and gently simmer over low heat. Remove from the stove top just before the milk commences to boil. Let cool and whip in a Vita-Mix total food machine or equivalent unit until the chiles are thoroughly pulverized. Next add the egg, oil, yogurt, brown sugar, and vanilla, and whip another minute. Set aside. Combine the flour, salt, and baking soda, and stir well. Add the wet mixture and stir until well blended. Mix in the oats. Add the cashews and raisins and continue to stir until the oats, nuts, and raisins are uniformly blended. Drop by teaspoonfuls onto an oiled cookie sheet and bake at 350° F. for about 15 minutes or until the cookies are light brown. Cool and store in a plastic container with a snap-on lid. Take a few with you when driving or working and nibble on them as needed to stay awake.

Salve for Sprains and Bruises

An ointment used in mainland China and Taiwan for treating athletic and work-related injuries such as sprains, bruises, and swollen painful joints is made with one part ground hot pepper and five parts Vaseline. Prepare by adding the ground hot pepper to the melted Vaseline, which is then mixed well and cooled until it congeals. This ointment is applied once daily, or once every two days, directly to the injured area.

In a 1965 report from a journal of traditional medicine from Zhejiang, 7 of 12 patients thus treated were cured and 3 improved, while 2 did not respond to this treatment. In the effective cases, 4–9 applications were usually used.

Energetic Recipes

Capsicum and paprika are known to increase energy levels within the body to a certain extent. Capsicum is included in some herbal energy products currently found on the market, such as Nature's Way Herbal Up sold in most health food stores nationwide. When I was in the Soviet Union in 1979 doing research, Dr. Venyamene Ponomaiyov, professor of chemistry and pharmacology at the Pyatigorsk Pharmaceutical Institute in Pyatigorsk, Georgia, informed me that he and some of his colleagues had discovered that cayenne pepper dramatically increased the intensity of electrical energy auras around the volunteers who used capsicum frequently in their diets. This finding indicates just how strength-promoting cayenne can be.

The three recipes that follow all use one or several kinds of the peppers cited in this section. They are designed to not only satisfy hunger by filling you up, but also to give you extra energy and vitality for an active life-style.

STUFFED BELL PEPPERS

Needed: 6 medium green bell peppers; 3 cups savory Spanish rice; 1 cup tomato sauce. Wash and core peppers but don't discard the seed cores. Instead finely dice them and mix them in with the savory Spanish rice. Steam peppers for 20 minutes. Then fill each one with 1/2 cup of this rice mixture. Place in a casserole and top with tomato sauce. Bake at 350° F. for 45 minutes, or until tender. Serves 6.

SAVORY SPANISH RICE

Needed: 1 medium chopped onion; 1 small chopped green pepper 1/4 lb. lean ground round; 1/2 cup raw brown rice; 1 clove minced garlic; 2 cups chopped tomatoes; 1 bay leaf; 1/8 tsp. cayenne pepper. Sauté onion and green pepper until tender. Brown beef and drain. Combine all ingredients. Bring to a boil, stir and reduce heat. Simmer 45–60 minutes or until rice is tender. Stir often to prevent sticking. Remove bay leaf before serving.

MILD TOMATO SAUCE

Needed: 1 cup canned tomato sauce; 1 tbsp. apple cider vinegar; 2 tsp. Worcesersthire sauce; 1 tbsp. finely diced onion; 3/4 tbsp. paprika; 1/2 tsp. finely diced garlic clove. Combine and bring to a boil. Turn heat down and simmer about 8 minutes. Serve over bell peppers stuffed with Spanish rice.

Hors D'Oeuvres from Hell

Many college kids are into doing the craziest stunts to either prove their bravado or stupidity or maybe a little bit of both. Some years ago several of my students, who were also on the football team, came up with a recipe that beat anything I'd ever run across. The fraternity to which they belonged threw a big party and most of the frat brothers, so I was informed, partook of these devilish little inventions. I was told later on that most of these guys had to take their nourishment in liquid form through a straw and postpone French kissing their dates for several days after their stunt.

Before I give you their basic recipe, let me mention that a batch of pastry dough is necessary. This can either be bought from the frozen food section in a local supermarket or else made according to these simple instructions.

Needed: 1 1/2 cups flour, 1/8 tsp. salt, 1 4 oz. stick of butter (avoid margarine at all costs), and 1/3 cup goat milk.

Mix the flour and salt together. Work the butter into the flour mixture. Add the goat's milk and mix only until the dough comes together and can be formed easily into a ball. Wrap in oiled wax paper and refrigerate 2 hours before rolling out.

Heavy whipping cream (that hasn't been whipped) can be substituted for the goat's milk. This will at least reduce some of the heat; to maximize it, though, substitute grapefruit or pineapple juice for the milk or cream.

HORS d'OUEVRES FROM HELL

Needed: one batch of dough; 1 12 oz. jar of whole Habañero peppers; 1 12 oz. jar of whole jalapeño peppers; 1 8 oz. package of regular cream cheese; 1/2 cup nonfat yogurt; 1 oz. fresh cashew nuts, finely chopped; 1 oz. fresh pistachio nuts, finely chopped; 1/4 tsp. each sea salt and coarsely ground black pepper; about 1/4 cup mineral water; about 1/4 cup buttermilk.

Make the pastry dough. Wear rubber gloves and be sure to work in a *well-ventilated* kitchen (with all the doors and windows open). If you don't you'll be s-o-r-r-y! Pour off the vinegar in which the habañeros and jalapeños are packed and discard it. Rinse the peppers well 3 or 4 times in water, remove and discard the stems, but leave the seeds intact if the ultimate pain and misery is desired. Then set both kinds of peppers aside on a clean dish towel to dry.

Next mix the cream cheese and sour cream together. Add both types of chopped nuts, salt and pepper and stir until well blended. Using your gloved fingers, stuff the habañeros and jalapeños with this mixture. Then roll the pastry dough out very thin, and use a Mason fruit jar lid to cut it into circles about 6" in diameter. Place one stuffed habañero and jalapeño apiece within each circle. Then with your gloved finger, apply a light coating of ice water to the edge of the dough, neatly fold over, and seal. Prick an air hole or two in the top of each tart with a toothpick.

Place the tarts on a greased cookie sheet and bake at 450° F. until the crust is golden brown (this takes about 12 to 16 minutes). About 7 minutes before they are finished, brush the top of each tart lightly with milk; feed any remaining milk to the neighborhood cats. Remove tarts from the oven and permit to cool for 3 hours before daring to serve. Makes about four dozen fiendishly delightful hors d'ouevres, which are better nibbled at by the timid and prudent, than wolfed down by the reckless and foolish. These tarts, indeed, live up to their name and are, pardon my English, quite "hot as hell"!

Increase Your Knowledge of the Capsicums

There are a number of items with information on chile peppers in print. I recommend just a few, however, which I believe bring together the best data in an enjoyable and comprehensible form. First, there is an article by Richard McCourt, "Some Like It Hot" in the August 1991 issue of *Discover* magazine (pp. 48–49). It neatly summarizes the scientific work with chiles. Then there are these two books for expanded information on the subject.

Jean Andrews, *Peppers: The Domesticated Capsicums* (Austin: University of Texas Press, 1984).

Amal Naj, *Peppers: A Story of Hot Pursuits* (New York: Alfred A. Knopf, 1992).

There is also a magazine devoted exclusively to these "volcanic vegetables": *Chile Pepper*, P.O. Box 4278, Albuquerque, New Mexico 87196.

One can also become a member of the Chile Institute by writing to Box 3003, Mexico State University, Las Cruces, New Mexico 88003, (505) 646-3028.

Persimmon
(Diospyros virginiana)

Brief Description

Most of the persimmons grown for the U.S. market are an Oriental type—the tomato-shaped, bright-orange fruit known as "kaki." Many imagine them to be extremely sour, but in fact these fruits can be quite palatable—somewhat astringent, but rich and sweet. The key is ripeness: commercially grown persimmons are picked and marketed unripe because of their extreme perishability; they must be held at room temperature until quite soft for the flavor to develop.

Persimmons come in two varieties: the acorn-shaped type is called the Hachiya, and the tomato-shaped variety, the Fuyu. With the Hachiya it's especially important to wait until the fruit is so ripe that the skin begins to wrinkle before you eat it. The Fuyu, however, is ready to eat when it's still firm.

Tea for Hangover, Seafood Poisoning

Raw persimmon when combined with a little horehound herb is excellent for relieving the throbbing headaches accompanying an alcoholic hangover. Both the fruit and the herb are also good as an antidote in tea form to reduce the symptoms of poisoning encountered when eating any kind of spoiled seafood such as sushi or oysters, for instance.

Bring 1 pint of water to a boil. Add 1/2 cup of coarsely chopped, unpeeled, ripe persimmon and 1 1/2 tbsp. fresh or dried, coarsely cut horehound herb. Cover with a lid, remove from the heat, and steep 40 minutes. Strain and drink the entire contents while still lukewarm.

Wonderful Astringent

Willie Lena, a Seminole town chief residing in Wewoka, Oklahoma, mentioned in 1983 that his tribe often made a tea of the fruit to stop diarrhea. Six near-ripe persimmons were cut into sections and steeped in 3 cups of boiling water, covered, for 20 minutes before straining. Drink 2 cups in a 4-hour period for chronic diarrhea. The fresh juice of the fruit may also be taken, but the tea seems to work better.

Other Native Americans of the early nineteenth-century employed persimmon juice for cleansing gangrenous leg ulcers and wounds, to stop bloody bowel discharges, and to wash out an infant's mouth to cure thrush, a yeast infection caused by *Candida albicans*. The previously mentioned tea could also be used as a vaginal douche to eradicate this yeast infection as well.

The juice of one ripe persimmon mixed with 3 1/2 tbsp. of warm water makes an excellent gargle for sore throat brought on by the common cold and influenza. In Thailand the ripe fruits are valuable for getting rid of intestinal worms, particularly hookworms.

A Pair of Delightful Recipes

Persimmon-Raspberry Yogurt Parfait

Needed: 2 ripe persimmons; 1 tbsp. brown sugar; 2 cups vanilla-flavored low-fat yogurt; 1 cup fresh or frozen raspberries, thawed; 1 cup low-fat granola without raisins.

Cut each persimmon into 4 wedges; peel these wedges, using the fingers or a small paring knife. Cut each wedge into 4 wedges, then set aside. Combine the brown sugar and yogurt in a small bowl; stir until well mixed. Then spoon 1/4 cup of the yogurt mixture into each of 4 (8 oz.) dessert glasses; top with 4 persimmon wedges, 2 tbsp. raspberries, and 2 tbsp. granola. Repeat the layers, ending with the granola. Serve these parfaits immediately; Yield: 4 servings.

Persimmon-Cilantro Salsa

Needed: 1/4 cup chopped fresh cilantro; 1 1/2 tbsp. minced red onion; 1 tbsp. fresh lime juice; 1 tsp. minced jalapeño pepper; 2 ripe persimmons, peeled and coarsely chopped.

Combine all ingredients in a bowl, and stir well. Cover and chill. Serve with pork, turkey, or duck. Yields 3/4 cup.

Petunia

(See under ORNAMENTAL FLOWERS)

Pineapple

(See under TROPICAL FRUITS)

Plums and Prunes
(Prunus domestica)

Brief Description

Plums are the most diverse and widely distributed of all stone fruits, with varieties suitable to almost any climatic condition; in fact, they are grown on every continent except Antarctica. Most commercially grown plums are descendants of either European or Japanese varieties. The European plums, oval or round in shape, include all the purple to black varieties (such as the El Dorado) as well as the smaller, greenish-yellow, richly flavored Green Gage. The small, bluish-black Damson is prized for jams and preserves. The larger, dark-purple Stanley is usually eaten as fresh fruit. The Japanese varieties are larger, with yellow to red skins and juicy flesh—like the crimson Santa Rosa. Native American varieties are seldom grown commercially, but they have been extensively hybridized with Japanese plums to increase their hardiness.

Prunes are the firm-fleshed variety of plums with a high enough sugar content to permit drying without fermentation around the pit.

Helps Heal Cold Sores

Irritating sores in the mouth like cold or canker sores may be effectively treated by taking 2 tbsp. of any fresh plum juice and swishing the same around inside for a couple of minutes before swallowing. Or for a particularly bad sore, soak a cotton wad with some fresh plum juice and then press against the sore and hold in place with either the tongue, jaw, or a little bit of peanut butter smeared to one side.

An Undeniably Fine Laxative

Prunes have long had the reputation of being an outstanding laxative agent for constipated bowels. A therapeutic confection can be made by combining 5–6 pitted prunes, the same number of fresh figs, 1 tsp. of coriander seeds, and a tablespoon or so of powdered psyllium seed in a food blender and puréeing until smooth and somewhat stiff in consistency. The amount of psyllium to be used can be adjusted according to the thickness desired for this confection. Portion off into equal tablespoonful amounts on individual pieces of cut wax paper before wrapping and refrigerating. These keep up to a month and several may be taken at one time for chronic constipation. Results usually appear within a couple of hours *or less*.

Testimonial from Jimmy Durante

The stimulating effects of prune juice upon even the most stubborn form of constipation are legendary among those who've come drink it whenever they need an effective bowel movement. One such testimonial came from none other than comedian Jimmy Durante who died in 1980 at age 87. His self-deprecating humor about his long, Pinocchio-like nose, which he cleverly referred to as "my schnozz in front of me," won over the hearts of millions of Americans in the heyday of his success.

But while that woeful face, with the prodigious nose and sandpaper voice, may have been one of a kind, the health problem he continually suffered from certainly was very common. Early on, he went to several doctors for treatment, but all they gave him were little pills to take. Finally, he got so exasperated with their medications, he flushed everything down the toilet. He told his head TV writer Charles Isaacs (from whom I got this story) one time: "One pill's for one thing, another pill is for something else; hell, I sometimes wonder how the damn pill knows where to go anyway."

It was sometime in 1950 while considering making the transition from stage to television that he really became uptight in more ways than one. Because TV was such a new medium then, he didn't fully trust it. And he had little enthusiasm for doing a variety series. "Gotta memorize all that stuff," he growled, "and worse than that, I gotta read it before I memorize it."

Gloria Swanson, a film star, suggested he start drinking prune juice to help solve his evacuation dilemma, telling him just how much it had helped with her own constipation. He decided to give it a shot and was very pleased to see just how quickly it worked. After that he never failed to have at least one if not two glasses of prune juice every day to keeps his bowels regular. With this routine he started doing squatting and stretching exercises, too. When asked why he faithfully adhered to his prune juice and exercise regimen, Isaac recalled that he quipped, “It keeps me goilish figger!”

Very Moving Recipes

My appreciation goes to the folks at Better Homes & Gardens for letting me use this nifty little dessert found in their *Fresh Fruit & Vegetable Recipes* booklet.

Plums Poached in Wine

Needed: 1/3 cup brown sugar; 1/2 cup dry red wine; 1 tbsp. each lemon and lime juice; 2 whole cloves; 2″ stick cinnamon; 8 pitted and quartered medium plums; some vanilla-flavored yogurt or plain yogurt to which a little real vanilla flavor has been added.

In a saucepan combine sugar and 1/2 cup water. Cook and stir till sugar dissolves. Stir in wine, citrus juices, cloves and stick cinnamon. Gently stir plums into wine mixture. Bring wine mixture to boiling point. Reduce heat and simmer, covered, for 2 minutes. Remove from heat, then remove cloves and stick cinnamon. Serve plums warm, spooning some of the wine mixture over plums. Top each serving with a dollop of yogurt. Serves 4.

Prune Whip

Soak 2 cups dried and softened prunes overnight in 1 quart warm distilled water and then simmer on low heat the next day for 30 minutes Then mash 1 cup of these cooked and stoned prunes through a coarse sieve. Next beat 2 egg whites very stiff, adding 1/4 cup of sifted brown sugar and a pinch of seasalt afterwards. Fold the mashed prunes into the beaten whites. Bake in an oiled casserole 25 minutes in a slow oven (300° F.). Serve hot or cold with pure maple syrup.

Prunes Steeped in Tea

Needed: 1 lb. unpitted prunes; 2 tbsp. honey; 2 tsp. lemon juice; 2 3″ strips lemon zest; 3 jasmine or black currant tea bags.

Place the prunes, honey, lemon juice, and lemon zest in a medium-sized suacepan. Add 1 cup water and bring the mixture to a boil. Remove the pan from the heat, add the tea bags and let steep for 3 minutes. Remove the tea bags and let the prunes stand for a few more minutes. Serves 4.

POMEGRANATE
(Punica granatum)

Brief Description

The pomegranate grows wild as a shrub in its native southern Asia and in hot areas of the world. Under cultivation, it's trained as a tree to grow to 20 feet in height, being grown in Asia, the Mediterranean region, South America and the southern states of the United States. The slender, often spiny-tipped branches bear opposite, oblong, or oval-lanceolate, shiny leaves about 1–2″ long. One to five large, red or orange-red flowers grow together on the tips of the shoots. The brownish-yellow to red fruit, about the size of an orange, is a thick-skinned, several-celled, many-seeded berry; each seed is surrounded by red, acid pulp.

Seeds Expel Tapeworm

Dried pomegranate seeds have been used since time immemorial for getting rid of tapeworm, as incredibly long parasitic host which attaches itself to the intestinal walls of its host by means of spined or sucking structures. Constant hunger, large amount of food intake, yet persistent relative thinness are the most common symptoms.

DRYING POMRGRANATE SEEDS

Dry the seeds from 7–9 pomegranates either in the sun or in a low-set oven on a cookie sheet for 7 hours. Then crush into powder with a hammer or other heavy object. Take 1 tbsp. of powdered seed in a 6 oz. glass of unsweetened pineapple juice 3–4 times daily on an *empty* stomach. If this is done in conjunction with a mild food fast, it might work quicker in getting the tapeworm out of the system.

Exciting Recipes for Smooth Eating Pleasure

WATERCRESS SALAD WITH TANGELOS AND POMEGRANATES

Needed: 5 tangelos; 2 tbsp. minced fresh basil; 1/4 tsp. pepper; 1/8 tsp. salt; 2 tbsp. red wine vinegar; 1 tbsp. walnut or sesame oil; 8 cups loosely packed trimmed watercress; 1 cup pomegranate seeds.

Grate 2 tbsp. rind from tangelos. Squeeze 1 tangelo to extract 3 tbsp. juice; discard tangelo. Then combine tangelo rind, tangelo juice, basil, and the next 4 ingredients in a large bowl. Cover and let stand 2 hours. Peel and section remaining 4 tangelos; set aside. Stir the tangelo juice mixture; add the watercress, and toss gently. Place 1 cup watercress mixture on each 8 salad plates; arrange the tangelo sections and pomegranate seeds on top of the watercress mixture. Serve immediately. Serves 8.

Fruit Medley with Pomegranate Seeds

Needed: 2 oranges; 1 cup seedless green grapes, halved; 2 kiwifruit, peeled and sliced; 1 banana, peeled and sliced diagonally; 1 cup pomegranate seeds; 2 tbsp. orange-flavored liqueur.

Peel the oranges; cut each one in half lengthwise. Cut each half crosswise into thin slices. Combine orange slices and next 4 ingredients in a bowl. Drizzle with an orange-flavored liqueur such as Grand Marnier. Serves 5.

Grapefruit Sorbet with Pomegranate Seeds

Needed: $^1/_4$ cup sugar; 1 $^1/_2$ cups water; 3 cups grapefruit juice; 1 cup pomegranate seeds.

Combine the sugar and water in a saucepan. Bring to a boil over medium heat; cook 5 minutes. Combine the sugar mixture and grapefruit juice. Pour this mixture into the freezer can of an ice cream freezer; freeze according to manufacturer's instructions. Spoon into a freezer-safe container; cover and freeze for at least an hour. Top each serving with pomegranate seeds. Serves 8.

(See under SEEDS)

POTATOES

White Potato (Solanum tuberosum)

Sweet Potato (Ipomea batatas)

Yams (Dioscorea sativa, D. alata)

Brief Description

White potatoes were first cultivated by South American Incan Indians in the high Andes, and later taken to England in 1586 by Sir Francis Drake. From there their cultivation spread to Ireland, continental Europe, and finally, in 1719 to the American colonies.

Modern potatoes, the best of which are a far cry from the small, floury originals, fall into three general groups. New potatoes are the tender, thin-skinned ones usually harvested during late winter and early spring; they are used for boiling, creaming and potato salads. All-purpose potatoes like the Pontiac (red) can be boiled, mashed, baked, or fried. And the famous Idaho, or russet, is a popular baking potato.

Yams and sweet potatoes are often confused. Both are edible tubers, but they are from different plant families. Yams probably originated in West Africa, whereas sweet potatoes, like squash, are native New World vegetables. Sweet potatoes come in many varieties but are of two basic types: the dry fleshed, with rather mealy, pale-yellow flesh and the moist fleshed, with deep-yellow to orange-red flesh (often incorrectly called a yam).

The common name (*D. sativa*) and the 10 months yam (*D. alata*) are both widely cultivated throughout the South Pacific and have been known to reach weights of up to 100 lb. They can be baked, boiled, roasted, and fried or used raw in salads.

Potato Plaster for Relieving Pain and Inflammation

In 1987, I met a young pharmacy student from Nova Scotia, who had grown up in the potato country in nearby Maine. Over lunch one afternoon he shared with me some folk data passed on to his family by his late great-grandmother.

According to him, she made special potato plasters for reducing inflammations caused by bruises, sprains, burns, fractures, hemorrhoids, abscesses, appendicitis, arthritis, neuralgia, and eczema.

To make her special plaster, she would peel and grate ordinary potatoes and mix half of them with an equal amount of green vegetable leaves (cabbage, radish or spinach) that had been coarsely puréed in a food blender. To this wet mass was added about 10% white flour. Everything was then thoroughly mixed in a large pan by hand. Just enough ice-cold water (*never* warm) was slowly added to give the paste a wet, somewhat even, and thick consistency without being runny or lumpy.

His grandma then applied this potato plaster directly to the skin. On top of it she would place a clean cloth and secure it with a lengthy swath of bandage fashioned from strips of old bedsheets. She would require that the person being treated stay in a reclining position during the time that this plaster remained on, which usually averaged about 3 $^{1}/_{2}$ hours.

Quite often the plaster would have dried out, so she would apply some warm water on the dry mass until it became moist again, permitting easier removal without much pain and discomfort to the patient. Once the plaster was off, the skin would then be rinsed with some warm water.

Sometimes, he said, she might rub a little olive oil on the area to be treated to prevent or reduce any chronic itching which might take place while the plaster was on the skin.

He noted that she also used the same potato plasters for drawing out purulent matter from boils, abscesses, infected acne, carbuncles, infected cysts, and various types of tumors (benign, fibroid, and even malignant). In a couple of instances, the women she treated for breast tumors, he observed, had most of the cancerous matter gathered toward the surface from further inside the body, thanks to her potato plasters. These women then had surgeries to remove these malignant accumulations, but, interestingly enough, the doctors didn't have to operate as much as they might have done had these women never been treated by her first of all.

Maker of Tater Tots Had "Perfect" Gout Remedy

The man who put frozen french fries in American kitchens and Tater Tots in toddlers' mouths once shared with me what he terms his "perfect" remedy for gout. F. Nephi Grigg founded Ore-Ida Foods after buying a bankrupt frozen-food factory and discovering he needed vegetables year-round to keep his plant going.

Potatoes seemed the perfect vegetable plant for a plant in Idaho. By tinkering with spuds and preservatives, he discovered he could blanch raw fries in sugar water, and they would stay frozen perfectly until cooks popped them in the hot oil or the oven to cook. Thus, was the invention of frozen french fries born. They became an instant hit with the American public.

But the peeling and carving process to form the french fries created an interesting by-product—bits and pieces of potato. At first they were fed to cattle, but this became too expensive. So Grigg took the bits and pieces, added flour and seasoning, and pushed it all through a piece of 3/4" plywood with holes in it. Out the other side came his next invention, Tater Tots.

Grigg told me that whenever his gout would act up, he would simply take a plain, raw unpeeled potato, quarter it, boil it in 7 cups of water for 30 minutes, cool, strain the liquid, and drink several cups at a time. The pain subsided within minutes, he claimed. (Grigg died at age 81 in early January 1995 in Salt Lake City, Utah.)

Lose Weight Eating Potatoes

Potatoes themselves are not fattening. It's the toppings that have earned for them the undeserved distinction of being fattening. Common sense tells you that a potato dripping with butter, sour cream or rich gravy is heavy in calories and bound to add more inches to the waistline. But a plain spud is no more fattening than eating a pear—the potato itself is 99.9% fat free, with a whole half pound of baked potato containing *fewer than* 250 calories!

Nutritionists at the Home Economics Department of Douglas College, New Brunswick, New Jersey, demonstrated the potato can be included in a reducing diet with good results. In a carefully controlled study, students who followed a food plan containing potatoes lost an average of 14 pounds in 8 weeks—an ideal amount to lose in that period of time.

A Creole Grandmother's Remedies

A very old Creole lady by the name of Clothilde Rousseau gave me a number of very effective remedies for an assortment of health problems utilizing white potatoes.

"Back in the early 1930s I suffered from eye-strain. I washed one potato and then cut 6 slices the size of a quarter from it. These I laid on my eyes, 3 on either side, and tied a strip of cloth around my head to hold them in place. They were soothing and took away the pain and inflammation.

"For people who had neuralgia, I treated them by laying a baked potato against the side of their face or neck where it hurt them the most, and covered it with a towel to keep the heat in as long as possible. It worked everytime!

"For black-and-blue marks I found that potatoes worked a lot better than raw beefsteak ever did. I'd just make a simple poultice of grated raw potato and apply it to any bruises on the skin. After leaving it there for an hour or so the discoloration and tenderness usually had gone down quite a bit.

"When I was about 5 years old I suddenly came down with a bad case of gallstones. At first my momma thought it might be appendicitis 'cause it hurt my side so bad. But she consulted with an old Cajun healer who told her to peel some potatoes and make a broth of the peelings. This she did and gave me 4 to 5 cups a day for about a week. She usually checked my toilet waste and noticed that a number of the stones had started to come out. Pretty soon I was healed and never have been troubled with them since. I've used this remedy myself on many others who've been bothered with the same problem over the years and never knew it to once fail them.

"Down through the years I've had folks referred to me who couldn't always handle our spicy food. In every instance, I'd have the person just take a small piece of raw potato no bigger than my thumb and chew on it good before swallowing it. Relief generally came to most in a minute or two. For those folks who had serious ulcers I'd have them drink a cup of raw potato juice in warm water first thing every morning.

"A lady who lived next door to me once came over screaming for help because her youngin' had just poured boiling water all over himself, scalding his skin pretty good. I went over there every day for nearly a month and applied lots of grated raw potato wherever he was burned. He managed to come out of it okay and never had any signs of scars that I know if.

"Potatoes have the greatest drawing power to them that I ever saw in a vegetable. I've used potato slices on infected sores to help draw out the pus and infection when nothing else seemed to work. I doubt you'll find anything that works better for abscesses and wounds than this. It really helps to get out the rotten stuff so they can heal faster."

Potatoes Reduce Heart Strokes

In Ireland men consume more than 170% more potaotes than their Irish-American counterparts do. Irish men ate 267 grams of starch daily (mostly from spuds), while second-generation Irish-Americans in Massachusetts consumed only 116 grams of starch daily (with only a small portion coming from potatoes). Statistics indicated that only 29% of all deaths in Ireland for the late 1960s in men aged 45–64 years were the result of ischemic heart disease, compared with a whopping 42% of all deaths for the same period in second-generation Boston men.

An analysis of fresh potatoes showed that they rank as the richest source of vitamin C out of a number of fresh vegetables examined. Researchers concluded that such vitamin C–rich foods offer some protection against coronary heart disease. A final study reported that large amounts of potato given in experimental diets significantly reduced serum cholesterol levels in humans. And Dr. Elizabeth Barrett-Connor of the University of California at San Diego found that men and women out of a group of 859 between the ages of 50–79 years, with low-potassium diets were 3–5 times as likely to die after a stroke than those who had higher potassium diets. Among foods recommended to cut strokes significantly was "a small potato with 350 milligrams of potassium." Thus it would appear that potatoes are beneficial for the heart when consumed without fat and either baked or boiled.

Starch May Prevent Suicide

Suicide is often in the news these days, particularly among our young people. A San Diego State University history professor, Howard Kushner, is inclined to think that societies which consume significant carbohydrates have much lower incidence of suicide than those cultures which do not. In a paper published in the Summer 1985 issue of the *Journal of Interdisciplinary History*, Dr. Kushner postulates that potatoes and pasta could be keeping untold numbers of Irish and Italians from killing themselves, whereas the Danes, Germans and Austrians, who consume less of these carbohydrates have higher suicide rates as a result.

Swedish and American experiments included extracting spinal fluid from 30 volunteers who had attempted suicide in order to determine the level of an important brain chemical called serotonin. A large number had significantly lower levels of serotonin than did a control group of normal individuals. Furthermore, those with less serotonin were more violence-prone too.

Serotonin is a protein that transports messages to brain cells. It is produced in the body by the amino acid tryptophan, which must be supplied by diet. But tryptophan encounters stiff competition from other amino acids when it attempts to cross the barrier from the bloodstream into the brain. However, when carbohydrates like potatoes or pasta are consumed, insulin is secreted, and this then lowers the level of competing amino acids, so that more tryptophan can get through. Once it reaches the area of the brain, sufficient serotonin can be produced, which helps to lower depression and anxiety somewhat.

So it seems relatively safe to say that potatoes should figure frequently in the diets of those who may be experiencing mental and emotional instabilities of some sort as a reasonable precaution against potential suicide.

More Rapid and Frequent Bowel Movements

According to the September 1977 *Irish Journal of Medical Science*, frequent consumption of *un*peeled spuds can do wonders for cleansing the intestines. Twenty-five of 48 volunteers maintained a daily intake of potatoes approximating to 2.2 lb. for a minimum of 10 weeks and maximum of 20 weeks. The average consumption was slightly under 2 lb. and contained an estimated $^1/_7$ th of an ounce of crude fiber. A significant decrease in the length of time it took food to go through the intestinal tract and a reduction in colon-rectal pressures was evident. A significant increase in stool weights was also noted as well.

Yams Remove Heavy Metals from Body

Yams and sweet potatoes contain simple peptide substances called phytochelatins that can bind heavy metals like cadmium, copper, mercury and lead and thus participate in metal detoxification of body tissue. These metal-chelating compounds interact with the mineral sulphur to achieve this. In light of this it's interesting to note that doctors in the Soviet Union include potatoes, yams, and cabbage (all sulphur-rich vegetables) in a special prophylactic diet for factory workers constantly exposed to toxic chemical occupational environments. When one considers that most of our big cities have a terrible smog problem, it only seems prudent for us to eat more of these three vegetables ourselves so that we don't accumulate and become ill from the heavy metals in the air we breathe each day.

Sweet potato and yam are also good for helping to remove any kind of foreign object accidentally swallowed by a small child or a mentally retarded adult (such as safety pins, needles, coins, thumbtacks, and the like). Feeding the individual some boiled, baked, or steamed sweet potato or yam, will help to coat the object and safely expel it from the body through the colon later on. Ripe, mashed bananas work just as well, I might add.

Cooking Creatively

The following recipes are simple and easy to prepare. They offer some "quickie" meals if served alone that are nutritious and good tasting as well.

O'BRIEN POTATOES

Start with Potatoes Lyonnaise, using sliced ($^1/_8$) or diced, medium-sized boiled potatoes, peeled, browned in melted butter, combined with sautéed onion and chopped parsley and seasoned to taste. When the potatoes are ready, pop in 1/4 cup chopped green pepper, 1/4 cup chopped pimento, and 1/4 cup heavy cream. Cook the cream down over medium heat. Blend by shaking pan as the dish cooks. Serve in a hot dish.

Apricot Sweet Potato

Needed: 1 baked sweet potato; 4 chopped, dried apricot halves; 2 tsp. maple syrup; 1/2 tsp. lemon juice; 1/2 tsp. lime juice; 2 tbsp. applesauce; some toasted almonds to garnish (optional).

Cut about 1/3 off the top of the cooked potato. Scoop out pulp. Reserve shell and keep warm. Mix potato pulp with remaining ingredients. Stuff into the reserved potato shell. Serve hot.

Potato Latkes

Needed: 3 tsp. canola oil; about 5 Idaho russet potatoes, peeled; 3/4 cup finely chopped red onion; 1/4 cup all-purpose white flour; 1 tsp. salt; 1/4 tsp. freshly ground black pepper; 1 large egg, lightly beaten; 1 large egg white, lightly beaten.

Set oven racks at middle and lower positions; preheat the oven to 450° F. Prepare 2 baking sheets by brushing each one with 1 tsp. of the canola oil. Using a hand grater, grate each potato. Place them in a large bowl and add the onions, flour, salt, and pepper; toss with 2 forks to mix well. Then add the egg, egg white, and the remaining 1 tsp. oil and toss to mix. Drop rounded tablespoonfuls of the potato mixture onto the prepared baking sheets and press lightly to form cakes. Bake for 12 minutes, or until they are golden brown on the bottom. Then flip the latkes over, switch the position of each baking sheet in the oven, and continue baking for another 6 minutes, or until golden brown. After they're done, transfer in a platter, arranging the latkes browned-side up, and serve. Latkes may be prepared and stored, covered, in the refrigerator overnight, if desired. Reheat them at 350° F. for 12 minutes. Makes roughly two dozen latkes.

Garlic Mashed Potatoes

What could be more satisfying than mashed potatoes? The challenge lies in making them rich and creamy without an excess of butter and cream. In this version, the potatoes are flavored with poached garlic, thinned with chicken stock, and enriched with a small amount of reduced-fat sour cream. This recipe can easily be doubled or tripled.

Needed: 5 Pontiac (red) potatoes, unpeeled and cut into chunks; 6 cloves garlic, peeled; 1 tsp. salt; $^2/_3$ cup defatted low-sodium chicken stock, heated; 2 tbsp. low-fat sour cream; pinches of black pepper and powdered nutmeg to taste.

Place the potatoes and garlic in a large saucepan and cover with cold water. Add 1 tsp. salt and bring to a boil. Cook, covered, over medium heat until the potatoes are tender, about 10 minutes. Drain the potatoes and return them to the pan. Shake the pan over low heat to dry the potatoes slightly. Remove the pan from the heat. Next mash the potatoes with a potato masher or hand-held electric mixer (don't use a food processor for this). Add enough hot chicken stock to make a smooth purée. Stir in sour cream and season with salt, pepper, and nutmeg. Serves 4.

Sweet Potato Pudding

Needed: 4 cups mashed cooked sweet potato; 1 $^1/_2$ cups sugar; 2 tsp. ground cinnamon; 2 tsp. grated orange rind; 1 tsp. salt; 1 tsp. ground ginger; $^1/_2$ tsp. ground cloves; 1 cup beaten egg; 2 (12 oz.) cans evaporated skimmed milk; vegetable cooking spray.

Combine sweet potato and the next 7 ingredients in a large bowl; beat at medium speed of a mixer until smooth. Add the milk and mix well. Pour the sweet potato mixture into a 2 quart casserole coated with some of the cooking spray. Bake at 375° F. for an hour or until a knife inserted near the center comes out clean. Let the pudding cool; cover and chill for a couple of hours. Serves 12.

Curried Sweet Potatoes

Needed: 8 medium sweet potatoes, peeled and cut into 1″ pieces; 1 tsp. salt to taste; 1 cup loosely packed dried apricots, cut into $^1/_4$″ slivers; $^1/_2$ cup raisins; 1 tbsp. canola oil; 1 medium onion, finely chopped; 2 tsp. mild curry powder, preferably Madras; freshly ground black pepper to taste.

Place the sweet potatoes in a large pot and add just enough cold water to cover by an inch. Add the salt and bring contents to a boil over high heat. Reduce the heat to medium and cook, uncovered, until tender but not mushy, maybe 10 minutes. Drain well. Meanwhile, in a small bowl, combine the apricots, raisins, and 1 cup boiling water; let sit until plumped up, for about 10 minutes. In a large wide pot, heat the oil over medium-high heat. Add the onions and cook, stirring often, until softened, for a couple of minutes. Then add the curry powder and cook, stirring, until fragrant, for another few minutes. Next add the cooked sweet potatoes, apricots, raisins, and the fruit-soaking liquid. Season with the salt and pepper. Stir gently over medium-low heat until warmed through. Serves 10.

PRUNES

(See under PLUMS)

PUMPKIN, SQUASH, AND GOURD

Pumpkin (Cucurbita pepo)
Squash (Summer) Cucurbita pepo
Squash (Winter) Cucurbita maxima
Squash (Zucchini) Cucurbita pepo
Gourd (Bottle) Lagenaria vulgaris

Brief Description

Pumpkin is the fabled jack-o-lantern of Halloween tradition and that very same object which the headless horseman threw at a frightened and fleeing Ichabod Crane with great fury and deadly accuracy, in the immortal tale by Washington Irving. Pumpkin is a variety of winter squash recognized by its smooth, round shape and hard-ribbed, orange-colored rind. For cooking purposes, the small sugar pumpkins averaging 7 lb. or so are best. But for scaring the wits out of young kids, varieties like the Big Max weighing 100 lb. or more are hard to beat.

Squashes originated in the New World and were introduced to the conquistadors by early Native Americans, who in turn carried these food plants back to Europe with them. Squash is divided into two basic groups: the quick-growing, tender-skinned "summer" squashes, which are harvested immature, and the larger, slower growing, hard-shelled "winter" squashes, harvested when fully mature. Summer squashes like yellow crooknecks, pattypans, and zucchini, are consumed whole. But winter squashes such as Hubbards, butternuts, acorns, and sugar pumpkins have inedible skins which must first be removed before they can be eaten. However, they are usually tastier and more nutritious than the summer varieties.

The name pumpkin goes back to the Greek word "pepon," meaning ripe or mellow. In time the early French had it down to "poupon" and, having been nasalized into "poumpon," entered the King's English as "pompion," to which was later added the diminutive "kin" ending. Squash, on the other hand, is derived solely from the Algonquin Indian word "askoot asquash," which translated means "eaten green."

There are many kinds of gourds, serving a variety of uses, from domestic to ornamental; including inedible varieties that look like waxy, warted apples, lemons, and tangerines, all the fascinating produce of the same plant. During classical times and the Middle Ages a woody rind edible gourd, probably one of the bottle gourds, was used as bottles and cooking and eating implements.

Cooling Effect on Inflammation

A number of remedies have already been given in this book for sunburn and burns in general. One of the quickest things I've ever found for immediate relief from the intense pain is to cover the burn with ice-cold mashed pumpkin, either freshly cooked or from the can, providing both have been refrigerated overnight. This handy remedy was first brought to my attention by an Indian woman named Sally "Big Thighs" Henderson, who resided on the Navajo Reservation in northern Arizona. That, by the way, is her honest-to-goodness name!

Tea for Any Kidney Problem

Among the Cherokee Indians, pumpkin and squash seeds were valuable for treating edema, gout, kidney stones, urinary burning, and difficult urination. A handful of the seeds were crushed and then added to a quart of boiling water, covered, and simmered on low heat for about 20 minutes, then permitted to steep away from the heat for an additional half an hour. Several cups of the strained liquid were drunk each day as needed until the desired relief was obtained.

Leaves Reduce Sprains

A very useful application of the leaves of pumpkin, squash or gourd may be found in the treatment of sprains, bruises, torn ligaments, and the like. Certain herbal practitioners in Jamaica, who also dabble in black magic and voodoo, will often utilize the fresh leaves for these purposes. The picked leaves are first pounded with a hammer or small round stone in order to macerate them a little, before being bound on a sprain or dislocation of some sort. They help to take the swelling down quite a bit and seem to hasten the healing process when other internal remedies for inflammation are used.

Recently several people have benefited from a variation of this same remedy that I've recommended to them for minor fractures, lower backache and tendonitis. First, some special chamomile cream called CamoCare was gently rubbed on the injured areas, after which either a pumpkin or squash leaf poultice was then applied and left for several hours with very good results. The CamoCare cream is available from most health food stores or may be obtained from Abkit, Inc., of New York City.

Fevers and Diarrhea Go Away

Other very popular uses for the leaves is to reduce fevers and stop diarrhea.

Some Jamaican voodoo witchdoctors will make a tea from pumpkin or squash leaves or use equal parts of both. About 2 quarts of water are brought to a boil, after which a couple of double-handfuls of leaves are snipped into the pot with a pair of scissors or shears. The pot is then covered, removed from the flame, and permitted to steep for about 35 minutes. A cup at a time is administered every couple of hours until the fever or diarrhea stops.

Keeping the Prostate in Great Shape

As men get older, body organs tend to wear out. One of the most frequent of these to break down is the prostate gland, which often requires surgery to correct. But now there is a much easier and safer way to keep this organ functioning as it should. A homemade syrup can be taken each day to assure a man past his prime that his prostate won't fail him later on in life.

First, shell and bruise with a hammer 6 tbsp. each of 6 ripe pumpkin and squash seeds. Then put them through a meat grinder so as to render finely ground. Next mixed in equal parts of blackstrap molasses and dark honey or just enough of both to form a nice, thick syrup. Finally, flavor with a little powdered cardamom and cinnamon and the juice from half a lemon. Take 1 tbsp. every morning before breakfast or at least three times a week for sure. This same syrup, I might add, is also another excellent way to take the seeds for expelling tapeworms and roundworms.

Migraines and Earaches Disappear

In certain parts of India and Europe the scraped pulp of fresh pumpkin or yellow and orange squash is applied to the forehead and temples as a cooling application to relieve intensely splitting headaches. And the same grated pulp is also applied to the sides of the face, neck and throat to relieve neuralgia or to draw out the purulent matter in ripe boils. This same pulp is likewise excellent for burns.

Throughout the Philippines, native practitioners often squeeze the sap from fresh pumpkin stems into their patients' ears to help relieve earaches. A friend of mine tried this once when he got water into his inner ear after swimming. About an hour later he reported the misery had cleared up.

Possible Cancer Preventative

A symposium on nutrition and cancer was held at the University of Adelaide in Australia in November 1978. One of the featured speakers was Dr. Takeshi Hirayama of the National Cancer Research Institute in Tokyo, Japan. In his address, Dr. Hirayama revealed his research findings in his remarks on green-yellow vegetables. He consistently found *lower* risks of many kinds of cancer in those who frequently consumed pumpkin, squashes, carrots, broccoli, and green bell peppers than in those who ate below-average amounts of them. Highlights of his interesting paper were reported in the March 10, 1979, issue of *The Medical Journal of Australia*.

An Effective Cure for Tapeworms

Speaking from personal experience, I can attest to the benefits of either pumpkin or squash seeds for getting rid of tapeworm. Even now I can recall having to chew very thoroughly each day for 5 1/2 days a cup of dried pumpkin seeds from the health food store when I was just 13 years old.

For several years prior to this, I had been eating voraciously, but never gained a single pound of weight. At first everyone thought it was just the "growing boy" syndrome, which every tall, lanky kid goes through during his teenage years. In time, however, several naturopathic doctors who examined me (my folks never believed in regular M.D.s) confirmed that it was a severe case of tapeworm, which was robbing me of the proper nourishment I should have been getting from the tons of food I was shoveling down.

Well, they recommended the pumpkin seeds, which apparently worked quite well within just a short period of time. A succession of different bowel movements on my part discharged chunks and sections of what had been a pretty long parasitic worm attached to the walls of my intestine. Various estimates were made as to its overall length, ranging from 20 feet to well over 45 feet. For myself, however, I never bothered keeping track of these statistics. I was just glad to get the ordeal over with and back to a more normal diet. In the course of time I filled out very nicely. I would recommend grinding the seeds up into a powder and serving them 1/2 cup of powder at a time in 1 1/2 cups of apple sauce, or mixed in with some carrot juice in a food blender as a vegetable shake in order to make them more palatable. Some have even suggested making a tea out of the seeds, but I have never found this to work as effectively as taking the seeds straight.

A folk healer friend of mine from Merida on the Yucatan Peninsula shared with me some years back on old Mayan remedy for expelling intestinal worms of any kind that he had found never failed once in those of his patients to whom he had prescribed it in times past.

On an empty stomach 2 tbsp. of castor oil were first ingested. The next day $^1/_2$ cup of shelled and powdered seeds of pumpkin or squash mix with a little water were taken, followed by 1 cup of goat's milk. Then some two hours later, another 2 tbsp. of castor oil were ingested.

Helps Heal Scratches and Minor Wounds

The Zuni Indians of Arizona have long relied upon the seeds and flowers of pumpkin and squashes for healing cactus scratches and minor wounds. The dried seeds are first ground into a fine powder, after which freshly picked blossoms are gently crushed and added. Just enough water is then mixed in with them to make a smooth, even paste, which is then applied directly on any kind of skin injuries with good results.

WILLARD SCOTT'S LOVE AFFAIR WITH SQUASH

America's favorite meteorologist, Willard Scott of NBC-TV's "Today" show, considers squash one of his favorite vegetables. This lovable and wacky weatherman is also pretty adept at cooking as well. According to Scott, one of the tastiest zucchini dishes he's ever tried is a casserole made with peppers, celery, and cream. "You put that on crackers, and mm-mm," the forecaster rhapsodized.

Here is a slightly modified version of his scrumptious apple-squash bran muffins.

Needed: 1 $^1/_2$ cups All-Bran cereal; 1 $^1/_2$ to 1 $^2/_3$ cups canned goat milk; $^1/_3$ cup corn oil; 1 large egg; 1 $^1/_4$ cups all-purpose flour; $^1/_2$ cup brown sugar; 3 tsp. baking powder; $^1/_2$ tsp. salt; 1 apple, peeled, cored, and diced; $^1/_2$ medium acorn squash, peeled, prebaked for 20 minutes and diced; 1 $^1/_2$ handfuls of raisins; $^1/_2$ handful chopped, pitted dates; $^1/_2$ handful finely chopped walnut meats; 1 tsp. ground cinnamon to taste.

Beat together cereal, milk, oil, and egg with an electric mixer. In a separate bowl, mix flour, sugar, baking powder, and salt. Add dry ingredients to bran mixture and stir until well-mixed. Then stir in apple, squash, raisins, dates, walnuts, and cinnamon. Pour into greased muffin tins and bake at 400° F. almost half an hour or until the muffins have risen and are golden. Makes a dozen delicious muffins.

A Pair of Fantastic Squash Soups

Pumpkins and squash make great soups. But from my own experience, squash makes the better version because it adds a depth of enhancing flavor that pumpkin can't quite seem to match. Here are two delicious but very different squash soup recipes. The first is a basic American preparation; the second one has unmistakable Middle Eastern characteristics.

CREAMY SQUASH SOUP

Needed: 2 cups chopped, peeled potatoes; 2 cups chopped, peeled acorn squash; $^1/_2$ cup chopped onion; 1 $^1/_2$ tsp. snipped fresh marjoram or $^1/_2$ tsp. crushed, dried marjoram; $^3/_4$ tsp. instant chicken bouillon granules; 1 minced clove garlic; 2 cups milk.

In a large saucepan combine spuds, squash, onion, marjoram, bouillon granules, garlic, 1 cup water and $^1/_8$ tsp. pepper. Bring to boiling. Reduce heat and simmer, covered, about 20 minutes or until vegetables are tender. Transfer about half of the vegetable mixture to a blender container or food processor bowl. Cover and blend or process until smooth. Repeat with remaining vegetable mixture. Return all to saucepan. Stir in milk and heat through, but *do not* boil. Lightly season with a dash of cardamom just before serving. Makes 10 servings.

MIDDLE EASTERN SQUASH SOUP

Needed: 2 medium onions, diced; 3 garlic cloves, minced; 1 medium carrot, diced; 1 rib celery, diced; 1 tsp. whole cumin seed; $^1/_4$ tsp. whole cardamom seed; 1 tbsp. olive oil; $^1/_2$ cup Marsala wine; 3 cups vegetable stock; 3 cups squash puree; $^1/_2$ cup cream; $^1/_2$ tsp. Madras curry powder; $^1/_4$ tsp. garam masala; granulated kelp to taste. Note: Garam masala is the Indian equivalent of curry powder. It is a premixed powder consisting of several spicy ingredients. A typical garam masala might include 2 oz. ground coriander seed, 2 oz. ground chile pepper, and a pinch of ground black pepper. Granulated kelp is a seaweed and can be obtained in most health food stores.

Sauté the first half-dozen ingredients in the olive oil. Then add the wine to deglaze. Add the stock and squash purée and bring to simmer over medium heat. Then, add the cream and spices. Simmer for 5 minutes and serve. Makes 7 1/2 cups.

"No-Bake" Squash Pie for Certain Food Allergies

Imagine making a pie that requires no baking in the oven *and* is free of dairy and egg products. For those allergic to such foods, this will come as a real taste treat and surprise!

VEGAN SQUASH PIE

Needed: 1/2 cup dried currants; 1/2 cup dates, chopped; 1/4 cup hazelnut liqueur; 1 cup coconut milk; 1/2 cup brown rice syrup; 1 tsp. cornstarch; 1 tbsp. agar; 10 oz. package extra firm silken tofu; 1/4 cup squash puree; 1/2 tsp. cinnamon; 1/4 tsp. cardamom; 1/4 tsp. nutmeg.

Simmer the first 7 ingredients together in stainless steel pan on the stove for 10 minutes. Then blend with the remaining ingredients. Pour into a prepared pie crust. Refrigerate for about 3 1/2 hours. Serve 6–8 hungry people.

A Wonderful Native American Bread

The Mohicans (also spelled Mohegan) were once a powerful North American Indian tribe residing in the states of Connecticut and New York who became immortalized in James Fenimore Cooper's classic novel, *Last of the Mohicans*. Although they were wiped out as a nation through the cruelty of the white man, several of their unique recipes survive in pioneer diaries and journals kept during that period. Here is one of them, which has been slightly modified for twentieth-century cooking methods.

MOHICAN PUMPKIN-SQUASH BREAD

Needed: 3/4 cups each of chopped, peeled banana squash and pumpkin; 1/4 cup olive oil; 1 1/2 cups all-purpose flour; 1 tsp. ground cinnamon; 1/2 tsp. baking soda; 1/2 tsp. cardamom; 1/4 tsp. ground cloves; 2 slightly beaten egg whites; $1/2$ cup blackstrap molasses; $1/4$ cup pure maple syrup; $1/4$ cup brown sugar; $1/4$ cup chopped, pitted dates; some liquid lecithin from any local health food store.

Put squash and pumpkin in a large saucepan, with about $2/3$" of water. Cook, covered, for about 25 minutes or until sufficiently tender. Then drain. Place squash and pumpkin with olive oil in a food blender and mix until smooth and even in consistency. Then stir together the flour, cinnamon, baking soda, cardamom, and cloves. In a medium mixing bowl stir together the egg whites, molasses, maple syrup, sugar, and squash-pumpkin mixture. Then add flour mixture. Next fold in the dates. Thoroughly rub the insides of an 8" × 4" × 2" loaf pan with liquid lecithin. Transfer the batter to this prepared pan. Bake in a 350° F. oven for 50 minutes or until sufficiently done. Remove from the pan and cool thoroughly on a wire rack.

One loaf generally yields a little over a dozen servings.

PURSLANE
(Portulaca olerocea)

Brief Description

Purslane is a very common weed found on moist sandy soils, but there are garden varieties which have been developed by French gardeners for salad purposes. These varieties are more popular in Europe than in this country. The small plant has fleshy, watery stems that are reddish in appearance and crisp, succulent, smooth, fleshy rounded leaves which are the size of squirrels' ears. The plant grows flat on the ground unless crowded. It produces very small, white to pinkish flowers. The improved varieties tend to have bigger leaves and usually grow more upright than the wild forms. The flavor is rather mild and faintly acidic.

The scarcity of purslane in the United States and Canada makes it unlikely to ever be used very much as a salad or cooked vegetable, which is a pity as it was a favorite summer dish of our ancestors. Where it becomes available it should be used right away while still crisp and fresh. It is a good source of iron, but also, like sorrel and rhubarb, has a high oxalic acid content and shouldn't be eaten too frequently. Oxalic acid inhibits the absorption of calcium and magnesium.

Urinary Problems Cleared Up

King Henry VIII (1491–1547) of Great Britain was known for his famous gluttony and the number of wives he violently dispatched into the next world when they ceased to please him. Lesser known, though, was his keen amateur interest in botanical medicine. According to medical historian Charles Lawall in his *Four Thousands Years of Pharmacy* (Philadelphia: J.B. Lippincott, 1927; p. 210), Henry was "a great dabbler in physic and offered medical advice on all occasions which presented themselves . . . ." He found the actual preparation and compounding of individual formulas particularly fascinating, and undoubtedly had his own royal set of apothecary's equipment which accompanied him in his travels.

A manuscript I've examined in the British Museum, written in beautiful Tudor script, records a royal collection of 114 favorite recipes for "plastres" and "cataplasmes" for "balmes . . . waters, lotions and decoctions." Thirty-two of these are believed to have been composed by the king himself.

Many of the medicaments devised by Henry must have been for his own personal use. His strenuous sexual life seems to have brought its own special problems. One decoction, in particular, caught my attention. It called for the use of purslane "to coole and comfort the Kinges membre hee pissethe." In other words, Henry had apparently devised for himself a remedy using this particular herb which greatly alleviated "the payne of strangurye" or the slow and difficult discharge of urine due to spasm of the urethra and bladder.

The equivalent of a handful of cut fresh purslane leaves were put into what would amount to about 1 $^1/_2$ pints of boiling water, covered, and allowed to steep for 30 minutes. This tea was then "strayned" through a piece of muslin cloth and taken warm in several cupfuls at a time. It was reported that every time Henry resorted to this, "the Kinges highnes had his payne and heat taken awaye" whenever he urinated.

Relief for Gout

Another of Henry's special formulas involved a tea intended "to ease the payne and swelling about the ankles." Because of his excessive eating and drinking, Henry suffered from gout. But he found great relief by drinking a tea made from combining equal parts of purslane and chickweed herb together. Directions for making it have already been given in the previous example. Henry would drink several cupfuls of this tea about an hour following one of his traditional heavy meals.

In the event that purslane isn't readily obtainable, you can easily substitute domestic lettuce, raw cabbage, raw spinach, dandelion, or chickweed.

Recipes with an International Flavor

Greek Salad with Purslane and Spaghetti

Needed: 6 oz. spaghetti (a bunch about 2″ in diameter); 2 cups purslane flowers, leaves, and stems, washed, thoroughly dried with a heavy paper towel, and chopped medium fine; 1 medium clove garlic, finely chopped; 1 cup feta cheese in 1/2″ cubes; 12 Greek olives, pitted and sliced; 4 tbsp. extra virgin olive oil; 3 tbsp. lime juice; 1/2 tsp. basil; 1/2 tsp. oregano; 1/2 tsp. salt; granulated kelp.

Cook the spaghetti, rinse under cold water, then drain and cool. In a large bowl combine the spaghetti, purslane, garlic, feta cheese, and Greek olives. In a smaller bowl combine the olive oil, lime juice, basil, oregano, salt, and kelp. Pour over the spaghetti and purslane. Mix. Serves 4.

Spanish Purslane Soup

Needed: 1 cucumber, peeled and coarsely chopped; 1 1/2 cups purslane, washed; 2 medium tomatoes peeled, and coarsely chopped; 1 medium onion, coarsely chopped; 1 small green bell pepper, washed, and coarsely chopped (leave seed pod intact); 1 large clove garlic, coarsely chopped; 1 cup tomato and mixed vegetable juice; 1/4 cup extra virgin olive oil; 1/4 cup red wine vinegar; 1/4 tsp. granulated kelp.

Needed: (for the croutons): 6 tbsp. extra virgin olive oil; 1 large garlic clove, finely chopped, 1/4 tsp. each oregano, basil, and granulated kelp; 6 cups French-style white bread cut into 1/2″ cubes.

In a Vita-Mix whole food machine or equivalent unit, purée the cucumber, purslane, tomatoes, onion, green bell pepper, garlic, juice, olive oil, vinegar, and kelp. This should probably be done in several batches so as to assure complete puréeing. Chill in the refrigerator for at least 2 hours.

Shortly before eating the soup make the croutons. In a 10″ to 12″ skillet over medium heat, warm the oil. Then add the garlic, oregano, basil, and kelp. Stir and add the pieces of bread. Monitoring for any signs of burning, brown the bread until crisp and golden, stirring frequently. Serve the soup and croutons separately. This recipe is modified from the famous soup *gazpacho* served in restaurants in Madrid, Seville, and other cities throughout Spain. It is absolutely delicious. Serves 3–4.

R

Red Radish
(Raphanus sativus)

Daikon Radish
(Raphanus sativus longipinnatus)

Brief Description

Radishes have been around since the days of Moses. The Egyptian pharaohs included them as standard rations, along with garlic, leeks, onions and cucumbers, for the several hundred thousand Hebrew slaves who constructed their mighty pyramids for them.

In the United States, the cherry-sized red radish is the most common variety sold, but radishes come in all shapes, sizes, and colors, including an intriguing black one. Popular varieties besides the Scarlet Globe and Cherry Belle (both globular red radishes) are French Breakfast (an elongated, white-tipped red radish), White Icicle (long and mild tasting), and the favorite of all Japanese, the daikon (a long, sharp-tasting white radish).

Helps in Digesting Starchy Foods

Throughout much of Europe, radishes are frequently consumed with bread and breakfast cereals. And the Japanese really like the daikon variety to accompany virtually all rich dishes. In both instances, we find that radishes seem to help in the digestion of starchy foods, such as grains, pastas, and potatoes. This is due, in part, to the presence of a special digestive enzyme called diatase, which occurs in the daikon radish in large amounts. So with those meals you're apt to consume that are loaded with starches, be sure to include some raw radishes as well for easier digestion.

Removing Hard Fat Deposits

Several years ago when I was in Tokyo attending an herbal medicine symposium, a Japanese colleague called to my attention an article in the *Asahi News* (Japan's largest circulated daily newspaper). This scientist friend of mine translated it into English for my benefit. The story dealt with a remedy employed by *kanpo* doctors in the eastern hills of Kyoto. These doctors, while practicing alternative folk medicine, were all regular M.D.s and had previously graduated from the prestigious Kyoto University School of Medicine.

The remedy employed for reducing and eliminating solidified deposits of hard fat imbedded in body tissue, consisted of a vegetable drink made with carrot and white daikon radish. Equal portions of grated carrot and daikon (1 tbsp. each) were added to 2 cups of water with 7 drops of soy sauce, 1 tsp. lemon juice, and a pinch of kelp and allowed to boil for 5 minutes. The broth was later strained and 1 cup prescribed twice daily, in the morning and again at night.

Stops Chronic Coughing and Raging Fevers

An old Chinese remedy calls for a handful of chopped, pithy roots from an old radish plant gone to seed to be boiled in a quart of water with a little pork for 40 minutes. Several cups of this lukewarm broth are then drunk each day for relieving spasmodic coughing, reducing high fever, and creating an appetite in those recuperating from recent illness.

Controls Diarrhea

Where diarrhea may persist and nothing else is readily available to stop it, some radishes will do the job effectively. In a food blender thoroughly mix together a handful of chopped red radishes, 1 cup of cold milk, and 1/2 tsp. Kingsford cornstarch. Drink the entire amount slowly. This concoction should stop the runs in less than an hour. Repeat again in 4 hours, if needed.

Prevents Gallstones and Kidney Stones

Good remedial management of existing stones in the gall bladder or kidneys, or to prevent their occurrence, entails making a daily drink in your food blender consisting of 2 chopped red radishes and 1/2 cup of red wine. This mixture may be taken twice a day for difficult urination too.

An old English remedy for stones has been used with very good success for several centuries now. The expressed juice of white daikon or black Spanish radishes is given in increasing doses from 1–2 cups daily. These 1–2 cupfuls are continued for 2–3 weeks. Then the dose is decreased until 1/2 cupfuls are taken 3 times weekly for nearly another month. The treatment may be repeated by taking 1 cupful at the beginning, then 1/2 daily and later 1/2 every other day.

Deodorant for Feet and Underarms

As a nutritious vegetable and wonderful medicinal, radishes also offer a third usefulness in the form of a toiletry for offensive body odor. But for this a juicer is needed. The juice from about 2 dozen radishes may either be put into an empty lotion bottle with a squirt top or else into a bottle with a hand-spray on it. About 1/4 tsp. of glycerine should also be added to the juice before bottling it to preserve it longer, unless you intend to refrigerate the same, in which case no glycerine is required.

After your morning shower or bath, pour some of the radish juice into the palm of your hand and rub under each armpit. Or else just spray some beneath each arm and on the soles of your feet and in between the toes, rubbing it in good to afford several hours' protection against odor. Radish juice is also useful for bruises, frostbite, insect bites, and stings and minor burns.

Quick Remedy for Burns and Scalds

Throughout this book a number of reliable remedies for treating serious burns and scalds are offered. (See Table of Contents for complete listing.) Many of them, however, take time to prepare.

A really fast remedy, though, calls for a handful of cleaned and chopped radishes to be put into a food blender with some crushed ice and thoroughly puréed until a nice, thick, even mixture is produced. This is then applied directly onto a burn or scald, covered loosely with clean muslin, and taped to hold in place. This brings almost immediate relief from pain and slows down infection considerably. To speed healing even more, add some zinc tablets and 1 tsp. of vitamin E oil while mixing the radishes and ice in your food blender.

Radishes for Cancer Treatment

Fresh radishes and radish seeds have been employed in the treatment of cancer around the world. Such success has been carefully documented in a variety of scientific publications. Lily M. Perry's *Medicinal Plants of East and Southeast Asia*, published by the Massachusetts Institute of Technology Press in Cambridge, describes the use of the seeds made into a strong tea for reducing stomach cancer and their external application as a heated poultice for treating breast cancer in women.

Dr. Jonathan L. Hartwell, formerly of the National Cancer Institute in Bethesda, Maryland, carefully screened thousands of plants for potential anticancer activity. His comprehensive report appeared in different issues of the scientific journal, *Lloydia*. In the March 1969 number, Dr. Hartwell cited fomentations of radish juice in cooking oil as being useful for abdominal tumors, the entire plant itself cooked in wine and oil good for liver and spleen hardness and just the red radish itself boiled in red wine and honey for treating, as he described it, "cancer of the fleshy parts."

A thorough analysis of the possible anticancer constituents in daikon and red radish appeared in the September 1978 issue of *Agricultural & Biological Chemistry*. Among the many volatile sulphur components cited was methanethiol. This factor has an odor of rotten cabbage to it, also evolves from penicillin bread cultures and is used in the manufacturing of pesticides and fungicides due to its strong antibacterial properties. It's components such as this that make radishes very good for many different kinds of cancer. Certainly I'm not recommending that it alone be used, but am suggesting radishes be included with other sulphur-bearing vegetables and herbs like cabbage, kale, kohlrabi, brussels sprouts, mustard greens, watercress, garlic, and onion in a special dietary program to combat cancer nutritionally as well as medically.

Thyroid Condition Substantially Improved

Loyal and Ethel. N. are advanced in years and reside in Parksville, British Columbia, Canada. I've known them for a number of years now, ever since we met at a health lecture I once gave to a large crowd in the province. We've remained good friends ever since.

In the early part of 1994, Ethel wrote to inform me of a thyroid condition her husband had been troubled with for some time. Apparently he was suffering from hypothyroidism, wherein there is a diminished production of thyroid hormone; clinical manifestations include low metabolic rate, tendency toward weight gain, frequent drowsiness, slowness in reasoning and thinking abilities, muscle weakness, subnormal body temperature, occasional hoarseness, and some dryness and loss of scalp hair.

Because I had seen a kampo doctor in Japan use radish juice and a thyroid extract (an animal glandular) for the same thing with remarkable results, I recommended the same procedure to her.

In a letter to me dated November 17, 1994, Ethel replied:

> Your advice worked great. We osterize radishes with water adding carrots, zucchini, or comfrey leaves fresh from the garden for as long as the climate remains frost-free. Loyal drinks one cup of this per day and takes along with it two 50 mg. pills of Thyroxin from Parke-Davis [a pharmaceutical house]; they're said to be natural.
>
> Well, it's been about 7 months now, and the tests that Loyal has had since he started taking these things have indicated a reactivation of this thyroid gland. He looks and feels a hundred times better. Thank you from Ethel and Loyal, with our very best wishes for your continued successes. With love, Ethel N.

Some Salad Ideas

The Japanese have over 100 different ways to cook with daikon radish. Raw, it can be grated and eaten with fish or meat. It's added to miso (fermented bean paste) soup. It's used to make flowers for a garnish, and is that stringy white stuff that's put with sashimi (raw fish) in Japanese restaurants. It can be shredded with carrots and dressed with a sweet vinaigrette for a salad. Or it can be cut into chunks and put in stews. A nice characteristic of daikon is that while it will get soft when cooked it won't dissolve.

Radishes go especially well in salads. The following are interesting salad ideas for the fussy gourmet and health enthusiast.

RADISH-CABBAGE SLAW WITH SESAME-YOGURT DRESSING

Needed: 2 cups shredded cabbage (preferably a mixture of red and green cabbage); 1 cup grated carrot; 1 cup grated daikon radish; $^{2}/_{3}$ cup grated red radish; $^{3}/_{4}$ cup plain yogurt; 2 tsp. chopped fresh dill or 1 tsp. dried dill; 1 tsp. sesame seed oil; $^{1}/_{4}$ cup toasted sesame seeds; 1 $^{1}/_{2}$ tbsp. tamari.

In a large bowl, combine the cabbage, carrot and both kinds of radish. Stir in the yogurt, dill, sesame oil, sesame seeds and tamari. Taste and add more tamari if desired.

Latin Lover Salad

Needed: 1 head of romaine lettuce, torn into bite-sized chunks: 1 bunch of fresh, chopped coriander; 1 bunch of thinly sliced radishes and their leaves; 1 chopped green bell pepper (including its seeds); 2 peeled, pitted and sliced avocados; 2 sliced, ripe tomatoes; $^1/_2$ lb. shelled, de-veined and boiled shrimp; $^1/_3$ cup olive oil; 2 deseeded limes, cut in half; kelp to taste.

Arrange the vegetables and shrimp in a large bowl, and drizzle with the olive oil. Squeeze on the lime juice, and add kelp. Toss lightly. Serves 4–6. Note: The last two recipes have been adapted from the elegant book *Vegetables* by four San Francisco Bay Area writers and courtesy of the publisher, Chronicle Books.

Raisins

(See under GRAPES AND RAISINS)

Raspberry

(See under BERRIES)

CHINESE RHUBARB
(Rheum officinale)

GARDEN RHUBARB
(Rheum rhaponticum)

Brief Description

Species of rhubarb are denoted by their large and sturdy sizes and large leaves borne on thick petioles. These hardy perennials grow between 7 and 10 feet high, are native to southern Siberia, China, and India, and are widely cultivated elsewhere.

Chinese rhubarb is used more for medicinal purposes, while the garden variety is grown more for its edible stalks (petioles) and ornamental beauty.

Strengthens Tooth Enamel

Rhubarb is high in potassium and calcium with a lesser amount of phosphorus. These mineral salts, according to a Rochester, New York, dentist, occur in rhubarb juice and seem to coat tooth enamel with a thin protective film. Dr. Basil G. Bibby of the Eastman Dental Center believes that more frequent consumption of cooked rhubarb might be of some positive benefit in helping to reduce extensive decay.

Better still, a little bit of the expressed juice from fresh rhubarb stalks brushed on the teeth with a soft bristle brush or else rubbed on with some cotton balls every other day should coat the enamel with these protective minerals.

Shows Antitumor Value

Rhubarb has demonstrated some excellent tumor blocking abilities. For instance, the first supplement of Volume 20 of *Pharmacology* related that two of the laxative compounds in rhubarb, rhein and emodin, also blocked Ehrlich and mammary tumors in mice by 75% at the relatively high dose of 50 mg. per kilogram of body weight per day.

A 1984 issue of *Journal of Enthnopharmacology* reported that rhein and emodin inhibited the growth of malignant melanoma at a daily dosage of 50 mg. per kilogram of body weight. The percentages of inhibition were 76% for rhein and 73% for emodin. In certain parts of mainland China, rhubarb juice and rhubarb tea are used in the treatment of some forms of cancer with good success. About 1/2 cup of juice twice daily obtained by putting fresh stalks through a mechanical juicer, are administered to patients. More often, though, tea is made by simmering 2 cups of finely chopped stalks in 1 quart of boiling water, covered, for up to an hour. Afterward, the liquid is strained off and given to cancer victims in 1-cup amounts 2 to 3 times a day.

Relief for Psoriasis and Arthritis

The anthraquinones in rhubarb, besides exerting wonderful laxative action, also help to relieve the itchiness and pain accompanying psoriasis and arthritis. Combine 1 cup of chopped, slightly mashed rhubarb root, 1/2 cup of chopped, slightly mashed rhubarb stalk, 10 tbsp. powdered wide Oregon grape root, and 8 crushed zinc tablets (50 mg. size) in 3 cups of quality gin or rum. Put in a tightly sealed bottle and shake twice daily for 15 days.

Then strain the tincture through clean muslin cloth and add 1 1/4 cups of cool cabbage juice. Thoroughly stir to shake up until both liquids are well mixed. The vegetable juice may be obtained by simmering half a head of chopped or shredded green cabbage in 1 quart of boiling water until only half the amount (or 1 pint) remains. Strain and cool before mixing with the alcoholic tincture. Then put in bottle with a tight lid.

One level teaspoon of this tincture should be taken five times a day on an empty stomach. Not only will this help bring relief to psoriasis and arthritis, but also it will work equally as well for eczema, herpes, acne vulgaris, and hepatitis.

Great Laxative and Antidiarrheal

Two important compounds in Chinese rhubarb root, called sennosides E and F, exhibit the identical properties on the bowels as do sennosides A and B, which occur in another well-known laxative herb, senna. And when used in large doses, it will quickly remedy even the most obstinate form of constipation. But strange to say, it's also an astringent and will stop diarrhea when used in small amounts.

As many as 4–6 capsules of powdered Chinese rhubarb root may be necessary for chronic constipation, but a mere 2–3 capsules should be all that is necessary for clearing up diarrhea. Or a tea can be made by bringing a pint of water to a boil and adding 1 $^1/_2$ tbsp. of cut, dried rootstock for constipation or just 1 $^2/_3$ tsp. of rootstock for diarrhea.

Reduce heat and simmer for 3 minutes before removing to steep, covered, for an additional half an hour. One cup at a time may be taken for constipation, but only $^1/_4$ – $^1/_2$ cup for diarrhea.

Heals Digestive Tract Diseases

Some interesting clinical studies conducted with Chinese rhubarb emerged from the Central Hospital of Luwan District in Shanghai in the early 1980s. In the first study, some 890 cases (79% male) of upper digestive tract bleeding (57% were duodenal ulcers complicated by hemorrhaging) were treated with rhubarb either in powder, tablets or syrup. These were administered in 1 tsp. equivalents three times a day until the bleeding ceased, usually averaging only two days with most. A 97% success rate was achieved.

A random comparison between this single use of rhubarb and the combined treatment of Western medicine and Chinese herbs was made in other patients experiencing the same type of difficulties. Furthermore, six different combinations of rhubarb were also tested. In all tests made, the single use of rhubarb took the shortest time to stop bleeding, reduce fevers, and help patients towards quicker recovery than the others did. This action exhibited by rhubarb may be due to the presence of tannic acid, which constricts blood vessels.

In the next set of studies, 100 cases of acute inflammation of the pancreas (pancreatitis) and 10 cases of acute inflammation of the gallbladder (cholecystitis) were successfully treated with the equivalent of 4 tbsp. of a decoction of rhubarb between 5 and 10 times a day until full recovery was noticed in most of them. Related symptoms like abdominal pain, high fever, and jaundice usually cleared up within 5 days or less. To make a decoction for any of these problems, simmer 2 1/2 tbsp. of cut, dried Chinese rhubarb root in 1 1/2 quarts or 6 cups of boiling water, covered, for 40 minutes or until about half (3 cups) of the liquid remains. Strain this and take as previously directed for any of the foregoing maladies.

Lowers Dangerous Cholesterol

A liquid solution of rhubarb root was fed orally to normal and hyperlipidemic rabbits. Those with the elevated levels of cholesterol, triglycerides and lipoprotein experienced a significant *decrease* in all of these. Which suggests that a meal heavy in fats should be accompanied with a simple dessert of delicious cooked rhubarb to help control cholesterol.

Such can be made by washing and cutting into inch pieces about 7 cups (approximately 2 lb.) of rhubarb. Cook in a double boiler with a little water and 1/4 tsp. sea salt until nearly tender. Then add 1 1/4 cups of dark honey and continue cooking another 40 minutes or until done. It may also be cooked in a covered casserole in the oven using 1/2 cup of water, but with the same amounts of salt and honey. Bake at 350° F. for 50 minutes. A therapeutic dessert served with greasy or fatty foods should be about 1 1/2 cups of cooked rhubarb. Adding a little cardamom, 1/2 tsp. pure maple syrup and a touch of pure vanilla improves the flavor more for those who don't especially care for its puckering tartness.

Pines International makes a Rhubarb Juice Concentrate in powdered form, which can be added to fruit or vegatable juice and consumed that way. About one-quarter teaspoon of rhubarb juice powder can be added to an 8 oz. glass juice.

RUTABAGA AND TURNIP

(Brassica campestris rutabaga, B. campestris)

Brief Description

Rutabaga is thought to have originated in Sweden sometime during the Middle Ages. It is larger, coarser and more emphatically flavored than the turnip, of which it may be a mutant form. Rutabaga is more elongated than a turnip is and usually adorned with a slightly purple top. Most have yellow or orange flesh, but a few are white.

Throughout much of recorded history, turnip has occupied a lowly position on the gastronomic scale, being considered only as a food for peasants and livestock. Turnips come in many different varieties, but those grown here in America are generally the white-fleshed kind, with purple or green tops. This distinct-tasting vegetable goes well with braised beef or roast pork.

Reduces Cholesterol and Constipation

Both rutabaga and turnip are ideal for lowering serum cholesterol levels in the body, not to mention encouraging more frequent bowel movements. Half a cup of each, either cubed or sliced, or 1 cup of each mashed should be consumed at least once a week where the cholesterol count may be dangerously high or infrequent bowel movements persist.

A simple way to prepare either of them is to cut them into slices or cubes and boil in a small amount of water containing 1 tbsp. lemon juice and a sprinkle of kelp and melted butter, in a covered pot, for about 25 minutes.

Brooklyn Cure for the Common Cold

A health-minded friend of mine from Brooklyn, New York, Joel Bree, has developed what people in his neighborhood consider to be one of the very best remedies for treating the common cold and influenza, along with related symptoms such as high fever, swollen throat glands, ear infection, and runny nose.

"The first time I made this soup my two-year-old-daughter, Tziporah, ate about three bowls, which surprised all of us. She was better the next day and she has never gone back on antibiotics ever since

"I buy kosher chicken bottoms and put them in a pot with enough water to cover, leaving an inch above them. I cover with a lid and let this boil out until there's just a little over half of the liquid remaining. I also add a teaspoon of apple cider vinegar in order to leach or draw out all of the minerals in the meat.

"After the liquid has been boiled down to half the volume, I take the chicken out, but leave the broth in the pot. Next I peel and juice in my food extractor 3 garlic cloves, a large red onion, 1 sweet potato that's been cleaned and with the *skin intact*, 1 *un*peeled zucchini or yellow crookneck squash, 3 *un*peeled carrots, 1 *un*peeled parsnip, 1 *un*peeled turnip, and 1 *un*peeled rutabaga. These juices are then added to the broth.

"After this, I return the pot to the stove and cook on a medium setting just long enough to where it starts to boil. At this point, I remove it from the heat so as not to destroy any valuable vitamins and enzymes. I find that this brief reheating helps to take away most of the starchy taste that would otherwise linger and make the blend taste unpleasant. I highly recommend this old family remedy to anyone seeking relief from the miseries of colds, flus and fevers."

Dentifrice And Deodorant

Eating a raw turnip once a week is a good way to clean the teeth and massage the gums at the same time. Frequently doing this on a consistent basis has been known to help reduce dental plaque and tartar somewhat.

A Japanese delegate told me that turnip juice was one of the "best things to use for getting rid of goaty armpits," as he so bluntly put it. He showed me a small bottle of turnip juice which he carried everywhere with him.

Being curious to see if it actually worked, I asked for and received some to use. First I washed both armpits good, then briskly rubbed 1 teaspoon of the juice beneath each of them. And although the temperature was in the high 90s and the humidity factor about 65%, there was *virtually no odor* to the perspiration. I recommend this over commercial deodorants, which contain harmful amounts of aluminum that may cause skin cancer in time. And unlike them, turnip juice won't prevent the sweat glands from doing their normal tasks, but *will* keep body odor from occurring for up to 10 hours as a rule.

Removing the Discoloration of Bruises

From the same Japanese friend, I learned that either a turnip or daikon radish were very effective to reduce the swelling and eliminate the discoloration accompanying nasty-looking black-and-blue marks. Just grate 1/3 of a turnip, apply it directly to the bruise and leave on for half an hour. Repeat the process several days in a row as necessary. This works like an ice pack in some ways to reduce pain and swelling.

Culinary Quickies

You can make a delicious and nourishing soup by adding 2 tbsp. of chopped chives to 3 cups of canned goat milk in a food blender, together with 2 cups of mashed turnips and 1 cup of mashed rutabaga. Purée until smooth, then cook on low heat for 15 minutes, garnishing with chopped parsley and a dash of paprika.

Turnip greens can be steamed for 25 minutes with a little lemon and lime juice and a dash of kelp and tad of butter added for flavor. Serve with hard-boiled egg on the side.

Salad Burnet
(Pimpinella saxifrage)

Brief Description

In my occasional trips through the English countryside, I've discovered salad burnet growing in dry pastures and by the roadsides, especially on chalk and limestone grounds. It is quite nutritious to domesticated livestock and was extensively cultivated as a fodder crop on calcareous soils in former times. Cattle don't seem to like it as much as they do clover when fully grown, but sheep are very fond of it when the herbage is kept closely cropped. It has the advantage of keeping dry, barren pastures green all winter long, affording food for sheep when other greenery is hard to come by.

In the lovely herb gardens of earlier times, salad burnet always had its place. Sir Francis Bacon (1561–1626), the famous English philosopher and author, recommended that it be planted in alley ways along with wild thyme and water mint in order "to perfume the air most delightful," when it was "trodden on and crushed."

The tiny toothed bright green leaves, looking somewhat like strawberry leaves, have a sharp, nutty, cucumber flavor. They are still used quite a bit throughout England in salads and also in sauces and soups. Butters and cream cheese are sometimes flavored with salad burnet and it can also be infused in vinegar.

Tonic for Heart Problems

Rembert Dodoens, a sixteenth-century herbalist wrote in his *Herbal* that salad burnet was good for heart problems: "The leaves stiped in wine and dronken, doth comfort and rejoice the hart, and are good against the trembling and shaking of the same."

His suggestion was to take a handful of salad burnet leaves and steep them for awhile in strong wine which had been heated until it was hot. Approximately 2 cups per day on an empty stomach seems sufficient for alleviating pressure on the heart.

Wound Healer

Dodoens claimed that if salad burnet leaves were "made into powder and dronke with wine wherein iron hathe bene often quenched, and so doth the herbe alone, being but only holden in a man's hande," as some have written. Apparently the success to this remedy depends as much upon a certain mineral as it does the type of herb to be used. The use of cast-iron vessels for making herbal formulas was standard practice for that time, and Dodoens' suggestion was merely a reflection of that. However, he noted that "the herbe alone" could be applied directly to the wound with similar results of healing.

Salsify
(Tragopogon porrifolius)

Brief Description

Salsify also goes by the names of purple goat's beard oyster plant or vegetable oyster. Salsify itself is a corruption of the old Latin name *solsequium*. This was derived from the Latin words *sol* (sun) and *sequens* (following), meaning the flower that followed the course of the sun.

On the other hand, the Latin binomial originates from the Greek; *tragos* means "goat" and *pogon* translates into "beard." This cognomen comes from the seed heads, at the top of which matures a fluffy, beardlike tuft, which, with the entire flower looking like a huge ripe dandelion blossom some 2–3″ across, parachutes the seeds over the landscape in the late summer.

In fact, the most noticeable part of this tasty plant is the single-headed bloom, principally made up of large purple or yellow ray flowers something like a daisy, beneath which pointed, long, narrow, green, grasslike leaves borne on the flower axis generally reach out in some species farther than the violet or golden petals themselves. Ordinarily closing in the afternoon, these pretty wild flowers appear in June and July after the plant is two years old.

Thrusting vertically downward, the very long, thin, tapering roots resemble a skinny parsnip and, in fact, yield a parsniplike flavor after their second year of maturity. Others, however, say that the taste is more closely allied to that of oysters, hence its other common names.

The tender white flesh discolors when exposed to the air, so it is best to always scrape each root and drop it into a small mixture of vinegar and water. The roots can be parboiled, drained, cut to any desired length, and then cooked some more until tender. Or just scrub the roots and cook them in their skins, and peel them afterward. They are tasty enough with a little melted butter, kelp, and lemon juice sprinkled over them. Or they can be served cold as a salad. They go very well with shrimp, scallops, and, of course, oysters, wouldn't you know.

Poor Appetite Improved

Many Americans and English people don't fully appreciate salsify like other Europeans do. A traveler can visit a number of fine restaurants in France, Belgium, Luxemborg, Austria, Switzerland, and Germany and find this herb being served with various dishes. It is also sold in a number of markets in these countries from the fall to late spring.

A delicious soup for improving the appetites of those convalescing from illnesses can be made using either the young or older roots; if using the mature roots, be sure to boil them in several changes of water first.

SALSIFY CREAM SOUP

Needed: 1 bunch of salsify, washed, scraped, and cut into small pieces; 6 tbsp. butter; 1 medium onion, finely minced; 1 quart of white stock; 6 tbsp. rice; pinches of salt, pepper, and nutmeg to flavor; 8 tbsp. unwhipped cream; 2 well-beaten egg yolks.

Briefly sauté the onion in melted butter in a medium pot before adding the cut up salsify root. Continue cooking over medium heat for about 6 minutes, stirring constantly. Then add the white stock and rice. Cover with a lid, turn heat to low, and allow to cool until the rice is done and the salsify root is tender. Strain the liquid through a fine sieve, retaining the solids. Put the solids into a Vita-Mix whole food machine or similar blender. Turn on and purée thoroughly. Put the purée back into the same pot and add just enough of the reserved stock to give the mixture a smooth, creamy texture. (Don't add too much, however, or else you'll wind up with a soup of runny consistency.) Bring to a rolling boil on the stove, and sprinkle in the three seasonings. Reduce heat to low setting and add the cream and eggs; stir briskly with a wire whisk but don't let the soup boil again. Serves 2–3.

Doing Something for Dry Mouth

The mature salsify plant stems yield a milky cap. Cut 10″ lengths of these stems, and with lightly gloved hands commence pinching each of them from the top downward to push out the sap onto a plate. Let it set until completely coagulated, then chew a little piece of it to bring moisture back again to a dry mouth. I got this little remedy from the diary of a nineteenth-century frontiersman who scouted sometimes for General George Armstrong Custer before his final Indian campaign against Chief Sitting Bull, in which he (Custer) lost his life.

Seaweed

Brief Description

The term seaweed is most commonly applied to multicellular marine algae. Simpler forms consisting of one cell or just a few cells aren't generally considered seaweeds; these tiny plants help to make up plankton. The more highly developed types of seaweed usually have a basal disk, called a holdfast, and a frond of varying length and shape, which often resembles higher plants in having stemlike and leaflike parts. The simplest of these seaweeds are among the blue-green algae (division Cyanophyta) and green algae (division Chlorophyta), found nearest the shore in shallow waters and usually growing as threadlike filaments, irregular sheets, or branching fronds. The brown algae (division Phaeophyta), in which brown pigment masks the green of the chlorophyll, are the most numerous of the seaweeds and also happen to be the most common marine forms. They grow at depths of 50 to 75 feet. The red seaweeds (division Rhodophyta), many of them delicate and fern-like, are found at the greatest depths (100–200 feet); their red pigment enables them to absorb the blue and violet light present at those lower depths.

The largest of the green algae, *Ulva* (sea lettuce), grows to a ribbon or sheet 3 feet in length. In provides food for many sea creatures, and its broad surface releases a large amount of oxygen.

Fucus, called rockweed or bladderwrack, is a tough, leathery brown alga (though it often looks olive-green in appearance), which clings to rocks and has flattened, branched fronds buoyed by air bladders at the tips.

Seaweeds, especially species of the red algae *Porphyra* and *Chondrus*, form an important part of the diet and are farmed for food in China and Japan; other species (often called laver) are eaten in the British Isles and Iceland.

Commercial *agar* (vegetable gelatin) is obtained from species of red algae and is the most valuable sea *vegetable* product around.

Irish moss or carrageen (*Chondrus crispus*), a red alga, is one of the few seaweeds used commercially in the United States, where it is gathered chiefly by fishermen working in the Massachusetts Bay area. After being bleached in the sun the fronds contain a high proportion of gelatin, which is used for cooking, textile making, making cosmetics, and other purposes. In Japan, it is made into a shampoo to impart a gloss to the hair. Irish moss is common among the submerged rocks off the coasts of France and, naturally, Ireland.

The kelps generally include the many large brown seaweeds and are among the most familiar forms found on North American coasts. Some have fronds up to 200 feet long, and for this reason have sometimes been called "the redwoods of the ocean." Typical examples would include the Pacific Coast *Nereocystis* and *Macrocystis*, found also off the Cape of Good Hope. Common Atlantic species include *Laminaria* and *Agarum* (devil's apron). The kelps are an excellent source of iodine and potassium, and, to a lesser extent, other minerals. But more about their nutritional benefits later. When the kelp is burned, the soluble mineral compounds are removed from the ashes (also called kelp) by washing. They are used chiefly as chemical reagents and for dietary deficiencies in people and in livestock. Kelp is also a commercial source of potash, fertilizer, and, recently, herbal medicines and health food supplements from the plant's vitamin and mineral content. Kelp is available in most health food stores and nutrition centers in loose (powdered or granulated), liquid, tableted, or capsulated forms. Kelps are especially abundant in Japan, and various foods known as *kombu* are made from them.

The brown algae of the genus *Sargassum* are called gulfweed. They inhabit warm ocean regions and are commonly found floating in large patches in the Sargasso Sea and in the Gulf Stream. Gulfweed was observed by Christopher Columbus and his men from the decks of their three small ships. Although it was formerly believed to cover the entire Sargasso Sea, making navigation all but impossible, it has since been found to occur only in drifts. Numerous berrylike air sacs keep the branching plant afloat. The thick masses of gulfweed provide the environment for a distinctive and specialized group of marine forms, many of which are not found elsewhere.

Popular Drink for Lung, Kidney, and Bladder Problems

In some parts of the British Isles and France, a popular drink is made from Irish moss for problems affecting the upper respiratory system, the kidneys, and the bladder. First, a simple decoction is made by steeping $1/2$ oz. of the moss in *cold* water for 20 minutes and then boiling it in 3 pints of goat's milk or good water for 15 minutes. After which it is strained and $1/2$ tsp. powdered cocoa added, along with a pinch of powdered cinnamon, a dash of lemon juice, and a little honey. It is taken freely as needed throughout the day.

Aphrodisiacs from the Sea for Sexual Impotency

If you want to rev up your passion, race your love engine, and put your sexuality into high-screeching gear, then seaweeds are the things to try. Some of them are very rich in beta-carotene (a form of vitamin A). This nutrient is a potent antioxidant, meaning that it hunts down and destroys nasty free radicals which abound in the body. Think of these scavenger molecules like mean sharks zipping around in your cellular sea doing considerable mischief. But in addition to this, vitamin A provides better sexual health. These benefits include healthier sperm production, a stronger thyroid, and the making of progesterone.

Vitamin E is known as "the sex vitamin" among many nutritionists. It is the other antioxidant which keeps free radicals in check and prevents them from causing breakdown in body tissue or basic physiological functions. A few limited studies seem to suggest that vitamin E might help a person to maintain his or her sex drive. Research on animal models has shown that when present in sufficient levels, vitamin E prevents the shrinkage of testicles and ovaries. Scientists have also noticed that women who take vitamin E (about 400 I.U.) regularly report less vaginal soreness, dryness, and pain during sexual intercourse. Giant kelp (*Macrocystis pyrifera*) is "very high [in] vitamin E [and] vitamin A," according to Judith Cooper Madlener, the author of *The Seavegetable Book: Foraging and Cooking Seaweed* (New York: Clarkson N. Potter, 1977, p. 113).

Health experts state that vitamin C soaks up free radicals, which damage healthy cells, making people look and feel older. And staying young can have a positive effect on sexual energy. Also vitamin C can boost energy levels, which can be very important if aggressive and prolonged sexual activity is to be engaged in. Madlener notes that many of the seaweeds, such as giant kelp, have ample vitamin C.

Proteins are substances made of amino acids in peptide linkage. While "protein" is a well-recognized term, the term "amino acid" can be confusing to many consumers. For easy comprehension, therefore, think of amino acids as useful ammoniated vinegars. There are eight essential amino acids considered absolutely necessary for the body—lysine, leucine, isoleucine, methionine, phenylalanine, threonine, tryptophan, and valine. An individual will begin to die without ingesting these amino acids on a daily basis (although the gut flora or bacteria do provide small quantities of each of them). Now the best source for these essential amino acids are protein-rich foods, which are derived from plant and animal sources.

One of these essential amino acids, phenylalanine, is a precursor of the neurotransmitters that influence elevated moods and sexual arousal. Four sources extremely high in phenylalanine are wild game, pork, luncheon meat, and eggs. A fifth source are certain seaweeds, which have high protein content, meaning that they have a variety of most of the essential amino acids, including phenylalanine. According to Madlener, the following seaweeds should increase sexual desire and arousal if consumed frequently enough: fairies' butter (*Nostoc*, a type of blue-green algae), dulse (*Palmaria palmata*), and purple laver (*Porphyra umbilicalis*), among others.

When the trace element iodine is mentioned, people generally connect it with the thyroid gland, but usually don't connect it with any sexual functions. However, this nutrient is known to provide heightened stimulation to the endocrine glands, *especially* to the thyroid. If sufficient iodine is wanting in the diet, this gland can't produce enough thyroxine; without adequate amounts of this in the system, there is poor metabolism, a bad attitude, lousy skin tone, and a sex drive about as exciting as a block of wood. Women with underactive or overactive thyroid glands are likely to have irregular periods or severe premenstrual syndrome symptoms. But iodine-rich foods prove quite beneficial in regulating abnormal or ill effects of the menstrual cycle. Some wonderful sources of iodine include seaweeds such as nori (*Porphyra tenera*, used in sushi), wakame (*Undaria pinnatifida*, used in soups), and edible kelp.

All seaweeds will have some positive effect on health and libido, so *bon appetit* and enjoy the incredible power of such love foods. For more information on the seaweeds themselves, consult Judith C. Madlener's *The Seavegetable Book* (New York: Clarkson N.

Potter, 1977). For additional foods that put human sex into overdrive, consult Raymond McIlvena's *The Pleasure Quest: The Search for Aphrodisiacs* (Specific Press, 1988). Since enhanced sexuality takes a toll on the central nervous system, the reader may wish to check out H. L. Newbold's *Nutrition for Your Nerves* (New Canaan, CT: Keats, 1993). For ideas on the value of sex from a therapeutic dimension, Judith Sachs's book, *The Healing Power of Sex* (Englewood Cliffs, NJ: Prentice Hall, 1994) covers the subject in a sensitive way.

Fatigue Gone with Kyo-Green "Power Drink"

Credit must go to the Japanese for coming up with some unique energy-enhancing health products containing seaweed. One of the more popular dietary supplements for boosting strength and stamina is called Kyo-Green. It consists of concentrated juice of young barley and wheat grasses and brown rice combined with chlorella and kelp from the Pacific Ocean.

One level tablespoonful in an 8 oz. glass of water makes a refreshing drink with a nice emerald-green color that would make any person from Ireland downright proud. Athletes not only benefit from this, but so do everyday working people who soon find their energy dissipated due to the stresses around them. One glass of this "power drink" will give them an unbelievable surge of energy that can last for several hours.

Wakunaga of America, which distributes this product through health food stores, also has the world's premier garlic product called Kyolic. It comes in various forms, including liquid. I sometimes like to add a teaspoon of this liquid aged garlic extract to my glass of Kyo-Green; the two make a terrific duo that is hard to beat for quick and lasting energy!

Some Seaweed Soups

In the year 1842, the *Domestic Dictionary and Housekeepers' Manual* referred to Irish moss as "exceedingly nutritious and by no means disagreeable when made into soup with meat and other vegetables; it is, in fact, quite equal to the famed birds' nest soup of the Chinese."

Irish Moss Soup

Needed: $^1/_4$ cup dried Irish moss; 4 quarts cold water; 1 stalk celery with leaves, cut up; 6 carrots, cut up; 3–4 lb. stewing beef, cut up into small bite-sized pieces; 1 tbsp. sea salt; 2 tbsp. soy sauce; 1 tsp. cracked black pepper; $^1/_4$ tsp. crushed red pepper; 2 bay leaves; pinches of sage and thyme powders; 1 cup dry white wine.

Soak the Irish moss for 30 minutes in cold water to cover. Wash. Drain. Remove any foreign matter. Chop into florets about the size of thumbnails. Place all the soup ingredients, including the sea vegetable, in a deep pot. Simmer until tender for 2 $^1/_2$–3 hours. Serves 24.

Sargasso Sea Soup

Needed: $^1/_2$ cup fresh or dried Sargassum (use blades of uppermost fronds only); 8 cups chicken broth; $^1/_2$ cup dried nori sheets; $^1/_2$ cup dried wakame fronds; $^1/_2$ cup small dried shrimp; $^1/_2$ cup dried sardines; $^1/_2$ cup soy sauce; 2 tbsp. sesame oil. Note: Various species of *Porphyra* may be substituted; these include purple laver from Scotland, nori from Japan, kim from Korea, chi choy from China, and summer seaweed from the Pacific Northwest. You can find most of these seaweeds in specialty food stores catering to Chinese, Japanese, or Korean cultures as well as general Oriental food stores.

If fresh Sargassum is being used, first rinse it under cold water to remove all particles of sand. Boil the broth and blanch the sea vegetables in it to make them pliable. Remove the leaflike blades from the Sargassum. Chop the nori and wakame and add to the boiling broth. Add the dried shrimp and sardines. Add the soy sauce and sesame oil. Boil gently for about 2 hours or until everything is tender. Serves 8–10.

SEEDS

Poppy (Papaver somniferum)
Safflower (Carthamus tinctorius)
Sesame (Sesamum indicum)
Sunflower (Helianthus annuus)

Brief Description

Poppy seeds are grown in the temperate and subtropical regions of the world. In India, the seeds are mainly grown in the states of Madhya Pradesh, Uttar Pradesh, and Rajasthan. The seeds are a good source of protein and oil and have long been used in food preparations like curries, breads, sweets, and other confections. Ironically, this healthful food comes from the very same plant that the highly addictive, dangerous narcotic, opium is derived. This illegal and destructive substance is obtained from the milky sap and the urn-shaped seed capsules or poppy heads.

The safflower is an annual plant native to the Mediterranean region and also cultivated in the United States and Europe. It's botanically related to lettuce, sunflower, artichoke, chicory, and daisy. Safflower is a well-branched plant, varying in height from 1 to 4 feet with smooth white or light gray stems. The thistlelike inflorescence is made up of a dense mass of blossoms which might be white, yellow, orange or red. They are surrounded by numerous protective bracts formed by the prickly terminal leaves. The fruit or achenes are produced deep down inside the flower heads where they are safe from hungry birds. Each achene contains a single oblong seed, which, like the stems, is either white or light grey in color. Each seed contains 24–36% of its own weight of an oil used extensively in cooking.

Sesame can be grown in tropical or less warmer climates such as the southern United States. It's a very attractive annual, reaching heights of between 2 and 6 feet, with heavy, glistening stems supporting variable leaves. Those near the base are broad, fleshy, and tooth-edged while the upper ones are slender and smooth-edged. The flowers of sesame resemble those of the foxglove or digitalis plant. They are extremely beautiful, colored white with spots or a tinge of red, yellow, or lavender. The flowers are always self-fertilizing and produce their seeds in inch-long keel-shaped pods or capsules. The tiny seeds are flat and oval with one end pointed, weighing no more than $^1/_{10}$ of a grain each, and containing 50–60% of a fixed, semidrying oil. A single pressing and filtration yields a clear pale-yellow oil without any definite taste, but highly edible. A much darker, inferior oil is also produced by two further pressings, under intense pressure from the heated residue, but needs refining and deodorizing to make it usable.

The state flower of Kansas, the sunflower, grows everywhere almost like a weed in many instances. The scientific name of the genus, *Helianthos*, comes from two Greek nouns—*helios* for "sun" and *anthus* for "flower." The showy yellow flowers are not only like a primitive reproduction of the sun, but they turn on their own accord toward this luminous celestial body's radiations. The annuals and perennials are upright and mostly rhizomatous. With up to a couple of dozen golden ray petals emanating from yellow and reddish purple to brown central disks where the seeds later mature, the often solitary terminal flowers become masses of gold in the summertime, from several to sometimes more than 15 feet high. The differing stems and leaves are mostly hairy. The numerous species all have neutral ray flowers and flat and fertile disk blossoms which also bear chaff. Sunflower seeds in the *H. annus* number about 650 to an ounce and contain about 50% of a semidrying, pale yellow oil used in cooking and salad oils and in the making of margarine, cheese, and other dairy products.

Seeds Inhibit Cancer

Recent epidemiologic evidence seems to suggest that plant seeds may lower the risk for developing certain types of cancer generally associated with too much meat and fat consumptions. At least that's the conclusion drawn by Dr. Walter Troll, professor of environmental medicine at the New York University School of Medicine and reported in the May 27, 1983, *Journal of the American Medical Association* (*JAMA*).

Troll analyzed computer data from epidemiologic studies indicating that prostatic, breast, and colon cancer rates are considerably lower in populations whose diet is rich in "seed foods." He found that certain enzymes, common to all seeds, when injected into lab mice previously inoculated with melanoma cells, prevented the development of cancers in the mice. In a similarly inoculated control group without benefit of these seed enzymes, tumors rapidly developed.

Dr. Troll told a reporter from *JAMA*, "A prudent diet would be one in which from $^1/_3$ to $^1/_2$ of all protein consumed is provided by seeds." The *New England Journal of Medicine* for January 19, 1978 confirmed that frequent consumption of sunflower seed oil in Bulgaria, Romania, and the Soviet Union resulted in lower mortalities from malignant tumors than elsewhere. So seeds are of definite benefit when it comes to preventing cancer!

Safflower Oil Prevents Heart Disease

Today's number one killer disease in America is coronary heart disease. If you were to outline on graph paper the statistical increases of it in the last three decades alone, you would almost have a sharp vertical line shooting to the top, although slightly tilted to the right. Or, put another way, there have been *NO* decreases in heart disease for the past 30 years!

The use of hydrogenated cooking oils by the food industry is the cause. Way back in 1956, the respected medical journal, *The Lancet* (2:557), issued this dire warning: "The hydrogenation plants of our modern food industry may turn out to have contributed to the causation of a major disease!" How tragically prophetic they proved to be.

But to understand why this nation of ours has experienced such great leaps in heart disease, unparalleled to that of any other nation on earth, we need to briefly examine just what hydrogenation is all about.

First, the process of hydrogenation originated as a way of making low-cost soap from discarded animal fats. From there it eventually crept into the powerful food industry. Simply put, liquid oil or soft fat is purposely hardened at a high temperature and under intense pressure. Hydrogen is then bubbled through the oil in the presence of toxic metals like nickel, platinum or some other cancer-causing catalyst. The hydrogen atoms combine with the carbon atoms and the product becomes saturated or hardened.

Hydrogenation is generally done to cheap plant oils like palm and coconut or animal fats from which shortening and lard are made. After hydrogenation what you have is a bad-smelling and loathsome-looking grease that would be quite unacceptable to any normal human being. But the magical skills of clever food technologists are used to bleach, filter and deodorize this rank stuff with myriad chemicals into a pure, snowy white, odorless, tasteless, highly artificial fat.

So when you eat fried foods at your favorite fast-food place, you are, in effect, laying the groundwork for coronary heart disease. Take it from an expert. During those early years that I was getting my schooling as a medical anthropologist, I worked full time on the side for nearly seven years in the restaurant business and almost four more years in the funeral business.

In my job at the funeral home, I occasionally witnessed autopsies whenever I had to pick up a corpse from the medical examiner's office. Several of those who were cut open had been individuals whom I'd previously known from my restaurant experiences. And I knew each of them had been *heavy* consumers of deep-fried food, which was recalled when the doctor slicing them open let me take a look at their hearts. Sure enough, every one of them had hearts choked full and clogged up good with yellowish fat beyond belief. I can't begin to tell you what their livers and gallbladders looked like!

The best comparison I can make to eating anything deep fried is if you were to pour a lot of motor oil *into your gas tank*. In no time at all, the valves would begin to stick, the carburetor would become clogged, and the spark plugs refuse to ignite properly. In short, the vehicle would no longer function. Much the same thing takes place inside the human body whenever any deep-fried food is consumed.

For the sake of your health and to prevent getting heart disease, you should cease and desist in eating such greasy fare. Instead, you ought to seriously consider eating only those foods which have been either fried or baked in safflower, sesame seed, or sunflower seed oils. A recent Veterans Administration hospital dietary study showed that the average serum cholesterol was 17% higher on a diet rich in palm oil than on a diet containing equivalent amounts of a highly unsaturated safflower oil. This is by no means a trivial difference since a 10% drop in serum cholesterol would mean 24% less heart disease in the United States.

This is no joking matter! A report published in the February 1984 issue of the *American Journal of Clinical Nutrition*, for example, showed beyond a doubt that when hydrogenated fat was systematically fed to swine, it induced *far more* hardening of the arteries than swine fed other types of fat such as some of these seed oils.

Sesame and sunflower seed oils also play a valuable role as well in preventing coronary heart disease. Both oils contain a group of compounds called phytosterols, which researchers have found significantly lower blood serum cholesterol and do *not* build up any fatty plaque on the artery walls of the heart. In other words, a person could technically eat a lot of foods fried or baked with any of these three cooking oils and still have a relatively fat-free, squeaky clean heart.

Poor Blood Circulation to the Brain Improved

Years of dietary abuse, such as in frequently consuming foods cooked in or made with hydrogenated oils, can cause the blood to become somewhat thick and stagnant. Not only is overall circulation slowed down, but artery walls, especially those leading to the brain and heart, become narrower and occasionally clogged with clumps of thick blood and bacteria.

If you feel that your own state of blood is somewhat thick and sluggish and could possibly lead to cerebral vascular problems of some sort, then safflower oil may be just the thing for you. An intake of 1–2 tbsp. of the oil each day could very well improve a situation that's been going from bad to worse.

Difficult Wounds Healed

There are certain types of wounds which are very difficult to manage. These are wounds complicated with fractures and hard-to-heal surgical wounds, such as skin grafts or mediastinitis, a swelling of the tissue between the lungs that sometimes follows heart surgery.

Doctors in the Orthopedics Department of Tianjin Hospital in Tianjin, China, and others at the Hospital Bichat in Paris used several different ingredients individually, which if combined, appear to have a more diverse use. Of those substances employed by the Chinese physicians, only sesame seed oil (500 grams or 2 $^1/_2$ cups) and melted, filtered beeswax (90 grams or approximately 1/4 cup) added to the warmed oil, seem to have the most healing merits. I would, therefore, submit to the reader the following recipe for an effective wound remedy: 2 1/2 cups of sesame seed oil; 1/4 cup melted beeswax; 3/4 –1 1/4 cups white sugar; 3 tbsp. powdered bonemeal; and 2 tsp. crushed calcium supplement tablets in powdered form. Gently warm up the sesame seed oil on low heat, being careful that it doesn't smoke. Stir in the bonemeal and calcium powder. Then gradually add the sugar, stirring constantly with a wooden ladle or wire whip. Finally, turn in the melted beeswax and keep stirring until there is an even, smooth consistency to the mixture, but not too stiff or lumpy. If so, add a little more heated sesame seed oil until the desired thickness has been achieved. Pack on and into wounds when completely cooled and change several times each day.

Sesame Oil for Ear and Eye Problems

Here are some folk uses for sesame seed oil that I learned from an Egyptian pharmaceutical student. Warm sesame seed oil is sometimes used as eardrops when the ear is plugged by excessive and hardened earwax. The oil, he assured me, would soften the wax, so that it can be washed out easier. The best way to warm the oil, he suggested, was to put a small amount in a glass jar (empty baby food jar, for example), then set it in a pan filled with about 2″ of boiling water for no more than a minute or so.

For eye problems that involve the accumulation of water in the eye, such as glaucoma, myopia and trachoma, the finest sesame oil should be used. First, about half a cup of the oil needs to be cooked until it is rather hot. This can be accomplished in the same manner as the above oil is heated for use in the ear. Only in this case, about 2 1/2″ of boiling water are put in the pan *and* a small plate put in the bottom in an inverted position. On top of this is placed the clean baby food jar containing the oil. The pan is returned to the stove and medium heat applied to keep the water boiling. When the oil is very hot to the touch, remove from the pan and strain through several layers of gauze. Allow to thoroughly cool until the oil is comfortably warm. This can be tested by simply dropping a little bit on the tip of the tongue. Then, using an eyedropper, put 3 drops of this oil into each eye just before retiring for the night.

Getting Rid of Dandruff and Headaches

This same Egyptian colleague of mine also shared with me another important remedy using sesame oil for scalp problems and headaches. Grate two fresh ginger roots on the smaller holes of a hand-held grater so as to obtain a finer pulp. Enclose this in enough double-layered gauze or cheesecloth and press hard enough to extract 1–2 tsp. of juice. This small bag of grated ginger can be placed between two small blocks of wood, set in a large flat pan, and a carpenter's bench clamp then put over the wood and slowly turned to get the necessary juice out.

The juice is then mixed with 3 tbsp. of sesame seed oil and $^1/_2$ tsp. lemon juice. This sesame-ginger oil can then be rubbed with a wash cloth into the scalp by parting the hair with a comb every $^1/_2$". This should be done 3 times a week to get rid of dandruff and seborrhea and to keep hair from falling out. This same oil can be rubbed on the forehead with a piece of linen cloth to relieve a headache. It's also very good for relieving neuralgia and sciatica if rubbed directly on the sore areas.

Sesame Soup for Constipation

Noted research chemist and microbiologist Albert Y. Leung recalled that

> By far the most common uses of sesame seeds in Chinese homes are as nutrients, tonics and laxatives. All three effects can be obtained from a drink (perhaps more appropriately called a soup) made from sesame seeds and rice.

To prepare this soup, 11 parts of sesame seeds are soaked in water, together with a small amount of rice. After the sesame seeds and rice are well soaked and become tender, they are ground to a paste by running them through a small food grinder or nut mill of some sort. The resulting milk mixture is strained to remove the coarse particles and then diluted with a little more water and some honey before cooking on low heat until the consistency is somewhat syrupy. Two cups of this delicious soup usually clear up the most obstinate form of constipation within an hour or so.

Paste for Spider Bites, Sores, and Burns

An eighth-century herbalist from China gave an effective remedy for treating insect bites, especially painful spider and centipede bites, as well as minor burns and various types of skin sores. About 2–3 tbsp. of sesame seeds should be ground into a coarse powder and then made into a paste with the addition of a little water. This, in turn, is then applied to the afflicted area and left until the pain and swelling subsides somewhat. The seeds may be ground in a mortar with a pestle or finely crushed on a counter top with a heavy rolling pin or carefully pounded with the bottom of an empty Coke bottle or hammer.

In another herbal from seven centuries later, sores on the head and face were treated just by chewing the raw sesame seeds for a couple of minutes and then applying the resulting wet mash to them. I've had occasion to try this only once on a small pimple on the side of my neck right by the collar line of my shirt which was irritating the heck out of me during my travels abroad. I secured some sesame seeds in one place on my trip, masticated them well, and then applied this wet poultice to my neck and kept it in place with several extrawide bandages from my shaving kit. I pretty much forgot all about it until the next morning, when I noticed after removing the Band-Aids that the redness and swelling had disappeared and the pimple itself was almost entirely gone.

Breaking the Smoking Habit

Back in 1980–81, a medical doctor by the name of John M. Douglass discovered an effective way to quit smoking while working long hours as an internist at the large, plush Sunset Kaiser-Permanente Medical Center near Hollywood. "It's a terrific tool with which to stop smoking," he admitted.

"They are an excellent substitute for smoking because the seeds have a comparable effect on those eating them to that of tobacco on smokers," he reasoned. According to him, this is the way they work: Tobacco releases stored sugar (glycogen) from the liver and this perks up one's brain. Sunflower seeds provide calories that give the same mental lift.

Tobacco has a sedative effect that tends to calm a person down. Sunflower seeds, too, stabilize the nerves because they contain oils that are calming and B-complex vitamins that help nourish the nervous system. Tobacco increases the output of adrenal gland hormones which reduces the allergic reaction of smokers. Sunflower seeds do the same. These allergic reactions can be a major problem to individuals attempting to stop smoking.

Dr. Douglass said that he had seen patients develop respiratory problems when they stopped smoking and he thought this was likely due to allergic reactions that had been kept in check by the antiallergic effects of tobacco. "The sooner you start munching the sooner you will be an ex-smoker," he confided. "A few weeks can do the job. Only eat raw, shelled sunflower seeds. Stash several ounces of the seeds in your purse or pocket and every time you get the urge to light up, reach for a handful of seeds instead."

And if one gets tired of munching seeds all the time, then Dr. Douglass has accommodated the primal urge for variety in the form of sunflower wafers for smokers. This is how he told us to make them: Grind the seeds into fine meal. Moisten slightly to make a thick dough. Some people like to add raisins or miso (a salty paste made from soybeans). Pinch off small pieces and form them into half-dollar sized wafers. Put them on a screen and place on top of the refrigerator or some dry place for 3–4 days until the wafers are dry. "The resulting uncooked seed bread is filling, extremely nutritious and satisfying, and kills the taste for cigarettes just as raw sunflower seeds do," I remember Dr. Douglass saying.

Stops Ringing in the Ear (Tinnitus)

An old but effective Chinese remedy for reducing or removing altogether strange sounds and noises in the ear calls for drinking a decoction made out of the empty shells of sunflower seeds until the problem clears up. The seeds, themselves, are often added as well to the tea.

Bring 1 $^1/_2$ pints of water to a boil. Add 2 tbsp. *each* of crushed seeds and their empty shells. Cover, reduce heat, and simmer for 15 minutes. Then remove from heat and steep an additional half an hour. Drink 1 cup of lukewarm tea, after straining, every 4–6 hours.

Putting Good Recipes to Seed

The following several recipes involve some of the different seeds covered in this section. You'll find them to be healthy, wholesome, and delightful to eat.

POPPY SEED ROLLS

Needed: 2 tbsp. active dry yeast; 1/2 cup lukewarm water; 6 tbsp. honey; 2 1/2 cups lukewarm buttermilk; 1 tbsp. tamari (an Oriental-style soy sauce available in some Oriental food shops and health food stores); 1 1/2 cups wheat germ; 6 1/2 cups whole wheat flour; 6 tbsp. sesame seed or safflower oil; some poppy seeds.

Dissolve the yeast in the water in a large bowl, and stir in the honey. When the yeast is bubbly, add the buttermilk, tamari, wheat germ, and 2 cups of the flour. With a hand mixer on high speed, mix the wheat germ, flour, and yeast mixture for 5 minutes. Add 1/4 cup of the oil and the remaining wheat flour, 1 cup at a time, stirring after each addition. When the dough holds together, turn it out onto a floured surface and knead until smooth, adding only enough flour to keep the dough from sticking. Oil a large bowl or kettle, and turn the ball of dough around in the oil until it's coated. Cover the container and allow dough to rest in a warm place until doubled in bulk. Punch the dough down until it collapses, form again into a ball and cover, returning to a warm place. When the dough has doubled a second time, punch the air out again and turn the dough onto a floured surface.

Make small balls of dough, about 2" in diameter, and coat them with a little of the remaining 2 tbsp. of oil. Place the balls (12 should do) in the bottom of a lightly oiled, 8" round cake pan. Sprinkle generously with poppy seeds and set aside to rise. When the rolls have risen almost double in bulk, place them in a preheated 400° F. oven. After 15 minutes, turn the oven down to 350° F. and bake the rolls about 20 minutes longer, until they are browned and baked through. Remove from the oven and turn out on a cooling rack. The rolls will break apart easily for serving. Makes 1 dozen rolls.

SESAME-RICE CEREAL

Needed: 3/4 cup raw brown rice; 1 cup powdered milk; 3 1/2 cups water; 1 tsp. salt; 2 tbsp. whole sesame seeds briefly ground in food blender to make a meal; $^1/_2$ tbsp. Brewer's yeast; 1 tbsp. real vanilla. Toast rice in dry pan over medium heat, stirring until browned. Grind in blender, then toast powder again briefly in dry pan, stirring constantly. Combine milk powder and water with wire whisk. Put in heavy pan, add salt and boil. Add rice powder, stirring constantly. Lower heat and simmer, covered, about 10 minutes, or until cereal thickens. Toast sesame meal in dry pan over medium heat, stirring constantly for 1 minute or so and add along with yeast to cereal. Stir in sesame meal and yeast. Add vanilla flavor and stir again. Serve with canned or fresh goat's milk and dark honey. Serves about 5.

Lip-Smackin' Appetizer

One of the most delicious appetizers I believe I've ever tasted in my many trips abroad, was the *tahina* dish served to me in a small, out-of-the-way cafe in Doha, the capital of the tiny Arab oil kingdom of Qatar in midsummer 1980. Tahina is an incredibly delicious paste made from the finely ground seeds of sesame. It can usually be obtained in certain gourmet food shops carrying Mideastern delicacies.

What I was served though was tahina mixed with a little dark red port wine and flavored with finely minced garlic and a hint of lime juice. With a saucer of this came chunks of dark rye bread, which I used for dunking purposes. My waiter also served me a glass of arrack (a rumlike alcohol) to go with my snack. So pleasing was this specially prepared tahina to my palate that I didn't even bother ordering dinner afterward, but made a meal just of this simple and extremely tasty fare!

McSun Burgers

Here's a completely vegetarian burger that not even the people working beneath those golden arches of McRaunchydom could ever dream up. Preheat oven to 350° F. *Needed:* $^1/_2$ cup grated raw carrots; $^1/_2$ cup finely chopped celery; 2 tbsp. chopped onion; 1 tbsp. chopped parsley; 1 tbsp. chopped green pepper; 1 beaten egg; 1 tbsp. oil; 1/4 cup tomato juice; 1 cup ground sunflower seeds; 2 tbsp. wheat germ; $^1/_2$ tsp. salt; $^1/_4$ tsp. basil.

Combine ingredients and shape into patties. Arrange in an oiled, shallow baking dish. Bake in a moderate oven until brown on top; turn patties and bake until brown. Allow about 15 minutes of baking for each side. Serves 4. I'm indebted to Rodale Press for these recipes from their *Rodale Cookbook* and *Natural Healing Cookbook*.

SNAPDRAGON

(See under ORNAMENTAL PLANTS)

Sorrel
(Rumex acetosa, R. scutatus)

Brief Description

Of the two types of sorrel cultivated for use as vegetables, garden sorrel (*R. acetosa*) is an indigenous perennial commonly found throughout much of Europe and Asia in damp meadows, along roadsides and by shorelines. It is a slender plant about 2 feet high, with juicy stems and leaves, and whorled spies of reddish-green flowers, which yield their lovely colors during the summer months of June and July.

French sorrel (*R. scutatus*) is preferred for kitchen use in soups by French chefs more than garden sorrel; the leaves are the principal ingredient of the famous *Soupe aux herbes*. This type of sorrel is also employed in ragouts and fricassees. It is distinguished from the other sorrel by the shape of its leaves, which are cordate-hastate, very succulent, fleshy, and brittle. Put another way, they resemble spinach leaves.

The name sorrel is derived from an old Teutonic word for "sour." It is a member of the dock family to which rhubarb belongs. The dock family is similar in many ways to that of the beets and spinach. Sorrel has a refreshing, somewhat bitter, and sour taste. This is due to the presence of oxalic acid, the compound that tickles the palate and gives the herb its inimitable scent.

Sorrel should always be cooked for a minimum time in order to preserve its fresh flavor, and it should be chopped only with a stainless-steel knife and never cooked in an iron pan. Iron with sorrel goes black and makes a nasty metallic taste. It can be used raw in salads, but the leaves should be very young and tender. It is excellent as a purée melted in butter for a few minutes after being chopped raw or blanched or lightly cooked in boiling salt water. This puree goes well with veal, pork, fish, and eggs (especially in omelettes). Sorrel is also a superb sour flavoring for soups made with chicken stock, milk, lentils, tomatoes, and cucumbers. Sorrel is also an important ingredient of the herb mixtures used to stuff or cook with fish of many kinds ranging from sole and salmon to coarse fish and eels. Last of all, it is the basis for old English sour green sauces to be served with pork or goose.

Delicious Soup that Cures Persistent Headaches

I've always been of the philosophy that "food should also be your medicine." Here is a delicious soup *formula*—I call it that instead of a recipe because of its great medical benefits—that was devised in 1940 by a British herbalist named Mary Thorne Quelch. She often prescribed it to those of her patients who suffered from "persistent headaches." Her instructions were simple: "Eat a bowl of this soup every day for a month *with* a slice of dark brown bread and your headaches will be gone for good!"

Here is her food *formula* given *exactly* as she wrote it down over a half-century ago. "Well wash and carefully pick a quarter of a pound of sorrel and a small lettuce. Chop all rather fine and put into a stewpan with an ounce of butter, margarine, or other cooking fat. Keep it over a slow fire for about ten minutes, stirring all the time. Then add a pint and a half of milk, slightly thickened with flour, or water in which practically any vegetable has been boiled. Simmer for half an hour. Remove from the fire, season with pepper and salt to taste, and add two tablespoonsful of tarragon vinegar. If eggs are obtainable [remember this was during World War II when some food staples were scarce] the yolks of two, beaten up with a little milk, should be added. Heat again thoroughly, stirring all the time, but be careful it does not boil after the eggs have been added."

Bear in mind that while some of the words in the foregoing quote may seem a little strange to us, they were a formal part of people's expressions in those dangerous times of war-ravaged London. "Well wash" was typical of those speech elements I'm referring to.

Mrs. Quelch also gave a short *recipe* for boiled sorrel. "Pick three pounds of sorrel from the stalks and wash thoroughly. Put it into a saucepan with a very little water—there should be only just enough to cover the bottom. Simmer for half an hour, turning the herb frequently and pressing it down well to make sure it is all equally cooked. When it is quite tender, strain and chop finely. Then return to the saucepan with the liquor that was pressed from it. Season with pepper and salt, add an ounce of butter or margarine. Stir over a slow fire for about ten minutes, and serve." (Since margarine isn't good for the body, I advise using just butter instead.)

French Remedy for Constipation, Hemorrhoids, Worms, Sore Throat, Fever Blisters, "Cold" Sores, Fevers, Abscesses, Acne, Herpes, Ulcers, and Food Poisoning

Maurice Mességué was one of Europe's greatest herbalists with an incredible healing career that spanned the 1940s to the 1980s. He was born, as he loved to tell it, "on Sunday, December 14, 1921, at 4:30 in the afternoon, under Sagittarius, and by chance at Calayrac-St. Cricq (Lot et Garonne . . .)"

He came into a family with a long and rich healing tradition—his father, grandfather, and both great-grandfathers were all skilled herbalists. "The Mességués have lived on the same land for over 450 years and have always had this special knowledge of plants," he once stated.

Sorrel was one of those herbs he came to reply on and use extensively in his healing practice for a wide variety of health complaints. Mességué used the leaves in two different types of decoctions. For internal use he advised: "Put a handful of fresh leaves into a litre (1 3/4 pints) of water; boil for 10–15 minutes; drink a cupful a day." For external applications only, he suggested: "Put two handfuls of leaves into a litre (1 3/4 pints) of hot water; steep 12 minutes"; bathe skin with it or apply a wet pack, as deemed necessary.

Maurice Chevalier, the famous French singer and film actor who died at the remarkable age of 94 (1888–1972), came to Mességué one time with severe constipation; he was ordered to drink sorrel tea and after just two cups taken that evening and early the next morning, he experienced his first really good bowel movement in *two weeks*. The director had to stop in the middle of an important film scene Chevalier was starring in, while the actor spent 45 minutes in *le lavatorie* doing more than just sitting and twiddling his thumbs!

Mességué's career really took off when he started lecturing on the wonderful healing properties to be found in medicinal and food plants such as sorrel. "Within a few months," he said, "I had spoken in more than twenty towns, from Antibes to Paris, from Rouen to Bordeaux, from Toulouse to Lyon." Medical doctors in attendance were, at first, somewhat skeptical of this upstart young herbalist. But when some of them began to refer a few of their worst cases to him and discovered the health miracles he was working with herbs, they became convinced that here was a fellow who truly understood the healing arts. Moreover, his fame quickly grew and spread to other parts of Europe as well.

In time he acquired enough money to afford a spacious "little villa" (as he modestly called it) at the very upscale Cap d'Aïl. "My little villa was surrounded by British Press Lords," he chuckled. He was soon introduced to Lord William Maxwell Aitken Beaverbrook (1879–1964), the British financier, statesman, and newspaper owner. The first baron had, shall we say, a slight problem *inside* his *derrière*, which Mességué *very carefully* took care of. In trying to explain it to a perfect stranger, Mességué—ever the gentleman—searched hard in his mind for just the right words to use. Put delicately, sorrel packs are dandy for hemorrhoids, even when the treatments were for a stuffy, rigid Britisher of pomp and considerable wealth.

Beaverbrook was so pleased with the results he got that he referred one of his cronies to Mességué. "I liked Winston Churchill from the first," he stated in accented English. "He was a man with an extraordinary, overwhelming personality, and I was completely won over by his simplicity, his outspokenness, his humor, and his way of taking himself seriously and at the same time making fun of himself. We were soon on familiar terms, and I even invited him round for lunch. Churchill's eyes were moist with the affection of a gourmand for good food, and his cheeks were like two little round apples. I had never seen his nose look so small or his smile so broad."

According to Mességué, this famous British prime minister of World War II (1874–1965) "was not terribly fond of doctors." He taught Mességué that well-known proverb: "An apple a day keeps the doctor away," but then added his own little rendition at the end: "Especially if you're a good shot!" In speaking to Mességué of his personal physician, Lord Moran, Churchill remarked: "He's marvelous, he really works very hard at the job of poisoning your life in order to save it. Thanks to him I know what I'll probably die of: B-O-R-E-D-O-M!"

He never fully committed himself to Churchill's *exact* problem, for which he used both decoctions of sorrel, but enough was given for me to form an educated guess: genital herpes, "cold" sores, and fever blisters.

There were other equally fascinating world figures that this man treated in his lifetime. Jean Cocteau (1889–1963), the renowned French writer, visual artist, and filmmaker, benefited greatly from sorrel for an abscess that had formed on an unspecified part of his body. King Farouk of Egypt (after whom the fictional college Farouk U. was sarcastically named) was, in Mességué's own words, "*the* most spoiled, indulged man in the world." Sorrel cured him of a bad case of indigestion, which later turned out to be suspicious food poisoning.

These and numerous other episodes attest to the incredible medical and nutritional benefits of sorrel. No wonder it had been one of Mességué's favorite herbs for so many decades. Sorrel is available in some parts of North America, but not everywhere. In the event you can't locate sorrel, here are some reasonable substitutes for it; I suggest using *double* the quantity, however; raw spinach, raw matured dandelion, raw purslane, raw chickweed, raw endive or romaine, prickly lettuce, or raw mustard greens.

SOYBEAN

(See under BEANS)

SPINACH
(Spinacia oleracea)

Brief Description

As a cultivated plant, spinach originated in or near Persia and later reached Spain by way of the invading Moors around 1100–1200 A.D. Spinach can be either smooth-leafed or, more commonly, of the crinkle leafed "Savoy" type. It can also be round seeded or prickly seeded. The prickly seed varieties, once thought to be the hardiest, are often grown in the United States; whereas, the newer smooth-seeded kinds—thought by many to be superior in flavor—are more popular in Europe. Fresh, young, tender leaves are delicious in salads or can be steamed with a little water until tender. This is the vegetable of purported strength, which has been forever immortalized by that colorful cartoon character, Popeye the Sailor Man.

Potent Cancer Inhibitor

Recently spinach has been linked as a good food for helping to cut the risks of getting cancer. First there was Dr. Chiu-Nan Lai's study published in a 1980 issue of *Mutation Research* (Vol. 77, p. 248). In her report, Dr. Lai of the University of Texas System Cancer Center, M. D. Anderson Hospital and Tumor Institute in Houston, found that the high histidine content in spinach, along with cabbage, parsley, mustard green, and broccoli, exhibited definite antimutagenic activities. That is to say these vegetables keep normal body cells from undergoing mutation, thereby becoming cancerous within a short time.

In August of that same year (1980), another report appeared in *Revista Brasilica de Biologia*, in which it was observed that extracts of spinach which had either been heated at 100°C. for 10 minutes or else freeze-dried, successfully inhibited Ehrlich tumor growth in 2 1/2-month-old male and female Swiss albino mice between 86–95% of the time. Then the November 13, 1986, issue of the *New England Journal of Medicine* linked the beta-carotene in dark-green leafy vegetables like spinach, kale and collards with reducing the risks of getting lung cancer. Based on such scientific evidence as this, it seems advisable to include this simple vegetable in our diets more often.

Antidiabetes Drink

For those with either diabetes mellitus (adults) or juvenile onset diabetes (young people), there is some nutritional benefit to be expected from spinach, since it contains manganese—a trace element important to those who are diabetics. Several handfuls of spinach should be well washed and pulled apart, leaf from leaf. Put these into a saucepan with a little kelp, 1 tbsp. *each* of lemon and lime juice and 1 $^1/_2$ cups of water. Simmer very slowly on very low heat for at least an hour. Then strain out all of the juice through several layers of cheesecloth. Take $^1/_2$ cup before each meal by at least 30 minutes, twice daily.

Spinach May Prevent Macular Degeneration of the Eye

A recent scientific study strongly suggests that spinach and other leafy green vegetables may help ward off an eye disease that ranks as the leading cause of irreversible blindness in people age 65 and older.

In this disorder, called age-related macular degeneration, the macula—the part of the retina at the back of the eye—becomes progressively damaged. As time goes on, victims lose all central vision. Physicians have no cure and no way to prevent it.

Medical researchers have speculated that certain nutrients in foods might protect against or slow the progress of this eye disease. Johanna M. Seddon of the Massachusetts Eye and Ear Infirmary in Boston and her colleagues decided to test that theory.

The team started its investigation with 356 men and women, age 55 to 80, already diagnosed with advanced macular degeneration. A control group of 520 men and women in the same age range suffered from some other form of eye disease. The researchers questioned the volunteers extensively about their diet. From that survey, the team calculated the amount of specific nutrients and vitamins that each recruit regularly consumed.

The scientists discovered that people getting the heftiest dose of carotenoids (pigment molecules) in their diet had a 43% lower risk of advanced macular degeneration than people ingesting the least carotenoids.

The carotenoids—a family of yellow, orange, and red pigments—are found in fruits and vegetables, especially items like spinach. Beta-carotene, the best known of the group, has been associated with a lower risk of cancer and heart disease.

However, when Seddon and her colleagues zeroed in on the specific carotenoids protecting the macula, they got a surprise: beta-carotene wasn't the star of this story. The other pigments, lutein and zeaxanthin, accounted for the risk reduction.

The beauty of that finding is that it makes biological sense. Lutein and zeaxanthin form the yellow pigment in the macula of the eye. Researchers believe they may prevent damage to the eye by filtering out visible blue light.

Seddon's study of spinach nutrients for the prevention of this blinding disease, along with an accompanying editorial by Susan E. Hankison of Bringham and Women's Hospital in Boston, appeared in the November 9, 1994, *Journal of The American Medical Association.*

Simple Recipes

Here is a couple of ways to prepare spinach without having to do a lot of work. They may be served as part of a meal or simple snacks by themselves.

Creamed Spinach

Needed: Approximately 1 1/2 cups raw spinach; 1 minced clove of garlic; 1/2 tsp. tamari (available from some Oriental food shops and health food stores); 2 tbsp. yogurt; 1/2 tsp. kelp; 1 tsp. lime juice; 1 tbsp. grated Parmesan cheese.

Wash the spinach, remove large stems, and shake excess water from leaves. Reserve stems for making soup stock. Place spinach, along with garlic, kelp, tamari, and lime juice in a large saucepan and steam in the small amount of water that clings to the leaves. When spinach is limp and has turned a deep green, remove from heat. Place spinach, drained if necessary, in a blender with yogurt and cheese. Process on low speed until spinach is puréed. Serves 2.

Cold Spinach Soup

Needed: 1 $^{1}/_{2}$ cups yogurt; 1 cup chopped spinach; 2 chopped scallions; $^{1}/_{4}$ cup chopped parsley; $^{3}/_{4}$ tsp. *each* lemon and lime juice; 1 tsp. tamari; $^{1}/_{2}$ tsp. kelp; 1 mashed garlic clove; dash of paprika for garnish.

Put the yogurt in a blender and then slowly add the other ingredients, with the blender set on medium speed. Blend until reduced to soupy consistency. Some of the spinach will be totally pulverized, while much of it will still be in small bits. The combination of spinach, parsley, and scallions give a real zing to this soup. If it's a bit too zingy for your taste buds, dilute the flavor with some more yogurt. Finally, garnish each bowl with a sprinkle of paprika. Other variations are to add $^{1}/_{2}$ tsp. of fresh grated ginger root when blending everything up, or replace the paprika with a single cross-sectional slice of sweet red pepper instead.

Squash

(See under PUMPKIN, SQUASH, AND GOURD)

Stinging Nettle

(See under NETTLE)

Strawberry

(See under BERRIES)

Sunflower

(See under SEEDS)

Swiss Chard

(See under BEETS)

Tamarind
(Tamarindus indica)

Brief Description

This is an evergreen tree which grows in tropical climates, reaching up to 80 feet or more with a shaggy, brownish-gray trunk that often is 25 feet in diameter. The alternate, even-pinnate leaves have from 20–40 small, opposite, oblong leaflets, and the pale yellow flowers have petals with red veins, growing in racemes at the ends of the branches. The fruit is a cinnamon-colored oblong pod, 3–8″ long, with a thin, brittle shell enclosing a soft, brownish, aciduous pulp.

An Expeditious Laxative and Fever Coolant

The principal and widespread use of the ripe, sweet-sour, stringy pulp throughout the Americas and Caribbean is as a laxative. The ripe pulp contains between 10.86 and 15.23% of tartaric acid, the constituent frequently found in some baking powders that is believed to be responsible for stimulating bowel movements. Generally, only one ripe fruit is needed for mild constipation problems.

In the West Indies, the ripe pulp from two fruits (minus their square, glossy, brown, hard seeds) is blended either by hand or mechanically with two cups of ice-cold water and 1 tsp. of white sugar (an equivalent amount of honey may be substituted instead). This, then, is administered to those who are burning up with fevers and within a relatively short time has cooled the body temperature down by several degrees.

Baking Powder Alternative

Toward the end of the section devoted to Grapes and Raisins, a recipe is given for making baking powder. Those who have access to ripe tamarinds may want to substitute the two parts of cream of tartar with four parts of the fresh fruit juice. This should be done *only* at the time any baking is intended. Combine the one part each of baking soda and cornstarch first, then slowly stir in the tamarind juice, mixing everything thoroughly. Then add this to the rest of your recipe that calls for a specific amount of baking powder (which may be less than what you have to work with). This unusual type of homemade baking powder really gives breadstuffs an original and unique flavor.

Jelly Making in the Tropics

Throughout the Philippines the brownish pulp and seeds of ripe tamarinds are often made into balls of less than fist size and sold in the public, open-air marketplaces under the local name of "tamarindo." A number of Filipino housewives make jams, jellies, sweets, and sherbet drinks from them, some of which I've tasted and found exceptionally delicious. This is because tamarind contains nearly 7% pectin, which is an important ingredient in making jellies and jams (see under Peach, Pear, and Quince for more details).

TANGERINE
(Citrus reticulata)

Brief Description

Tangerine is the generic name for a number of small citrus varieties with fragrant, loose, orange skin. They also go by the name of mandarin oranges. These are the fruits of memorable childhood Christmases for many of us, one of the healthier stocking-stuffers our parents gave us.

The tree has a slender trunk and grows upward to about 10 feet. The leaves are evergreen, alternate, and rather aromatic. Each leaf is somewhat leathery and dotted with oil glands, which account for the odor. Flowers are quite fragrant, white, and five petaled.

Ripe tangerines yield a nice scent and can be orange or red-orange in hue. The loose skin is dotted with oil glands. Many fibers loosely clasp the 8 to 15 easily separated segments containing the juice sacs and seeds. The center of the fruit is hollow.

Tangerine is named after the ancient walled Moorish town of Tangier in North Morocco; it sits on the Strait of Gibraltar and has a busy port. The fruit grows in abundance there.

Juices Makes a Wonderful Decongestant for Asthma and Bronchitis

Bothered with respiratory disorders? Consider drinking more tangerine juice. According to scientific research done back in 1965, not only is the fruit high in vitamin C, but it also has a large amount of the phenolic amine called synephrine. This is a well-known decongestant. A glass each morning is recommended to clear mucus out of the lungs.

Diabetes Helped with Tea Made from the Peel

Tangerine peel has a reputation in Venezuela as being very useful for sugar diabetes. The peel of 3 fruits is boiled for 10 minutes in a quart of water. The decoction is kept in the refrigerator, *un*strained, and taken each day.

A man I know in Caracas told me that he and six of his friends, who are diabetics, have been taking this tea for quite some time now. In the beginning, Julio Havalar told me that his blood sugar level was 171 mg.%. Medication the doctors gave him didn't help. So he started drinking this tangerine peel tea every day and claimed that his blood sugar level dropped to 94 mg.%.

Combat Thrush/Yeast Infection, Worms, and Dysentery the Easy Way

Julio said his wife rubbed some of the fruit juice on the tongues of their infant children to combat thrush (a form of yeast infection). He said the fungus cleared right up.

Both of them drink glassfuls of tangerine juice when they suspect they have intestinal parasites; he said the worms never fail to come out when they do this.

A cousin of his had dysentery and took the tangerine juice every day, as well as eating the whole fruit. He got better in four days.

TAPIOCA
(Manihot esculenta)

Brief Description

The commercial tapioca pearls commonly used in puddings come from the starchy long, thick, tuberous roots of a half-woody shrub that grows anywhere from 3 to 12 feet high. These dark-brown, fat roots contain not only solid white starchy insides as potatoes do, but also considerable milky latex as well.

Poultice for Skin Eruptions

In Venezuela, Trinidad, and elsewhere throughout the Americas, the dried, powdered starch of the root is used as a highly effective poultice on erysipelas, eczema, whitlows, boils, carbuncles, abscesses, and herpes lesions. A similar poultice can be made by you at home for the same purposes. Combine 1 tbsp. of raw, peeled, and grated potato with $^2/_3$ tsp. of granulated, quick-cooking tapioca (*un*cooked, however) in just a tiny bit of water to make a nice, even, sticky paste. Spread this on several layers of gauze cut into a small square. Affix this to the skin eruption and secure in place with adhesive tape. Change every day until purulent matter is drawn out of the sore and healing ensues.

Helps Heal Stomach Ulcers and Colitis

In São Paulo, Brazil, some doctors and local folk healers employ cooked tapioca for helping to heal peptic ulcers and colitis. Bring 1 cup of water to a boil, adding 2 tbsp. of Knox unflavored gelatin until thoroughly dissolved by vigorous stirring. Then add 4–6 average-sized ice cubes to the solution. When cool, slowly add this mixture to 8 stiffly beaten egg whites. Then add 1 cup pure maple syrup, a pinch of cardamom, and 4 tbsp. of granulated, quick-cooking tapioca. Stir until the mixture becomes firm, then eat a dish of it. The rest may be refrigerated for consumption later. This simple pudding, if consumed regularly, has been known to promote healing of the entire intestinal tract within a short time.

Soothes Sore Breasts During Nursing

A common trend today among many women from all economic levels of society is to nurse their own infants, instead of bottle-feeding them ready-made preparations such as Similac, for instance. Usually women in the lower working and poverty classes who are used to having and feeding many children are seldom ever troubled with inflamed breasts and sore nipples. But middle- and upper-class mothers, many of whom are career oriented, apparently seem to be more frequently afflicted with these conditions when nursing their own, some for the very first time as a matter of fact.

Now there is a useful remedy from Argentina for these very things. Mix some granulated, quick-cooking tapioca in milk, according to the instructions on the package or box. Be sure what you're making has the consistency of cooked cream-o'-wheat cereal. When it reaches this stage, ladle out with a wooden spoon onto several layers of gauze or a clean piece of muslin cloth. Then apply this poultice while still somewhat hot to one or both inflamed breasts, covering it with a bath or hand towel to retain the heat for as long as possible. Repeat as often as needed and change when it has become cold.

Enema for Infant Diarrhea

In Venezuela, a little granulated, quick-cooking tapioca (but *un*cooked) is thoroughly dissolved in some water and employed as an enema in cases of diarrhea, especially in newborn infants. In the latter case, a bulb syringe would be used.

Delicious Pudding

This recipe can be made without orange juice by replacing the juice with a half cup of skim milk. Then add any fresh fruit in place of the orange segments. I'm indebted to the author and publishers of *Recipes to Lower Your Fat Thermostat* for the use of this delicious pudding recipe.

Tapioca Pudding

Needed: 1 $^1/_2$ cups skim milk; 1 tbsp. brown sugar; 3 tbsp. quick-cooking tapioca; 2 egg whites; $^1/_2$ cup orange juice; $^1/_2$ tsp. pure vanilla; either $^1/_2$ cup diced orange segments or $^1/_2$ cup pitted and quartered dates. Combine milk, sugar and tapioca in a saucepan. Let stand 5 minutes. Beat eggs till stiff. Set aside. Bring tapioca mixture to a boil over medium heat, stirring constantly. Add orange juice. Cool, stirring occasionally. Add vanilla. Fold in egg whites and diced orange segments or dates. Chill and serve. Makes about 6 servings. When dates are used, omit the sugar and orange segments. Add dates to boiling tapioca mixture *before* orange juice is added. Cook about 6 minutes. Add orange juice and proceed as above.

TEA
(Camellia sinesis)

Brief Description

Tea drinking in some Asian countries has evolved into a delicate art, much as wine sampling or tasting has done in France. There are connoisseurs of finely brewed tea who can tell what type of water was used, what kind of utensils were involved, and the approximate conditions under which a particular tea is made. In mainland China, some teas are so incredibly strong to the palate that they're served in (literally) thimble-sized cups.

Relieves Migraines

Black and green teas both contain caffeine (1–5%). Since caffeine constricts the blood vessels in the head, it's able to calm the pain caused when they throb and swell. During my 1980 visit to the People's Republic of China, I noticed a number of traditional medical hospitals administer black tea to their patients suffering from migraine headaches.

In the Soochow Chinese Traditional Medical Hospital in Soochow, there was a remarkable 92% recovery rate for migraines. To 1 cup of hot water, simply add 2 teabags of black tea and steep for 20 minutes until very strong. Then drink while still very warm, but not so hot as to burn the tongue or injure the inside of the mouth. "Works every time!" I was told by my hosts in China.

Great Infection Fighter

Throughout mainland China I found black tea being used in numerous hospitals and clinics to treat all kinds of infection and inflammation of the stomach, lower intestines, colon and the liver with recovery rates averaging 83–100% for the majority of patients treated. A strong cup of warm tea was given four to five times daily for these various infectious diseases.

Reduced Atherosclerosis

University of California scientists have discovered that tea drinkers experience a lot less hardening of the arteries than coffee drinkers do. It seems that the caffeine in coffee is bound with some heavy oils, which tend to elevate serum cholesterol levels. But not so with either dark or green teas. In fact, it's believed that the caffeine content in both teas may actually help to cut cholesterol. Besides this, the tea's polyphenols act in concert with vitamins C and P that help strengthen the blood vessel walls of the heart more.

Tea Removes Dental Plaque

During 1983 and 1984 dental scientists at Washington University in St. Louis conducted a series of experiments proving that black tea definitely inhibits the growth of decay-causing bacteria common to plaque buildup on the teeth. This is probably due to the high natural fluoride content found in both teas.

A 1977 dental study conducted in Taiwan showed that 50 weaning rats given cavity-producing foods like white bread, white sugar, and carbonated soft drinks, but also given green or black teas, had anywhere from half to three-fourths *less* cavities than those rodents which didn't receive any tea. Brew a Lipton Brisk Tea bag in a cup of hot water for at least 6 minutes to get the maximum removal of fluoride. Also squeezing the tea bag before discarding helps. Use either tea as a good mouth wash and dental rinse after every meal of sweets. The April 1986 issue of *Dentistry* encouraged people to drink more tea to reduce cavities and plaque buildup.

Celebrated Food Expert Clears Up Hoarseness with a Cure of Her Own

In the first week of January 1995, *The Wilmington* (Delaware) *News Journal* carried an interesting story about 83-year-old Julia Child, author of the definitive *Mastering the Art of French Cooking*.

It was on a Sunday morning and fog and humidity had taken their toll on her vocal cords. The culinary grande dame's distinctive voice kept breaking up in take after take of the opening shots for "Good Morning America's" visit to the Hotel du Pont in downtown Wilmington. Child was unable to get past her opening line.

She told members of the anxious TV crew that "my hoarseness just isn't normal." After the fourth try, she reluctantly admitted with an apology, "I still have that frog." Shaking her head and narrowing her eyes, she worked desperately to clear her throat, but to no avail.

Finally, turning to hotel manager Jacques Amblard, she ordered some tea. A makeup artist on the set rushed off to fetch some lozenges as a backup measure in case the other failed. The program producer stood with hands on hips, frowning at the hotel lobby's ornate ceiling.

But television's most famous chef wasn't the least bit worried. Once the tea arrived, she added $^1/_2$ tsp. of lemon juice and 1 tsp. honey and slowly sipped the hot liquid. After she had finished, she set the empty cup aside and remarked, "Let's give it a few more trys, shall we?"

By now her voice had assumed the firmness fans have come to expect. And she was correct; in fewer than six takes, her voice was again clear and distinct and she executed her just-written lines perfectly.

She told reporter Valerie Helmbreck afterward that "this remedy will always take care of hoarseness, no matter how bad it gets. I've depended on hot black or green tea with a little lemon and honey for many years to keep my voice in good shape for critical moments like this."

TOMATO
(Lycopersicon esculentum)

Brief Description

The earliest tomatoes were harvested by the Incas of the Andes, but later carried to Europe by the Spanish conquistadors. These small, yellow fruits were about the size of today's cherry tomatoes. But fear and ignorance, to a certain extent, kept them from becoming very popular, especially in France and England. Even when reintroduced to America, they were still thought of as being poisonous, just like other members of the deadly nightshade group happen to be. The Creoles in New Orleans finally brought tomatoes into the kitchen in 1812, but another half-century had to pass before other sections of the country got up enough courage and curiosity to try them.

But even then the tomato's troubles weren't over. Botanically, they are really fruits, which confused many people because they are most often used like vegetables. Finally, it took, believe it or not, a ruling by the Supreme Court in 1893 to reclassify the tomato as an official vegetable, even though botanically speaking, it isn't and never will be!

Varieties of tomatoes available today include the large, all-purpose beefsteak types; the oval plum variety, used chiefly for cooking purposes; the small, tasty cherry tomato, often served in salads; and the large, yellow or orange, low-acid tomatoes.

Looking for the Best Tomato

In the United States, more tomatoes are consumed than any other single fruit or vegetable. They far outdistance oranges, potatoes, and lettuce, to say nothing of such popular staples as peas, carrots, and bananas. But as the editors of *Reader's Digest* have noted, "Today's thick-skinned commercial tomatoes, usually picked green and 'ripened' with ethylene gas, have nothing to recommend them beyond the dubious values of the marketplace: ease of shipment and long shelf life."

"Such tomatoes are really only symbols of tomatoes," griped the food editor of the *Los Angeles Herald Examiner*. "They are tasteless, have a mealy texture, and a lower vitamin content than fully grown tomatoes that have ripened on the vine. To distinguish a vine-ripened tomato from a gas-ripened one (without tasting it), smell it. Gassed tomatoes are odorless, and vine-ripened ones have a telltale tomato fragrance about them."

As you've probably guessed it by now, the very best tomatoes to look for are the succulent, vine-ripened kind bought in season from local farms, outdoor produce stands, or truck markets or simply grown in your own backyard. Nothing can compare with them in flavor and nutritional goodness!

Dried Tomato Prevents Diarrhea

Researchers at the Animal Nutrition Lab at Cornell University reported in the April 19, 1940, issue of *Science* that tomato pomace (dried residues from tomato juice) stabilized and gave form and consistency to the feces of dogs, foxes, and minks when it was added to their diets. Before these improved mixtures, the looseness of these animals' stools almost bordered on diarrhea. But by adding to the diet a quantity of ground, dried tomato pomace (seeds, skin and minor pulp) equal to 5% of the wet ration, their feces assumed more solid and uniform consistency.

A similar preparation can be made at home to improve the form and consistency of your own stool. Choose firm, just-ripe tomatoes. Wash and remove stems. Slice them into 1/2" thick slices with an even thickness throughout (not like the salad-sliced wedges, however). Place them on plastic trays, not metallic due to their high acid content, if you intend to dry them in a dehydrator for 2 hours at 155° F. and another 9 hours at 125° F. Or line a flat cookie sheet with foil and put them on if you intend to dry them in your oven instead for the same length of time at the identical temperatures just cited.

By the time they're done, they will be very thin and crisp. Check around the peel for any remaining moisture beads. When cool, store in airtight containers. Four to five slices equals a fresh tomato, depending on the size, of course. You can also put the dried slices in the blender and make a nice tomato powder. Figure about a dozen dried slices for enough tomato powder to stir in 2/3 cup of cool water to drink for mild diarrhea. This amount can be increased, as needed.

For a really effective antidiarrheal preparation combine equal parts of dried tomato powder and dried apple powder, having rendered them into this state by way of your food blender. Plan on 2 level tbsp. (one of each) in 1 1/4 cups of warm water to take care of the problem nicely.

To make dried apples, cut *un*peeled apples into slices or rings about 1/4″ thick. Pretreat them by dipping slices in lime juice before drying. Dry at 155° F. for 2 hours, then at 125° F. for about 3 hours more in your dehydrator or oven. Test for dryness by removing a slice, allowing it to cool, then cutting it in half. No moisture beads should appear when squeezed. Pieces that aren't quite dry tend to feel cooler, as well as moist. Remove the slices as they dry, allow them to cool, then store in airtight containers.

Dried tomato slices can also be topped with a slice of slightly melted cheese for a tasty snack or used in soup. Apple and tomato slices may be rehydrated for culinary purposes by covering them with boiling water and steeping for 20 minutes in a pan with a lid on top.

Vine-Ripened Tomatoes for Hypertension

Clinical evidence cited earlier in this work indicates that potassium has a very positive influence upon the kidneys, whereby high blood pressure can be substantially reduced in many instances. Now just one jumbo tomato approximately 3″ in diameter and weighing some 7 oz. contains nearly 450 mg. of potassium. Imagine then if two such good-sized tomatoes were consumed every other day in the form of a delicious vegetable cocktail, just how much potassium the body would be getting.

But wait, there's more: if 1 tsp. each of ground tarragon, paprika, ground turmeric, and ground basil and 1 tbsp. of lemon juice, along with $^{2}/_{3}$ cup of spring or distilled water were added with the two ripe tomatoes and thoroughly blended, you would have a zesty, lip-smacking beverage that would contain almost 1,200 mg. of straight potassium *and only* about 15 mg. of sodium. This is an incredible ratio of 80:1 and potent therapy for helping to bring hypertension under control!

Turning Sunburn into a Nice Tan

Paul Neinast, owner and operator of Dallas' leading beauty salon, came up with a nifty idea for turning a relatively painful sunburn into a modestly decent tan. He takes peeled tomato slices, soaks them in buttermilk, then applies them directly to the skin. They not only help to relieve the pain, but also close up the pores and turn the burn into some kind of a tan. Another way he has used them is to make a purée out of *thinly* peeled tomatoes with a little bit of buttermilk added, but not enough to make it too runny. This, then, is spread over sunburned skin to give a more even and slightly darker tan.

Heals Festering Wounds and Sores

In Papua, New Guinea, some Stone Age tribes still rely on the pounded leaves of wild tomato plants in the form of poultices as a means of helping to heal old wounds and sores. A couple from Rhode Island shared similar minor experiences of their own with me in this regard. He had a badly festering forefinger that kept oozing out a lot of pus and blood. So he took a slice of fresh tomato and wrapped it around the finger, holding the same in place with some adhesive tape. He changed this a couple of times a day and within 2–3 days the infection had all cleared up.

His wife, on the other hand, had been trying out some new high heels recently purchased at an expensive department store in downtown Providence. But they were too small for her and kept pinching her big toe. Soon an excruciating pain developed. She just put a slice of tomato over it, then some gauze, and finally taped it down. *In less than a day*, the pain had entirely gone! Such is the power of tomatoes.

Building Strength in Place of Fatigue

One of the common symptoms seen so frequently in hypoglycemics or those who have low blood sugar is their constant fatigue and apparent lack of energy. Well it seems that tomatoes can play a vital role in rebuilding their strength. Doctors at Tohoku University in Sendai, Japan reported that fresh tomato juice was extremely effective in accelerating the glycogen (blood sugar) formation in normal rabbits by further stimulation of the liver.

Nor should it come as any surprise either that the next most important constituent of the tomato are various naturally occurring sugars, which account for some 50% of the total dry matter. Ripe tomatoes are especially high in glucose and particularly fructose. Interestingly enough, when tomatoes are grown in the shade, their sugar contents are drastically reduced.

Furthermore, for those who are athletically active in any way, they may find powdered tomato seeds worth taking as an ideal protein supplement. The amino acid profile of tomato seed protein, says the *CRC Critical Review in Fd. Science & Nutrition* for November 1981 is "similar to that of sunflower and soybean." Anyone fortunate enough to live near a cannery or soup factory can probably get all the free tomato seeds that their hearts desire.

The Health of the Liver

In the Soviet Union, many doctors prescribe tomatoes in the diets of factory workers exposed to toxic chemical occupational environments. One reason for this may be due to the fact that tomatoes contain two very important detoxifying trace elements, namely, chlorine and sulphur.

Natural chlorine helps to stimulate the liver in its function as a filter for body wastes and further assists this major organ in its efforts to remove toxic waste products from the system. Sulphur helps to protect the liver from cirrhosis and other debilitating conditions. In 100 grams of raw, edible tomato ($2\ ^2/_3$" diameter and $3\ ^1/_2$ oz. weight) there are 51 mg. of chlorine and 11 mg. of sulphur.

Limited clinical evidence also suggests that fresh juice from vine-ripened tomatoes can actually help the liver to regenerate or reproduce a part of itself if another portion has been destroyed or surgically removed. Thus we can see that tomatoes definitely help promote the health and well-being of one of the body's most important organs. "A tomato a day, keeps the liver in good stay!"

A tomato is also good to consume when eating too much animal fat in the form of butter, cheese, eggs, pork, beef, and many deep-fried foods. The tomato will usually help to dissolve this fat, thereby preventing hardening of the arteries.

Challenging Recipes

The following culinary uses for tomatoes should challenge your imagination and definitely tantalize your palate for sure.

Early New England Tomato Catsup

"The best sort of catsup is made from tomatoes," observed a very serious Mrs. Child in her 1832 *American Frugal Housewife*, as if to imply almost that catsup can be made from other types of vegetables as well. "The vegetables should be squeezed up in the hand, salt put to them, and set by for 24 hours. After being passed through a sieve, cloves, allspice, pepper, mace, garlic and whole mustard-seed should be added. It should be boiled down one-third and bottled after it is cool. No liquid is necessary, as the tomatoes are very juicy. A good deal of salt and spice is necessary to keep the catsup well. It is delicious with roast meat; and a cupful adds much to the richness of soup and chowder. The garlic should be taken out before it is bottled." Anyway that's pretty much how they made catsup in Boston in those days.

Tomato Fettucini

Needed: 7 oz. (approximately $^3/_4$ cup) tomato pasta; 6 tbsp. raw, chopped shallot bulbs; 8 slices of dried tomato (see this same section under "preventing diarrhea"); 1 tbsp. olive oil; 1 tbsp. melted butter; 4 tbsp. white wine; 10 olives; 2 slices of goat cheese; pinches of basil, rosemary, sage, tarragon, and parsley. Sauté shallots with butter and oil. Add sun-dried tomatoes, olives, and white wine. Reduce slightly and add herbs. Toss with tomato pasta and goat cheese.

Tomato Leather

A popular snack and much less expensive to make at home is tomato leather. Before you begin make sure the tomatoes are clean, with any bruises or spoiled areas removed. Cut them into $^1/_2$" chunks and leave the peel on. Purée the fruit in a blender until no large chunks remain. You can then add 1 tsp. lime juice per cup of tomato purée to help prevent it from darkening any.

Place the puréed fruit on lightly oiled plastic dehydrator sheets or foil-covered and lightly greased cookie sheets. Spread the purée out as evenly as possible, less than $^1/_4$" thick. Dry the fruit leather at 135° F. for 6–8 hours in a dehydrator or low-set oven until leathery and no longer sticky to the touch. Peel the leather off the trays or sheet pans and cut into serving size pieces. Place on clean plastic wrap, rolling the slices along with the plastic wrap, jelly-roll fashion to keep the tomatoes from sticking to themselves. Store the slices in a plastic bag in a cool, dark place. This same process can be used for different kinds of fruit as well; some, like bananas, pears, pineapples, watermelon and oranges, would need to be peeled, however, while others would not.

Enchilada Casserole with Tomato Sauce

The author is indebted to La Rene Gaunt and the publishers of her book, *Recipes to Lower Your Fat Thermostat* for letting him use their "out of this world" Mexican meal treat.

Needed: 1 chopped onion; 1 minced garlic clove; 1 chopped green bell pepper; $^1/_2$ cup sliced mushrooms; 1 recipe basic tomato sauce (see below); 1 tsp. chili powder; 1 $^1/_2$ cups cooked pinto beans; $^1/_2$ cup yogurt; 1 cup cottage cheese; 8 corn tortillas; $^1/_2$ cup grated Mozzarella cheese.

Sauté onion, garlic, green pepper, and mushrooms until onions are transparent. Add tomato sauce, chili powder and beans. Heat through. Stir yogurt and cottage cheese together. In a 1 $^1/_2$ quart casserole dish rubbed with liquid lecithin (available from any local health food store), put a layer of tortillas, a layer of sauce, sprinkle of Mozzarella cheese and a layer of yogurt mixture. Repeat until all ingredients are used, ending with a layer of sauce. Top with yogurt mixture. Bake at 350° F. for 15–20 minutes. Serve hot. Makes 8 servings.

Basic Tomato Sauce

This sauce has a wide range of uses. It can be poured over spaghetti, stuffed cabbage, stuffed green peppers, cooked rice and so forth.

Needed: 1 cup diced onion; 2 minced garlic cloves; 1 28 oz. can stewed tomatoes; 3 tbsp. tomato paste; 1/2 tbsp. kelp; 1/2 tsp. oregano; 1/2 tsp. basil; 1 diced green bell pepper (including the seed center); 1/2 cup mushrooms.

Tomato Versatility

One of the tomato's great virtues is its versatility. You can stew them using no water at all by dicing them and stirring over low heat. You can broil them under moderate heat for about 10 minutes, creating fresh taste sensations through judicious applications of toppings, or bake halves in a preheated 425° F. oven for 10–15 minutes. Here are a few ideas for toppings, from the kitchen of the California Fresh Tomato Advisory Board: (1) bread crumbs seasoned with a pinch of basil, sage, thyme, or lemon thyme; (2) grated Parmesan or cheddar cheese; (3) chopped green onions or chives and butter; (4) blue cheese crumbled in sour cream; (5) chopped smoked salmon in softened cream cheese.

Finally, you can stuff your tomatoes, but be careful of the ripeness and consistency. Leave them unpeeled, cut out the stem and core, then, starting at the center of the uncut end, divide into quarters, three-fourths of the way down. Pull apart gently and fill with tuna, chicken, or vegetable salad; cottage cheese; or avocados chunked or mashed. Or try a real treat by precooking chopped onion, cracked wheat, tomato pulp, chopped mint and raisins in olive oil, then stuff tomatoes with this mixture. They are then baked in a 350° F. oven for 30 minutes. This last version of Tomatoes à la Greque is an international favorite.

TROPICAL FRUITS

Carambola (Averrhoa carambola)
Guava (Psidium guajava)
Mango (Mangifera indica)
Mangosteen (Garcinia mangostana)
Papaya (Carica papaya)
Passion Fruit (Passiflora edulis)
Pineapple (Ananas comosus)

Brief Description

CARAMBOLA. This small shrub in the wood sorrel family is also called star fruit. It is an orange fleshy fruit in appearance. Carambola looks like nothing else you've ever seen in the way of a fruit: a glossy golden blimp with ribs, it forms star shapes when cross-cut. The edible skin encloses a juicy, slightly tangy flesh that suggests a milky blend of plums, apples, and grapes. The sourness is due to high amounts of oxalic acid. But cultivated varieties which abound in Florida are much sweeter in taste when eaten raw. The juice is often used to remove stains from clothing and other articles, because of the oxalic acid content. It is available from August into March, but mainly at specialty fruit markets. If refrigerated it can keep much longer than other tropical fruits, up to 2 weeks, if in good condition.

GUAVAS. Guavas are small, thin-skinned tropical fruits. They are often processed into jellies, jams, and preserves, but they can also be consumed fresh. The fruits are round to pear shaped, usually less than 3″ in diameter, with green or bright-yellow skins; some have a reddish blush. Ripe guavas have a musky, pungent odor. They contain small, hard seeds that may irritate the throat, but some of the newer varieties are relatively free from seeds. Guavas are sensitive to frost, which explains why the majority of them grown in this country are found only in California and Florida.

MANGO. These are the most luscious of all tropical fruits. However, when they aren't of good quality, the flesh can be disagreeably fibrous, with a flavor of turpentine. These highly perishable fruits also vary greatly in size—anywhere from 6 oz. to 4–5 lb.—and shape: they can be round, oval, pear, or kidney shaped, or even long and thin. The tough skin is usually dull green, with red and yellow areas that broaden as the fruit ripens. Some people have an allergic skin reaction to the fruit beneath the peel and must wear protective gloves when peeling the fruit. India still produces 80% of the world's crop of mangoes.

MANGOSTEEN. David G. Fairchild was a distinguished botanist. He organized the Office of Foreign Seed and Plant Introduction (SPI) for the Department of Agriculture in 1898. Fairchild was instrumental in the establishment of the Florida and California avocado industries. Between 1898 and 1912, he saw SPI bring in more than 30,000 distinct varieties and species of foreign plants. At its peak, in 1916, SPI was receiving 40 distinct plant forms a week from abroad.

Mangosteen from Malaysia and Indonesia was one of these in which Fairchild took a personal interest and considerable liking to. The mangosteen is not related to the mango. But I realized that the very first time I saw one of these unprepossessing little fruits in Kuala Lumpur, the capitol of Malaysia. It is hard to believe that this is one of the most incredibly delicious of all tropical fruits I've ever had the pleasure to sink my teeth into.

Mangosteen is the size of a small apple with a leathery, purplish-brown, inedible skin. The skin and pith are thick for the size of a fruit, and a deep, bright, staining pink. (It's not advisable to eat one of these ripe juice little fruits while wearing a white shirt; believe me, the stain will *never* totally come out!) The enclosed fruit itself consists of five pearly white, translucent segments in the center of which are two inedible stones. The flavor is exquisitely sweet and fragrant with a lively touch of acidity.

Fairchild established his own exotic garden (named after him) in Coconut Grove, Florida, where he later settled with his wife, Marian, who just happened to be the daughter of the famous Alexander Graham Bell, the inventor of the telephone. Fairchild believed that his beloved mangosteen was too rare and delicate a fruit for its beauty to be masked with heavy cream or sickly sweet syrup. So he instructed his wife to prepare the fruit without these additions; she cut around the top third of a mangosteen, lifted it off and scooped out the flesh with a small teaspoon.

Mangosteen is now grown throughout much of Southeast Asia and Central America (especially Honduras) and also in parts of the American Deep South.

PAPAYA. A native of the Caribbean, the papaya now grows abundantly throughout tropical America. The fruit is usually pear sized and has a central cavity filled with edible, pea-sized black seeds; the sweet, juicy flesh is rather bland, with a slight muskiness and a melonlike texture. Unripe papayas can be baked or boiled as vegetables, and the leaves, if attached, are often cooked as greens. Like mangoes, papayas also secrete a fluid that usually causes an allergic skin reaction in some people, thereby necessitating the wearing of rubber gloves while peeling the fruit.

PASSION FRUIT. True passion fruit is a smooth ovoid about 3″ in length. In its natural ripened state, it can have either a reddish-purple leathery skin or else a pale yellow one instead. The fruit is green, however, when unripe. It originated in Brazil, but is now cultivated in Hawaii and South America, particularly in Colombia.

By the time it reaches us it is often somewhat shriveled in appearance. An old wives' tale claims that the more wrinkles there are, the better they taste. But this isn't true; they just get older. The leathery skin encloses a thin reddish pith surrounding a white membranous lining which in turn encloses small edible crunchy seeds, each one set in a fragrant, intensely flavored, sweet-sour, and translucent pulp, dark greenish orange in color.

To get to the sweet pulp cut the top off the fruit like an egg and spoon out the interior. Both the pulp and seeds are edible. If you just want the pulp, which is excellent, simply rub through a fine sieve with a tablespoon of boiling water, which will help to remove the maximum amount of pulp from the seeds.

The juice can be used for marinating. It especially goes great with game like venison and pheasant. The marinade can also be added to the sauce or gravy of such items at the end of their cooking time.

PINEAPPLE. These plump, heavy fruits with fresh, green crown leaves, emit a fragrant aroma and have a very slight separation of the eyes or pips. Though the shell turns yellow as the fruit matures, pineapples do not ripen after harvest as some are inclined to think is the case. Columbus encountered the first pineapples on his second voyage to the New World on the tiny island of Guadeloupe. These fruits were not introduced to Hawaii until 1790 and it took until the early part of the twentieth-century for Hawaiian plantations to dominate the world market in this delicious fruit.

Guava Juice for Congestion

The Negritos of the Philippines have used ripe guava as a tonic for strengthening weak hearts. Among the Choco Indians of Panama, ripe guavas are consumed to overcome congestion of the lungs and throat. Fresh juice can be made by thoroughly mixing one coarsely chopped, ripe mango in 1 cup of crushed ice and $^{2}/_{3}$ cup of ice water. Sip slowly. Or 1 cup of canned mango juice may be taken at least 3 times a week for helping improve the heart. Those bothered with asthma, bronchitis and hayfever would benefit from drinking $^{1}/_{2}$ cup of *lukewarm*, canned mango juice with $^{1}/_{4}$ tsp. of lime juice and 4 drops of pure vanilla stirred in, twice a day during bouts with serious respiratory congestion. Drinking this just before retiring will enable an asthmatic to rest better without choking up so much during the night.

Guava Leaf Extract for Seizures

In parts of the West Indies, an alcoholic extract is made from the pounded leaves of guava plants for treating epileptic seizures and convulsions. The tincture is also rubbed into the spines of young children and young adults, besides being given internally.

If guava plants are available near you, then pick about 10 leaves, crush them lightly and snip into inch-square pieces. Put them into the bottom of a quart glass Mason fruit jar and cover with 1 cup vodka and 1 cup gin. Cap and permit the herbs to steep in this solution for 15 days, remembering to shake the jar twice daily. Then strain through a fine wire sieve and put in a clean pint jar with a ring lid. Store in a cool, dry place until needed.

Two teaspoonfuls of this alcoholic extract may be taken by young to middle-aged adults suffering from seizures and convulsions (half this amount for children under the age of 15). Some of the tincture should also be rubbed on the spine each day from the base of the neck to the tailbone.

Mango and Hypertension

On the island of Curacao, many residents drink a decoction of the semidried leaves for high blood pressure. Pick 1 $^1/_2$ large or 2 medium-sized leaves and dry halfway on a piece of clean cloth for only one day. While still somewhat moist, snip with shears into inch-size pieces and place in 1 quart of boiling water. Cover, remove from the stove and steep for an hour. Strain and drink 2 cups a day, 3 days in succession, as they do in Curacao. Then go for several days without the tea, before repeating the process all over again. This sedative action to the leaves may be due, in part, to eremophilene, which also occurs in the herbal tranquilizer, valerian.

A Proven Antidiabetic Remedy

A group of tannins called anthocyanidins are found in mango leaves. For over 20 years, according to a back volume of the French journal, *Plantes Médicinales et Phytothérapie* (Vol. 11, pp. 143–151, Suppl.), some European physicians have been using a watery extract (tea) of mango leaves to treat not only diabetes, but also blood vessel problems and eye complaints related to this disease. They help to slow down the progress of diabetic angiopathy (disease of the blood vessels due to diabetes). In fact, the definite improvement observed in diabetes is due primarily to the healing influence of these leafy compounds on the blood vessels in and around the pancreas.

Excellent results have also been obtained with a tea made from mango leaves in the treatment of diabetic retinopathy. This is a noninflammatory degenerative disease of the retina occurring in those who've had diabetes for many years. The condition is marked by small hemorrhage dots and very tiny enlargements of blood vessels within the retina itself, as well as sharply defined waxy exuded matter.

Make a tea according to the instructions previously given for hypertension. Drink 1 cup each day to which has been added 2 tbsp. of guava, mango, or papaya juice. This remedy in the same amount is also ideal for strengthening fragile blood vessels, treating purpura (hemorrhage spots in the skin) and varicose veins, and in the prevention of bleeding accidents through the use of anticoagulant drugs.

Chronic Diarrhea in Children Cleared Up

Chronic diarrhea is a frequent problem in many Third World countries, because of inadequate sanitation, lack of proper hygiene and good medical attention, and poor diet. Children are usually the victims and can die if this condition is left untreated for very long.

But the Malaysians and Filipinos have a ready cure for this problem. They will peel sections of the rind, mash them into a pulp, add a little white sugar for flavor, and then feed this to their sick children. An astringent substance within the rind clears up the diarrhea in a hurry.

Hernia Pain Disappears

Dr. Jesus Alvarez of Philippine General Hospital in Manila informed me in July 1994 when I was there on a fact-finding trip for new remedies that he had fed several hernia patients some mangosteen rind. He said that their excruciating pains diminished, but didn't repair their ruptures, which only surgery was able to correct.

Carambola Juice to Whiten Dentures

Annie Rice, aged 72, of Boca Raton, Florida, told me about a natural agent she used to keep her dentures nice and white. "When they get badly stained, just soak them in a glass of carambola juice," she claims. "It removes even the worst stains in less than 10 hours." Sometimes she dilutes the juice with $^1/_4$ amount of distilled water if she thinks the solution is too strong.

Gastrointestinal Problems Relieved

Ripe mango, papaya, and pineapple are all extremely useful for any disturbances of the G.I. tract. In the Philippines, ripe mangoes are employed to settle nervous and upset stomach and to stimulate the bowels in times of constipation. Papaya contains the remarkable digestive enzyme, papain, just the same as pineapple contains its own unique enzyme, bromelain.

Those who have had difficulty digesting certain starchy foods, (breads, cereals, and potatoes) or meats (beef, pork, chicken) have sometimes accompanied their meals with an 8 oz. glass of mango, papaya, or pineapple juice. Any one of these can bring quick relief to the heartburn common to rapid eating and poor chewing habits and heavy meals.

It may be interesting to note here that the proteolytic enzyme from papaya (papain) is employed a lot in the tendering of tough meat fibers by such food outlets as Sizzler, which is known for serving steaks at relatively cheap prices. They can afford to do this in most instances simply because they purchase lower grades of beef carcass—standard (young steers), commercial (mostly old cows) and utility (tough, old bulls)—which are then saturated with papain for certain lengths of time in order to make them seem more tender. Also, those who like to drink beer may find it interesting to know that papain is extensively used in stabilizing and chill-proofing beer. Which just goes to show that a fruit like papaya with an enzyme in it good for the healing of stomach ulcers, likewise has diverse applications in the giant food and beverage industries.

Help for Insomnia, Restlessness, Neuralgia, Muscle Spasms, Convulsions, and Epilepsy

Ripe passion fruit makes a wonderful sedative for jangled nerves. A cup of passion fruit juice 30 minutes before retiring can relieve muscle spasms, jumpy or restless nerves, and facial neuralgia. Frequent consumption of passion fruit and its juice help to minimize the occurrence of epileptic seizures and reduces the incidence of convulsions.

Papain for Insect Stings and Deadly Snakebites

A retired school nurse by the name of LaFonse Webber, who resided for years in Houston, Texas until her recent death, shared the following information with me during one of many lecture trips I've made to that city in the past.

> When I worked as a school nurse, both in the junior highs and local high schools here years ago, I always used to keep a bottle of Adolph's meat tenderizer in my desk for treating bee, wasp, and hornet stings, and also mosquito and horsefly bites that the kids would get during the spring and early fall. I never knew any case to fail when I used the stuff on them.
>
> The way I used to apply Adolph's was to mix a little of the tenderizer in $^1/_4$ – $^1/_2$ tsp. of water, which I'd then soak up with a cotton ball. This I'd place on the kid's sting or bite and have him or her hold it in place until some of the pain had subsided. I figure it took about 20 minutes for the tenderizer to soak into the skin where it could do some good.
>
> One time a big, heavy-built kid in one of the high schools here in Houston happened to step on a couple of honeybees feeding on some dandelions, while he was training on the grass. Boy! Did he ever holler. He was still moaning and groaning when his coach brought him limping to my office. For a guy with so much muscle and size to him, he sure didn't have much courage. Anyway, I mixed 2 tbsp. of Adolph's in a pint of warm water, which I then put into a small square pan and had the kid soak the bottom of his foot in this for awhile. In less than 15 minutes the pain went away and in another half hour or so half of the swelling had gone down too.

Jim Nelson, M.D., an allergist in Fort Wayne, Indiana, told the *Intermountain Reporter* (a publication of the U.S. Forest Service regional office in Ogden, Utah) once that the best way to treat a bee sting was first to "scrape off the stinger, then apply a paste of meat tenderizer and water ($^1/_4$ tsp. in 1–2 tsp. water) to destroy the venom." This also works well for insect bites too. And in Australia a similar method is used with meat tenderizer to treat jelly-fish and man-o'-war stings.

Now it is interesting to note that the proteolytic enzyme in papaya, called papain, is far more effective in the hydrolysis of protein substrates than related proteases such as trypsin and pepsin. Which helps to show its extreme importance in the successful treatment of poisonous snakebites from the likes of cobras, black mambas, and other deadly species. The *Journal of the American Medical Association* for December 22, 1975, pointed out that trypsin and similar enzymes like papain help to degrade the venom protein molecules injected into the blood stream.

In experimental animals, if a dose of trypsin followed injection of venom in less than 15 minutes, all animals survived. If the enzyme was injected 50 minutes after the snakebite, at least 50% of animals survived. Since the papain from papaya is superior to trypsin in this respect, it only stands to reason that it should be taken orally when an individual is accidentally bitten by a poisonous snake here in America, such as a rattlesnake or copperhead, for instance. Papaya tablets are available from some health food stores. Between 5 of them and about 1 level teaspoonful of Adolph's meat tenderizer dissolved in a cup of warm water, there should be enough papain ingested to begin neutralizing quite a bit of any such snake venom. First, swallow the tablets one by one, then take small gulps of the warm water and meat tenderizer solution.

Pineapple and Sports Injuries

Dwight McKee, M.D., a New England physician, has recommended the digestive enzyme bromelain, and raw pineapple in which it's found, for acute cases of tendonitis. "Sometimes just a day of eating nothing but raw pineapple will clear it up," he observed.

Another health practitioner, Dr. Robert W. Downs, a noted chiropractor who writes often for *Bestways* magazine, also believes very much in the therapeutic potential of bromelain. "It's going to be used by more and more doctors for younger individuals engaged in heavy athletics—such as jogging, running, and weight lifting—because it speeds up the healing process," he stated. "There have been studies suggesting that people who are damaging their tissues with athletics (especially the weekend athlete) can reduce the inflammation and speed the healing dramatically by the use of bromelain. In the future," he predicted, "I look for quality bromelain products to be one of the major supplements recommended for sports injuries."

One of the contestants in the Iron Man Triathlon held yearly in Hawaii with whom I spoke, confided that he ate nothing but raw, ripe pineapple and drank only pineapple juice several days before competition began in order to greatly minimize the pain and inflammation expected from such a grueling test of physical strength and human endurance.

Bromelain Thins the Blood

For those with thick blood and poor circulation, the rat poison compound called warfarin has usually been the standard drug prescribed for them by their physicians. But warfarin, coumadin, and related medications are toxic and can produce nasty side effects in the course of time. Now scientists have discovered that the bromelain in ripe pineapple and pineapple juice can also thin the blood and prevent blood clots from forming, not to mention helping increase circulation more.

One cup of pineapple juice daily or the inner ripe flesh of half a raw, ripe pineapple are recommended in place of harmful drugs for the above conditions. However, it should be noted that hemophiliacs and others with bleeding problems should avoid this fruit, since it is liable to only aggravate their situations more.

Pineapple Juice and Dental Surgery

The removal of impacted wisdom teeth from teenagers is a fairly common dental procedure in many parts of the country. But this type of operation is not without its drawbacks, namely, extreme swelling, black and blue marks, and considerable pain. But pineapple can change all of that in no time before and after surgery.

Prior to the operation, the patient should eat at least one can of pineapple chunks packed in its own juice for 15 days and drink one 6 oz. glass of unsweetened pineapple juice each day during this period as well. Following the operation, the patient should begin drinking two 8 oz. glasses of pineapple juice each day, one glass fresh and the other glass canned juice. Along with this should be taken one 50 mg. tablet of zinc, up to 3,500 mg. of vitamin C tablets, a couple of strong B-complex tablets, and about 750 mg. of bioflavonoids (unless they are included with the vitamin C supplement to be used). Ice packs may also be applied externally to the jaw as well.

The results will be nothing short of amazing. Swelling will be extremely minimal, pain will be practically nonexistent as such and there will be no black and blue marks. If fresh pineapple is readily available, then it should be the obvious first choice no doubt. By way of interest, the crushed leaves and petioles of mango are used in parts of Mexico to scrub the teeth with. This cleans them, hardens the gums, and relieves pyorrhea.

Getting Rid of Warts and Corns

Papaya and pineapple are very useful in getting rid of warts and corns. In Jamaica the milky latex of the green papaya fruit is slowly dripped onto warts several times each day for a week or less. This gummy residue shrivels them up and they soon fall off of their own accord. A slice of pineapple rubbed gently on a wart will remove it, but several applications are necessary before success results.

As a corn cure the pineapple is nearly always successful. Cut off a small, inch-square piece of the peel and bind it on the corn with *wide* adhesive tape, making sure that the inner side of the peel faces the corn. Leave it on all night and in the morning soak the foot in hot water. The corn will be easily removed. Some stubborn cases, however, may require 4–5 applications, but it nearly always works!

Eliminating Intestinal Worms

Both papaya and pineapple are excellent for removing intestinal parasites such as roundworm, hookworm, and the like. The gummy, milk sap in the unripe papaya as well as the little black seeds destroy such parasites by digesting them. If you're not up to eating some of the green, bitter fruit, then why not chew and swallow a piece of the leaf or a tablespoonful of the seeds after each meal? The seeds have a pungent taste to them, not unlike that of watercress or radishes.

In Venezuela and Colombia, coarsely cut, ripe, unpeeled pineapple fruit is steeped in 1 quart of boiling water for up to 3 hours. This can be accomplished by first putting the cut chunks of unpeeled fruit in a 2 quart fruit jar and then pouring the hot water over them. Seal with a ring type of lid while steeping this way. Strain the infusion and drink up to 4 cups a day in between meals and on an empty stomach in order to expel greater numbers of intestinal parasites.

Relieving an Aching Back

At one point or another in their lives, 8 of every 10 people on earth it's estimated, will suffer from this universal affliction. In the United States alone, as many as 75 million Americans have back problems. Of these 5 million are partly disabled, and 2 million are unable to work at all. Beyond the personal grief, back pain exacts a staggering social cost. In the United States 93 million workdays are lost each year because of back problems. Americans spend over $5 billion a year for tests and treatment by a dizzying array of back specialists, including orthopedists, osteopaths, physical therapists and chiropractors, to say nothing of self-styled gurus who promote every manner of cure imaginable.

Besides complicated back surgery, which can be very risky and only moderately successful, there is another alternative—the injection of an enzyme called chymopapain from ripe papayas. Developed in the early 1960s by orthopedist Lyman Smith of Elgin, Illinois, the treatment is designed to dissolve a ruptured disc's gelatinous pulp and eliminate the need for an operation.

The chymopapain is injected while the patient is under a light general anesthetic. Fluoroscopy is used to help guide the needles to the proper location. Two milliliters of fluid are inserted into the disc and allowed to remain for 4 minutes, after which the needles and fluid are removed. The fluid becomes cloudy, because during that 4 minutes, the enzyme has already begun dissolving the disc. Usually pain is alleviated within 48 hours and all symptoms are gone by six weeks.

A University of Wisconsin (Madison) study found that 83 of 114 patients who had received chymopapain for herniated discs had little or no back pain 3–6 years after the treatment. Now-a-days, though, because of FDA skepticism and a lot of "bad press," chymopapain is mostly used by Canadian physicians. Mark Brown, M.D., who was with the University of Miami School of Medicine in 1980, commented that 90% of his patients referred to Toronto doctors, came back feeling just fine and "didn't need anything else" to help them.

In 1983, some 70,000 treatments of chymopapain were given in the United States. But as of 1986 this figure had been drastically cut to the paltry sum of only 15,000. A couple of years ago, Smith

Laboratories, one of the makers of the drug, sent a form letter to over 22,000 surgeons warning them of several of its possible side effects. Among them are transverse myelitis (inflammation of the spinal cord), which can lead to muscle paralysis and analeptic shock.

Still, though, chymopapain remains an effective therapy for relieving back pain where surgery can and ought to be avoided. Using several herbs will help to reduce the risks associated with it. Just prior to and for sometime thereafter, an individual receiving chymopapain from his or her attending physician here or in Canada, should be drinking several cups of yarrow-chamomile tea each day *and* rubbing the spinal column with some CamoCare Cream from Abkit of New York. A good peppermint oil from any local health food store is likewise recommended for the spine too, rubbing a little in each day. To make the tea, bring 1 quart of water to a boil. Remove from the heat and add 1 1/2 tbsp. cut, dried yarrow and 1 1/2 tbsp. cut, dried chamomile. Stir a little before covering with a lid. Steep for 50 minutes before straining.

Cosmetic Applications

Paul Neinast employs papaya and pineapple in his renowned Dallas beauty salon. He prepares a papaya facial mask by puréeing a little bit of the peeled, ripe fruit (minus seed) in a blender. He claims that this "really helps to lift blackheads off the skin." And he finds that rubbing a chunk of peeled ripe pineapple segment over the skin not only neutralizes fatty acids, but also wipes up any greasy film which might be on the surface.

For a youthful, smoother complexion, try using a formula that includes green papaya concentrate, sunflower seed oil, and blackstrap molasses. Take one green papaya, if available in your area, and cut it up into 2" chunks. Put them and the milky latex in a food blender and purée until thick and even. Then add approximately 2 tbsp. of sunflower oil and 2–3 tbsp. of molasses and whip again until smooth. The consistency should just be thick enough to adequately spread on the face, forehead, throat, and neck each night without being runny or messy. A couple of tablespoons of whipping cream and 4 egg whites should also be added at the very last, but only blended for about 15 seconds and no longer.

Those who've had occasion to use this wonderful remedy report that it seems to work great on acne vulgaris and dried, old, scaly skin. There appears to be a rapid turnover of epidermal (surface) skin cells, which leads to a much younger, baby-soft type of skin for many. Some have even claimed that the papain present in this formula has actually lightened dark patches in their skin or nearly faded away existing freckles. But documented proof is wanting on these assertions, so they must be taken for just what they are—untested claims—until more solid evidence can be forthcoming in the near future.

Healing Wounds and Sores

The papain in papaya, because of its incredible ability to digest dead tissue without affecting the surrounding live tissue, has gained for itself the reputation of being a "biological scalpel."

This reminds me of an episode which occurred about a decade ago in a London hospital. A young doctor recently assigned to the medical staff from South Africa, was treating a male patient with a lingering infection from abdominal surgery. Antibiotics had proved virtually useless.

Remembering an old folk remedy, this physician purchased all of the ripe papayas at a local market. He instructed the nurses to slice some of the fruit in strips and judiciously place them over the patient's wound, leaving them there for about 5 hours before replacing with fresh ones. Needless to say the infection cleared up in a day or so and the patient was discharged shortly thereafter.

Cold sores around the lips and inside the mouth, cracked and furred tongue and inflamed tonsils can all be readily cleared up by sucking and chewing on several papaya tablets a couple of times each day until healed. In Brazil a piece of the fresh leaf is chewed for oral sores or else just tied onto a wound or external sore with good success.

Treating Inner Ear Infection

A really good oil for curing inner ear infection can be made by soaking 4–6 crushed, powdered papaya tablets, and 2 peeled, finely minced garlic cloves in 1 pint of virgin olive oil for 10 days. This can then be strained through several layers of gauze and rebottled in a dark, well-sealed container and stored in a cool, dry place until needed.

For a child under the age of 12, use about 2–3 drops, while teenagers and adults should be given up to 6 drops in whichever ear canal is infected. Do this once a day and repeat as often as needed until the infection clears up. It seems to work well in most cases.

Bathing Rashes

Some Indians of the Amazon Basin in Brazil boil the rind of ripe pineapple with rosemary, then frequently wash skin rashes and hemorrhoids with the concentrated decoction. In a quart of boiling water, simmer the cup rind of one pineapple with 1 tbsp. of dried rosemary, uncovered, for the space of 35 minutes. Besides washing eczema, dermatitis, psoriasis, jock itch, and diaper and poison ivy rashes with it, several cotton balls can also be saturated in this solution, the excess liquid squeezed out, and then inserted into the rectum on hemorrhoids. Fresh pineapple or powdered guava leaves can be put on ringworm with good results.

Preventing Nausea

For those experiencing nausea while flying or traveling on a ship or train, or for pregnant women with morning sickness, an 8 oz. glass of canned pineapple or papaya juice diluted from concentrate may be just the ticket for ending their unsettling discomforts.

Dealing with Foot Fungus

One of the niftiest ways to cope with athlete's foot or fungus around the fingernails is just to soak the hands and feet in some canned pineapple each day for about an hour. Wipe dry and sprinkle with a little Kingsford cornstarch. Rub this all over the feet and in between the toes, as well as on the hands and fingers.

Stimulating Milk in Dry Breasts

An effective way of promoting mammary secretion in more or less dry breasts, appeared in the *Ceylon Medical Journal* (Vol. 26, p. 105). Coarsely chop 1 green papaya and add it to 2 cups of boiling apple cider vinegar and 1 cup of water. Cover and simmer the same on low heat for half an hour. Remove and steep for an additional 15 minutes. Strain and take internally in 1 tbsp. doses 5 times a day or as needed.

A solid extract of the fruit of somewhat glutinous consistency can also be taken in place of the liquid extract. Thoroughly blend together one semiripe papaya cut into sections with seeds included and only enough water to make a thick purée. Use an electric blender for making this. A nursing mother should eat about 1 cup of this per day.

An alcoholic tincture can also be topically applied to the breasts if preferred. In a Mason quart fruit jar, soak half of a finely chopped green papaya and 4 finely chopped leaves in 2 cups of Jamaican rum for 10 days, being sure to shake the container a couple of times each day. Strain and take 1 tsp. 3 times daily, or as required.

There will be an abundant stimulation of milk production in most normal breasts. And even in the virgin breast a transient, limpid fluid will soon be evident.

Cooking in Paradise

Tropical fruits add certain distinctive flavors to otherwise dull dishes and really help to liven up so-so meals. Additionally they shorten both the cooking time and digestion of some foods, as recalled by noted research chemist, Albert Y. Leung:

> It is common knowledge among peoples of the tropics that cooking meat with a piece of green papaya or wrapped in papaya leaf will make it tender faster.

GUAVA FRUIT SLAW

Needed: 3 cups shredded cabbage; 1 peeled and sectioned orange; 1 cup halved, seedless red grapes; $^1/_2$ cup sliced celery; one 8 oz. carton orange yogurt; 1 small guava, chopped and with seeds removed. In a large salad bowl combine cabbage, orange sections, grapes, and celery. For dressing stir together yogurt and guava. Spread dressing over cabbage mixture. Cover and chill. Just before serving, toss salad gently. Serve on cabbage-lined plates, if desired. Makes 10 servings.

Mango Ice

Needed: 1 medium, very ripe mango; 2 tbsp. sugar; 1 egg white; 1 tbsp. sugar. Peel mango and cut flesh from seed. Cut mango into chunks. In a blender combine mango pieces and the 2 tbsp. sugar; cover and blend till nearly smooth. Set aside.

In a small mixer bowl beat egg white and the 1 tbsp. sugar with an electric mixer on medium speed till stiff peaks form. Lighten mango mixture by stirring in some of the egg white mixture. Fold mango mixture into remaining egg white mixture. Pour into an 8″ × 4″ × 2″ loaf pan. Cover surface with clear plastic wrap. Freeze about 3 hours or till firm. Scoop mixture with a small ice cream scoop or melon baller into dessert dishes. Makes 4 servings.

Both the guava fruit slaw and the mango ice recipes may be found in Better Homes & Gardens' *Fresh Fruit and Vegetable Recipes* and *Eating Healthy Cook Book*, which may be obtained from your local book store or directly from the publisher. I'm grateful to the publisher for their use here.

Baked Papaya Custard

This recipe is one of the favorite of restaurateur and chef, LaMont Burns of Encinitas, California, and may be found in his book, *Down Home Southern Cooking.* I'm grateful to his publisher, Doubleday & Co., Inc., for its use here.

Needed: 4 cups papaya pulp; 1 cup shredded coconut meat; 1 orange (pulp, juice, and grated rind); 4 eggs; 4 cups milk; 1 cup sugar. Preheat oven to 350° F. Arrange papaya, coconut, and orange in baking dish. Make a custard by beating together the eggs, milk, and sugar. Pour over the papaya. Insert knife in center of custard. If it comes out clean, it is done. If any of the milk clings to knife, bake longer. Serves 4–6.

PINEAPPLE-ALMOND TURKEY SALAD

Needed: 1 pineapple; 2 cups cooked, diced turkey (or chicken); $^1/_2$ cup Italian dressing; 1 tbsp. soy sauce; $^1/_2$ cup toasted, silvered almonds; 1 chopped green onion (including stem); 2 thinly sliced celery ribs; $^1/_2$ cup seedless grapes.

Cut pineapple into halves or quarters, lengthwise. Remove fruit, leaving a half-inch of shell. Drain shells on paper towels. Cut fruit into small chunks. Set aside 1 cupful for salad. Refrigerate remaining fruit to use in another recipe. Combine turkey, dressing, soy sauce, almonds, onion, celery, the reserved pineapple, and grapes. Toss gently to combine, making sure all ingredients are coated with dressing. Divide equally among pineapple shells. Chill for 1 hour. Serves 4.

Tropical Fruit Cocktail for Getting Well

A rather health-minded friend of mine who lives in Brooklyn has come up with a simple combination of fruit juices for getting well. Joel Bree insists that "this stuff is great for getting you or the kids well again after your family has been hit pretty hard with the flu."

First, take two medium oranges and ream the juice out of them by hand on a glass juicer. Next peel close to the skin one medium-sized, ripe pineapple and scrub but do *not* peel one ripe papaya. Run these through a juice extractor. Then mix these two juices with the hand-reamed orange juice and refrigerate. It keeps about two days before some more needs to be made. Drink this in the morning and early afternoon as much as you want.

Then wash but do *not* peel two large red apples. Cut them up and run through the extractor, core, seeds, skin, and all. Do the same with an equal volume of black grapes, seeds, and all. Mix the two together good. Drink this from midafternoon on into the evening as much as you want. I think anyone will agree after trying this regimen that these different juices really help to flush poisons out of the body, while giving you some badly needed energy and bounce at the same time.

Vita Drinks for Vitality

These two energy-revving drinks are courtesy of the Vita-Mix Corporation, maker of the world's finest juice and total food machine. (See the Appendix for more information.)

Pineapple, Coconut, and Yogurt Slurp

Needed: 1/2 cup fresh or canned pineapple, with juice; 1/4 cups lowfat pineapple yogurt; 2 tbsp. shredded coconut; 1/2 cup skim milk; 1/4 cup ice cubes.

Place all ingredients in the Vita-Mix whole food machine container in the order given. Secure the complete 2-part lid by locking under both tabs. Move the black speed control lever to "High." Lift the black lever to the "On" position and allow the machine to run for 15 seconds. Makes 2 1/4 cups.

Papaya-Mango-Pineapple Delight

Needed: 1/2 cup fresh or canned pineapple, with juice; 1/4 papaya, peeled; 1/2 mango, peeled; 1/2 cup pineapple juice; 1/2 cup ice cubes.

Place all the ingredients in the Vita-Mix whole food machine container in the order given. Secure the complete 2-part lid by locking under both tabs. Move the black speed control lever to "High." Lift the black lever to the "On" position and allow the machine to run for 15 seconds. Makes 1 3/4 cups.

Broiled Salmon with Passion Fruit Sauce

Colorful, festive, elegant, and quite simple to prepare, this recipe is a fine example of passion fruit's charm in savory dishes.

Needed: 2–3 passion fruits (to yield 1/4 cup juice); 1 tsp. cornstarch; 1/8 tsp. cayenne pepper; 1/4 tsp. salt; 1/4 cup fresh orange juice; 3 tbsp. diced red onion; 1 tbsp. light rum; 1/2 tsp. sugar; 1 1/4 lbs. skinless salmon fillet, cut into 4 equal portions; 3 tbsp. vegetable oil; 2 small pickling cucumbers, skin peeled in alternating strips, sliced; 1 bunch watercress, trimmed, washed and dried (about 2 cups); 1 tsp. balsamic vinegar.

Cut the tops from passion fruits, and scrape out all the pulp into the container of a Vita-Mix or similar food processor. Process the pulp until liquefied, then strain through a sieve to remove seeds. If you must do this by hand, just work the pulp through a sieve, pressing hard with the back of a wooden spoon to separate the pulp from the seeds. Set the juice and seeds aside.

In a small bowl, combine cornstarch, 1/8 tsp. cayenne pepper, and $^1/_4$ tsp. salt. Gradually stir in the orange juice, onions, rum, and $^1/_4$ cup of the passion fruit juice. Add sugar and additional cayenne pepper to taste. Set salmon fillets in a shallow dish and cover with the passion fruit mixture. Marinate in the refrigerator for 20–30 minutes, spooning liquid over occasionally.

Preheat the broiler and broiler pan. Remove the salmon from the marinade, reserving marinade. Brush the salmon and broiler pan with 2 tsp. of the oil. Broil the salmon until the flesh is opaque, about 6 minutes, basting once.

Meanwhile, combine the reserved marinade and 2 tbsp. water in a small saucepan; bring to a simmer over medium heat. Cook, stirring, until thickened slightly, about 1 minute. Remove from the heat and season to taste with salt and cayenne pepper. If you like, stir a spoonful or two of the passion fruit seeds back into the sauce for contrast and crunch.

Just before serving, toss cucumbers and watercress in a large bowl with balsamic vinegar, the remaining 1 tsp. oil, and salt to taste. Transfer the greens to a platter, set the salmon on top and spoon the sauce over the salmon. Serves 4.

Passion Fruit Sauce

Passion fruit provides concentrated tropical flavoring power, functioning as an essence that develops the tastes of other fruits. Stir up this simple sauce to transform all kinds of soft fruit, such as mango, papaya, melon, strawberries, blueberries, or raspberries.

Needed: 4 passion fruits; 2 tbsp. honey; fresh lime juice; 1–2 tbsp. light rum. Cut tops from the passion fruits, scrape out all pulp into the container of a Vita-Mix whole food machine, or similar piece of equipment. Process the pulp until completely liquefied; then strain through a sieve to remove the seeds. Blend the juice with honey and lime juice and rum if desired, to taste. Chill before serving. This sauce keeps in the refrigerator for 2 days. Makes $^2/_3$ cup sauce.

Carambola, Avocado and Wilted Spinach Salad

This makes an attractive salad that is pleasing to the eye, unusual, and easy to fix.

Needed: 1 tbsp. peanut or corn oil; 1 tsp. ground cumin; 2 tbsp. chopped red onion; 1/3 cup Japanese rice wine vinegar; 1 1/2 tbsp. brown sugar; 1 tsp. soy sauce; 1/4 tsp. red pepper flakes; 1 lb. fresh spinach, stemmed, washed and dried; 2 carambolas, thinly sliced; 1 avocado, peeled and sliced.

In a small heavy skillet, heat oil over medium heat, stir in cumin and cook for 30 seconds. Add onions and cook, stirring, until soft, about 2 minutes. Add the vinegar, brown sugar, soy sauce, and red pepper flakes, stirring to mix well.

In a large bowl, toss spinach with 3 tbsp. of the warm dressing. Divide the salad among four plates, arrange carambola and avocado slices over the top and drizzle with the remaining dressing. Serves 4.

TULIPS

(See under ORNAMENTAL PLANTS)

TURNIP

(See under RUTABAGA)

V

VANILLA
(Vanilla planifolia)

Brief Description

Liquid vanilla flavoring is obtained from the fully grown but unripe pods or beans found on large green-stemmed perennial herbaceous vines growing wild and extensively cultivated throughout the tropics (Mexico, Malagasy Republic, Comoros Islands, Tahiti, Indonesia, Seychelles, Tanzania, and Uganda, among others). Pollination is all done artificially, except in Mexico, where it's partly performed artificially and partly by certain hummingbirds and butterflies not found anywhere else in the world. Because of the high price of vanilla and the low cost of vanillin, vanilla extracts have been extensively adulterated. Vanillin is the major flavor component with over 150 other aroma chemicals being present as well.

Sexual Stimulant

In various Central and South American countries such as Mexico, Argentina and Venezuela an alcoholic extract of the dried pods is taken for increasing amorous desires. Generally 4–6 pods are steeped in 2 cups of tequila or imported cognac for 21 days in a well-stoppered flask or glass container of some kind. The bottle is shaken several times each day during this period of time. The tincture is then taken in doses of 10–20 drops 2–3 times a day, usually at night.

Calms Hysteria

Pure vanilla flavor fluid extract may be used to help calm hysteria and related emotional traumas. The best manner for successfully accomplishing this is to soak one or two cotton balls completely with the vanilla extract, lightly squeezing out any unnecessary excess fluid before placing them beneath the tongue, one on either side of that vertical fold which elevates the tongue as needed.

By positioning each cotton ball thusly, it permits the vanilla to slowly penetrate the smallest of the salivary glands called the sublingual gland. The vanilla then travels through the tiny blood vessels which empty into the sublingual and submental arteries. These last two arteries, in turn, hook up directly with the much larger internal jugular vein and the external and internal carotid arteries. Now the internal jugular vein and the internal carotid artery both supply blood directly to the brain, which suggests that this is the route by which vanilla travels in order to reach that part of the brain responsive to its calming effects in episodes of hysteria.

VIOLET

(See under ORNAMENTAL FLOWERS)

VINEGAR

Brief Description

Vinegar is among the oldest of foods and medicine known to humankind. It has been around at least since the time of Noah; its existence before the Flood is doubtful, since what few ancient records there are of the era never mention wine. Both vinegar and wine require fermentation, and it is believed by some Flood scholars that this bacterial action (as we know it) didn't exist in antediluvian times.

In simple chemical terms, vinegar is a sour liquid consisting mainly of acetic acid and water, produced by the action of bacteria on dilute solutions of ethyl alcohol derived from previous yeast fermentations. The coloring and flavoring are characteristic of the alcoholic liquor (as cider, beer, wine, fermented fruit juices, solutions of barley malt, hydrolyzed cereals, starches, or sugars) from which the vinegar is made.

Vinegar was widely employed throughout the ancient world. The Phoenicians, Egyptians, Greeks, and Romans relied on it for a variety of applications, from cooking to curing. The Greeks and Romans in particular were known to have kept quantities of vinegar in their cellars, especially prizing Egyptian vinegar. They used vinegar for cooking—steeping vegetables and marinating meat to tenderize and add flavor, making pickles, and preserving herbs, flowers, and vegetables in vinegar. Sweet-and-sour dishes, especially meat cooked in honey and vinegar and fruits and vegetables preserved in vinegar, were enjoyed in other ancient civilizations such as the early Chinese and Arabic cultures.

Bowls filled with vinegar were placed on the tables for dipping bread during meals, a use first mentioned in the Old Testament in the Book of Ruth, where it is noted that the reapers soaked their bread in vinegar to freshen it (see Ruth 2:14). Believe it or not, there is even a specific printing of the Bible that is known as the "Vinegar Bible." An Oxford printing of 1716 by John Baskett, the royal printer for the king of England, was so called because the Parable of the Vineyard (Luke 20:9–18) was printed as the Parable of the Vinegar.

At the time of Christ, Roman soldiers when in camp were issued with their rations a certain portion of a thin, sour wine called *acetum*, vinegar, both in the pure state and diluted with water. In the latter state it was termed *pasca* (see Pliny the Elder's *Historia Naturalis* 19:29 for more details). It was this type of vinegar which a kind-hearted Roman soldier soaked a sponge with, affixed it to the end of a long hyssop branch, and put it close to the mouth of the crucified Jesus in order to alleviate some of his burning thirst (John 19:29–30).

By the thirteenth century, a wide selection of vinegars—including those flavored with clove, chicory, fennel, ginger, truffle, raspberry, mustard, and garlic—was commonly sold by street vendors in Paris. In fact, it is from France that the name itself first appeared, being derived from *vin aigre* (or sour wine). Pepper vinegar was especially popular during the Middle Ages because wine that contained pepper was not taxed on importation into Paris. Wealthy French cooks in the Middle Ages went to the *saucier*—the vinegar maker for sauces—and seasonings made of vinegar, mustard, herbs, and spices.

It is safe to say that vinegar can be found in nearly 99% of all homes in America today. Until recent years, though, consumption was static and mainly limited to cider and distilled vinegars. But with increased interest in the many different kinds and flavors of vinegar over the last several decades, sales have risen 10%. The acidity and nutrient content of these vinegars varies. By law in the United States, vinegar must be at least 4% acidity, or 4 grams of acetic acid per 100 cubic centimeters. Other countries have different legal minimums for acidity. Sometimes the acidity is specified by the term *grain*, which refers to the amount of water dilution. Hence, a 40-grain vinegar implies a 4% acetic acid content.

Ancient "Gatorade" for Fatigue and Weakness

Vinegar as a beverage has a lengthy history. It has been claimed by some historians as part of the reason for the success of the Roman army. Spartianus, a Latin scholar, wrote that vinegar mixed with water was the drink of soldiers. This beverage is credited with helping them survive the rigors of battle as well as the various climates they encountered in Europe, Asia, and Africa. Undiluted vinegar not only proved more portable than wine, but also more invigorating.

Roman laborers working on huge public works projects such as the Coliseum, roadways, or long aqueducts managed to survive this grueling manual labor under a burning sun, with an improvement over what the military legions carried with them on their many foreign expeditions. To the aforementioned mixture of $^1/_4$ water and $^3/_4$ vinegar they would add a pinch of salt and sometimes the equivalent of $^1/_2$ teaspoon honey. This is the earliest version of the popular sports drink Gatorade developed at the University of Florida years ago which has become a big hit with athletes everywhere.

Four Thieves Vinegar

Among all the therapeutic vinegars, Four Thieves Vinegar has its own unique place in history. It is told that during one of the bubonic plague epidemics in Marseilles, France, four robbers were able to steal from the dead and dying without becoming sick by dousing themselves with this mixture. Upon being captured and hauled before the local authorities, they were given a chance to escape execution by sharing their "health secret" which had kept them from getting the plague.

Here for the very first time in print is their unpublished recipe. I am indebted to French medical historian Jacques Dupreau for discovering this in the archives of Marseilles.

FOUR THIEVES VINEGAR

Take 1 quart of red wine vinegar made from the *second* pressing of red grapes. To this add the following: 2 chopped garlic cloves; 2 chopped shallots; 1 tbsp. each powdered cinnamon, cloves, and nutmeg; 1 tsp. each rosemary, sage, and wormwood. Place in corked bottle; set in cool place for 7 days, shaking once a day.

Perfect Antidote to Food Poisoning

DeForest Clinton Jarvis, M.D., was born in 1881 and graduated from the University of Vermont Medical College. In 1909 he commenced practicing medicine in Barre, Vermont. During his long and illustrious medical career which spanned many decades he developed a keen interest in the folk medicine of Vermont. Not only did he collect a lot of remedies from folks he knew, but he also recorded the results of other remedies he recommended to many of his patients.

Apple cider vinegar was one of these. He used it for a lot of things. One of the more practical applications was as an effective antidote to food poisoning. Here is one example from Dr. Jarvis's files of just how well it worked.

"At a Shriners summer outing here in Vermont," he wrote, "lobster salad was served at dinner. Unfortunately the salad was spoiled and nineteen Barre people developed severe diarrhea, complicated in some cases by vomiting. One of the diners had taken precautions. As I had advised him when there was any chances that food might have become spoiled, he had brought along a little bottle of apple cider vinegar to the affair. At the outset of dinner he poured a liberal amount into his glass of water. It happened that he was particularly fond of lobster salad and he had two extra helpings. Whereas many of his table companions suffered bad effects from the spoiled lobster, the apple cider vinegar had so sterilized it in his digestive tract that nothing disagreeable happened to him."

A Simple Cure for Common Cold and Flu

Dr. Jarvis relied more upon natural remedies than he did prescription medications to treat ordinary maladies. The common cold and influenza were just two of these that could be easily solved with a little internal application of some apple cider vinegar.

Here's how he did it. Into an ordinary 8 fl. oz. drinking glass he measured 50 tsp. apple cider vinegar. Every 2 hours he would have his patients take 2 tbsp. of the vinegar. After the contents of the first glass had been emptied, another would be prepared in the same manner and the dose increased to 1 tbsp. every waking hour. A third glassful would be prepared, if necessary, but diluted with 15 tsp. of water. One tbsp. every 45 minutes would be swallowed.

Dr. Jarvis mentioned, however, that his experience taught him that the correct medicinal dosage was pretty much an individual thing. "Some people have found that the dose for them was one teaspoonful of the apple cider vinegar to a glass of water," he wrote. "Others say they pour the vinegar into an ordinary glass to the measure of the width of one finger, then fill the glass with water. Still others reported, variously, that two fingers and three fingers of vinegar in the water were enough for them."

He advised that "if for any reason apple cider vinegar isn't accepted by the body, try taking a small glass of apple juice instead."

Some Other Uses of Apple Cider Vinegar in Vermont Folk Medicine

Dr. Jarvis kept in his extensive medical records hundreds of other observed uses for apple cider vinegar. The best of these have been gleaned from this inventory and are presented in the paragraphs that follow.

JOINT STIFFNESS/MUSCLE SPASM or *"CHARLEY HORSE."* Whip up the yolk of an egg with a fork; mix in with it 1 tbsp. each of genuine turpentine and apple cider vinegar. Apply this mixture to the skin surface, rubbing well in a clockwise motion to bring immediate relief.

RASH. Use equal parts of apple cider vinegar and water. Dab on the afflicted part and allow to dry on the skin. Apply often throughout the course of the day.

SHINGLES. Apply apple cider vinegar, just as it comes from the bottle, to the skin area where the shingles are located, four times during the day and three times at night if you are still awake. The itching and burning sensation in the skin will soon leave once the vinegar has been applied. The shingles are apt to heal more quickly this way.

NIGHT SWEAT. If the skin surface of the body is given a cupped-palm hand bath of apple cider vinegar at bedtime, the night sweats will be prevented.

BURNS. Undiluted apple cider vinegar straight from the bottle should be gently applied to a burn on the skin surface; it will quickly remove all pain and soreness.

SHRINKING VARICOSE VEINS. Apply apple cider vinegar from the bottle by means of the cupped-hands treatment or else sponge it on the skin. Do this twice daily for a month. Besides this, take 2 tsp. of vinegar in a glass of water twice daily for the same period.

IMPETIGO. Technically, impetigo is a staphylococcus or streptococcus infection of the skin. Anybody of any age may catch it, but children seem to be especially susceptible. In treating this problem, use apple cider vinegar straight from the bottle. Dip the forefinger in it and apply to each affected part of the skin. Application should be six times a day, beginning in the morning and at different intervals throughout the day to bedtime. If this procedure is faithfully allowed the impetigo should be gone in four days or less.

RINGWORM. Children are more easily affected with this fungal problem than adults; boys are affected 6 to 9 times more often than girls. It can also be transmitted from household pets to children. Apply apple cider vinegar with the fingers or 3 cotton swabs taped together to the ringworm area at least 6 times a day. Start the treatment in the morning and complete it by bedtime.

Facial Tonics

Genevieve S. is a cosmetologist at a well-known beauty salon in Philadelphia. She uses cosmetic vinegars all the time with her clientele. "I find they're able to close pores and preserve or restore the skin's natural acidity," she said. "Vinegars keep both oily and dry complexions soft and fresh. Vinegar tonics are much better for the skin than most commercial tonics because the latter usually contain drying alcohol, which can be hell on the face. Vinegar tonics also combat the ravages of alkaline soaps and makeup. Depending on the herbs infused in the vinegar, tonics can tone, heal, soothe, or soften the skin of both men and women. They also help hold acne at bay, improve blood circulation, reduce broken spider veins and capillaries, smooth out wrinkles, and bleach out age spots and freckles."

She uses vinegar as a facial skin tonic by first diluting it with 6 parts of distilled water or rose water and then spritzing it on her clients' faces after washing. Or else she'll just apply the mixture with cotton balls, rubbing in a circular motion.

The instructions for making her basic cosmetic vinegar for face, scalp, or body are given here. She kindly shared them with me. Combine together equal parts ($^1/_2$ cup each) fresh or (1 tbsp. each) dried rosemary, sage, tarragon, basil, peppermint, and lavender. Pour 5 cups apple cider vinegar or wine vinegar over this mixture and let set in cool, dry place for about 1 $^1/_2$ weeks, then strain.

Next, mix together $^1/_4$ oz. gum camphor, $^1/_2$ oz. gum benzoin, and 2 $^1/_2$ tbsp. vodka until totally dissolved. Stir into the vinegar solution, cover, and let stand for 1 week. Strain, bottle, cap tightly and store in cool, dry place. Use as needed.

WATERCRESS
(Nasturtium officinale)

Brief Description

Watercress is a perennial plant which thrives in clear, cold water and is found in ditches and streams everywhere. It is cultivated for its leaves, which are principally used as salad greens or garnishes. Connected to a creeping rootstock, the hollow, branching stem, 1–2 feet in length, generally extends with its leaves above the water. The smooth, somewhat fleshy, dark green leaves are odd-pinnate with 1–4 pairs of small, oblong or roundish leaflets.

Healing Mouth Sores

One of the most popular remedies among Chinese residing in Hong Kong and Canton, China, is a watercress soup used to treat canker sores on the tongue or lips, blisters in the mouth, swollen gums, bad teeth, and foul breath. There is no specific amount called for, but generally for one person, about 1/2 lb. each of cut watercress and chopped carrots are cooked in 2 quarts of water. The liquid is boiled down slowly to 1/3 or 1/4 of the original fluid volume and then the soup is consumed with the vegetables intact. It's also good for hot flashes when consumed cold.

Relieves Headaches

Watercress forms the basis of an excellent remedy for headaches brought on by some kind of sickness or general nervousness. A handful of watercress, having been first washed thoroughly, is then put into a clean quart fruit jar and 2 cups of boiling apple cider vinegar added. After the solution becomes cold several hours later, it is strained and rebottled for use later on. When the headache occurs, a clean handkerchief should be dipped in the vinegar, wrung out, and laid over the eyebrows and forehead. A wash cloth may also be used in place of a hanky.

Soothes Skin Ailments

An old and effective cure for eczema and dermatitis consisted of an infusion of watercress. A large panful should be thoroughly washed and put into a stainless steel saucepan that has just enough cold water added to cover the cress with. Bring this to a boil, then reduce heat and simmer slowly until quite tender. Strain through muslin cloth or several layers of gauze material and refrigerate.

The afflicted part should be bathed with this infusion often. It's better to use a piece of soft linen for this purpose. This infusion is excellent for roughness of the skin due to frequent exposure to the wind, sun and cold weather.

Diuretic and Expectorant

Watercress tea or juice is valuable for eliminating accumulated fluids in body tissue such as in gout and for clearing mucus congestion from the lungs. To make the tea steep 1 tbsp. chopped fresh cress in 1 cup boiling water for 20 minutes, then strain and drink. Fresh juice can be easily obtained from an electric juicer, but should be combined with some carrot or tomato juice before drinking.

WATERCRESS SOUP

Needed: 1 quart chicken stock; 1 tbsp. honey; $^1/_2$ tbsp. blackstrap molasses; pinch of kelp; 1 bunch watercress; water to cover stems; pinch of cardamon.

Heat stock and season with honey, molasses, kelp and cardamon. Before untying the bunch of purchased watercress, cut off stems. Wash stems and boil for about 10 minutes. Drain and add the cooking water to the stock. Cooked stems by themselves can be eaten when seasoned with a little tamari soy sauce.

Wash tops of watercress (flowers) and add to boiling stock just before serving. Boil soup 3 minutes, no longer, as you don't want to lose the emerald green color. Be sure to cook uncovered after you add the watercress or it will darken. Yields 4 cups.

Cress and Onion Dish

Watercress with Browned Onion

Needed: 2 tbsp. butter; 1 small to medium onion, finely chopped; 1 tsp. salt; 3 cups watercress, washed (about 4 $^1/_2$ oz.).

In a small saucepan or skillet, over low heat, melt the butter, and when the foam subsides add the onion. Sauté for about 20 minutes, stirring frequently, until browned. The onions should be dark brown but definitely not burned. Mix in the salt and set aside. Place the watercress in enough boiling water to cover. Over moderate heat cook, covered, for about 15 minutes until tender. Drain thoroughly and chop medium fine. Mix the watercress with the sauteed onion. Serves 4.

Super Salad

Feta, Pear, and Watercress Salad

Needed: 1 $^1/_2$ cups tightly packed torn red leaf lettuce; $^1/_2$ cup tightly packed trimmed watercress; $^1/_2$ cup red pear, thinly sliced; 2 tbsp. crumbled feta cheese; 1 tbsp. balsamic vinegar; 1 tsp. distilled water; 1 tsp. walnut oil or olive oil; 1 tsp. Dijon mustard; dash of liquid Kyolic aged garlic extract (obtained from your local health food store); a dash of dried oregano.

Divide the leaf lettuce, watercress, and pear between 2 salad plates, and sprinkle with cheese. Combine the vinegar and next 5 ingredients in a bowl, and stir well. Drizzle evenly over salads. Yield: 2 servings.

Zucchini

(See under PUMPKIN, SQUASH, AND GOURD)

With the zucchini comes the close of this informative work. Of all the some two dozen odd books on health, nutrition, and herbs that I've penned over the years, I believe that this project has been the most exciting, challenging and grandest undertaking I've ever attempted. Yet for me, it's also probably been the most personally satisfying because of the enormous amount of self-care data presented for the very first time on such a wide range of illnesses under one cover.

Nothing that I'm currently aware of even comes close to it. And I'm stating this more as an observed fact of truth than as a self-administered pat on the back.

Finally, let it be said that although we may never have the pleasure of meeting each other in this life, yet hopefully, some day you, the reader, and I, the author, will be able to meet together in another dimension where time stands still, peace abounds, love surrounds, health is for all and there are no more tomorrows—only an eternal period where the soul may finally rest and the spirit can be of good cheer!

APPENDIX

MISCELLANEOUS

Dried Fruit by Mail

These sources have any kind of dried fruit you want—choose from dried kiwifruit to bing cherries. Call for catalogs:

American Spoon Foods Inc., 1-800/222-5886.

Frieda's by Mail, 1-800/241-1771.

King Arthur Flour Baker's Catalogue, 1-800/827-6836.

L'Esprit de Campagne, 703/955-1014.

Melissa's/World Variety Produce Inc., 1-800/468-7111.

The Wooden Spoon Catalog, 1-800/431-2207.

Walnut Acres, 1-800/433-3998.

Quality Products and Services

Great American Natural Products, Inc., 4121 16th Street North, St. Petersburg, FL 33703, 800/323-4372

Aqua-Vite; Ginger-Up, Green Onion-Garlic Oil, Nettle Scalp Oil/Nettle Shampoo, Super Weight Loss Tea, Avocado Healing Salve (2 oz. jar), Thyroid Seaweed Formula. Also many dried fruits/vegetables.

Old Amish Herbs, 4141 Iris St. North, St. Petersburg, FL 33703 Cabbage Compound, Carrot Concoction, Cranberry Cleansed, Okra Farm-Lax, Fig Paste, Night Nip, Artichoke/Garlic Combination, Oat "Anti-Smoking" Extract, Oil Amish Digestive "Bitters" with Rhubarb, Cholester-Low with Apple Fiber, Digest-Eze with Pineapple Enzyme, Horseradish Sinus-Stop.

Pines International, Inc., POB 1107, Lawrence, KS 66044, 800/642-PINE. Beet Root Juice Concentrate; Rhubarb Root Juice Concentrate.

Vita-Mix Corp., 8615 Usher Road, Cleveland, OH 44138, 1-800-VITA-MIX (1-800-848-2649). The Vita-Mix Whole Food Machine/Juicer is "America's very best food processor and juicer," according to many industry experts. Vita-Mix also sells Neova Stainless Steel/Waterless Cookware, superior cooking utensils for very efficient cooking.

Wakunaga of America Co., Ltd., 23501 Madero, Mission Viejo, CA 92691, 714/855-2776. Makers of "the world's most preferred and best researched garlic," Kyolic Aged Garlic Extract products and Kyo-Green. For free samples and product information, call 1-800-825-7888. For general information on garlic, call the Garlic Information Hotline at 1-800-330-5922. (This is a public service jointly sponsored by Wakunaga of America and The New York Hospital-Cornell Medical Center.)

Rex's Wheat Germ Oil

The most pure and unrefined wheat germ oil available anywhere. Intended primarily as a nutritional supplement for use in domestic pets and livestock. Used by breeders and veterinarians for conditioning of skin and coat in animals, as well as during pregnancy, parturition and lactation. But for years I've recommended it over all other vitamin E brands currently sold in health food stores or nutrition centers. Our research center carries it as a service to readers, since it is difficult for ordinary consumers to buy it on their own. Send $65 to the name and address given below.

Utah Prime Times

A monthly 81,000-circulated senior citizen newspaper. The paper is edited by Dr. John Heinerman. Annual mail subscription is $20.

The Healing Benefits of Garlic

The most complete and up-to-date book ever written on the world's oldest and most frequently used spice. Entitled *From Pharoahs to Pharmacists: The Healing Benefit of Garlic*, Dr. Heinerman, the author, explores in detail both the ancient as well as modern history and uses of garlic. The book retails for $25.00

To order any of these, send the appropriate amounts (in U.S. funds) to:

Anthropological Research Center
P.O. Box 11471
Salt Lake City, UT 84147
801/521-8824

Index